Complex Locations

RGS-IBG Book Series

Published

Complex Locations: Women's Geographical Work in the UK 1850–1970
Avril Maddrell

Value Chain Struggles: Institutions and Governance in the Plantation Districts of South India
Jeff Neilson and Bill Pritchard

Arsenic Pollution: A Global Synthesis
Peter Ravenscroft, Hugh Brammer and Keith Richards

Queer Visibilities: Space, Identity and Interaction in Cape Town
Andrew Tucker

Resistance, Space and Political Identities: The Making of Counter-Global Networks
David Featherstone

Mental Health and Social Space: Towards Inclusionary Geographies?
Hester Parr

Climate and Society in Colonial Mexico: A Study in Vulnerability
Georgina H. Endfield

Geochemical Sediments and Landscapes
Edited by David J. Nash and Sue J. McLaren

Driving Spaces: A Cultural-Historical Geography of England's M1 Motorway
Peter Merriman

Badlands of the Republic: Space, Politics and Urban Policy
Mustafa Dikeç

Geomorphology of Upland Peat: Erosion, Form and Landscape Change
Martin Evans and Jeff Warburton

Spaces of Colonialism: Delhi's Urban Governmentalities
Stephen Legg

People/States/Territories
Rhys Jones

Publics and the City
Kurt Iveson

After the Three Italies: Wealth, Inequality and Industrial Change
Mick Dunford and Lidia Greco

Putting Workfare in Place
Peter Sunley, Ron Martin and Corinne Nativel

Domicile and Diaspora
Alison Blunt

Geographies and Moralities
Edited by Roger Lee and David M. Smith

Military Geographies
Rachel Woodward

A New Deal for Transport?
Edited by Iain Docherty and Jon Shaw

Geographies of British Modernity
Edited by David Gilbert, David Matless and Brian Short

Lost Geographies of Power
John Allen

Globalizing South China
Carolyn L. Cartier

Geomorphological Processes and Landscape Change: Britain in the Last 1000 Years
Edited by David L. Higgitt and E. Mark Lee

Forthcoming

Aerial Geographies: Mobilities, Subjects, Spaces
Peter Adey

Politicizing Consumption: Making the Global Self in an Unequal World
Clive Barnett, Nick Clarke, Paul Cloke and Alice Malpass

Living Through Decline: Surviving in the Places of the Post-Industrial Economy
Huw Beynon and Ray Hudson

Swept-Up Lives? Re-envisaging 'the Homeless City'
Paul Cloke, Sarah Johnsen and Jon May

Millionaire Migrants: Trans-Pacific Life Lines
David Ley

In the Nature of Landscape: Cultural Geography on the Norfolk Broads
David Matless

Transnational Learning: Knowledge, Development and the North-South Divide
Colin McFarlane

Domesticating Neo-Liberalism: Social Exclusion and Spaces of Economic Practice in Post Socialism
Adrian Smith, Alison Stenning, Alena Rochovská and Dariusz Świątek

State, Science and the Skies: Governmentalities of the British Atmosphere
Mark Whitehead

Complex Locations

Women's Geographical Work in the UK 1850–1970

Avril Maddrell

WILEY-BLACKWELL

A John Wiley & Sons, Ltd., Publication

For Bill, Sam and Breesha

Contents

List of Figures

List of Tables

Preface

During my DPhil research I started to study women geographers working before 1918, including writers such as Mary Somerville, Isabella Bird and Marion Newbigin, and educationalists such as Nora MacMunn and Joan Reynolds, and their relationship to geographical societies, including the 1892–3 RGS women's membership debate. This convinced me that there was a bigger 'story' to tell about women's geographical work before second-wave feminism, not least because the popular perception of my own generation, based on the most widely read histories of the discipline, was that women had contributed little to the discipline before the 1970s and 1980s. For me it became a political project to repopulate the historiography of geography with women's work and I was delighted when my long-standing proposal finally found a home in the relaunched RGS-IBG/Blackwell series.

My confidence about a quick completion of the manuscript proved false, as the primary and secondary materials were much richer than I had anticipated. This was to the ultimate benefit of the book, but not the publication schedule! I have found it engrossing and fascinating as the lives and work of individual women geographers took shape, and connections emerged in relation to each other, the wider discipline and the socioeconomic and political milieu. While some women were well known in their day and their work easily traceable, others were serendipitous 'discoveries' (as with most 'discoveries', there all the time and usually known to others). 'Unearthing' perhaps gives an appropriate sense of the archaeological nature of some of the historiographical work (see Foucault 1972) and the thrill of finding even shards which tell us something new and enlightening.

Complex Locations lived up to its name as I organised the structure of the book. The chapters focus on geographical societies, travellers, educationalists and academics and are organised in broad chronological order. By following this format the early chapters address topics such as women's entry to geographical societies and travel writing, which have been discussed in detail elsewhere but are nonetheless important for a rounded discussion of women's geographical work. The bulk of new research is found from Chapter Five onwards. For the most part the women studied were relatively easy to place, but there were numerous anomalies, where their chronology and/or multi-faceted careers challenged neat divisions. Having initially organised the academics under the headings of three successive 'generations', late in the day I realised that while a human reproductive generation might amount to my neat categories of about 20 years, academic reproduction, from new undergraduate to employable PhD holder can occur in as little as six years and some early academics were employed without higher degrees, making for much quicker production of successive academic generations. Thus the academic chapters have been divided into four broad periods reflecting wider socioeconomic and political periods in the UK: the 'first generation'; the interwar years; the war and immediate post-war years; and 1950 to 1970. Marion Newbigin proved difficult to place, geographical societies perhaps the most obvious location, but that chapter didn't allow enough space to discuss her influential writing; her dominance of a single chapter reflects the limitations of organisational structures (notably chapter length) rather than her pre-eminence within the discipline.

I started the book knowing I would have to be selective of women travellers and teachers of geography, but expected to be able to include all of the women academics working in geography 1900–70. In the event, this has not been possible, given (a) the number of women and (b) my decision to analyse the production and reception of individuals' work in relation to personal biographical, disciplinary and wider contexts. I have been able to include most women academic geographers working before 1950, but have not been able to include/analyse in detail the work of all those working 1950–70. Some notable geographers beginning their careers at the very end of the 1960s are flagged briefly in Chapter Nine, but require a fuller treatment elsewhere. The selection made is intended to be representative of a geographical spread across the country and a range of research and teaching interests within the discipline, as well as departmental roles. While some chapters may be too long to read at one sitting I hope the reader will be tolerant of this, in the knowledge that this reflects the desire to be as inclusive as possible and to give the women 'voice' through the use of publications, archive sources and interview material. My hope is that the book will be read in a variety of ways, from those who wish to read it cover to cover, to those who read selective chapters or biographies; either (any) way I hope it will lead to further critical engagement with the complex locations of women's geographical work in the historiography of geography.

Avril Maddrell 2009

Series Editors' Preface

The RGS-IBG Book Series only publishes work of the highest international standing. Its emphasis is on distinctive new developments in human and physical geography, although it is also open to contributions from cognate disciplines whose interests overlap with those of geographers. The Series places strong emphasis on theoretically-informed and empirically-strong texts. Reflecting the vibrant and diverse theoretical and empirical agendas that characterize the contemporary discipline, contributions are expected to inform, challenge and stimulate the reader. Overall, the RGS-IBG Book Series seeks to promote scholarly publications that leave an intellectual mark and change the way readers think about particular issues, methods or theories.

For details on how to submit a proposal please visit:
www.rgsbookseries.com

Kevin Ward
University of Manchester, UK

Joanna Bullard
Loughborough University, UK

RGS-IBG Book Series Editors

Acknowledgements

I have a great many people to acknowledge: those who have supported and assisted my research and those who have supported me in the long process of that research and writing the book.

I am grateful to the British Academy for a small grant funding a tranche of archive work 2004–5, and to Professors Morag Bell and Alison Blunt who supported my application. Although my research on women's geographical work dates back to my doctoral research in the early 1990s, this BA grant was pivotal to progressing the book. Alison Blunt has been a resolute source of encouragement for this project over the years, including when there was little other academic interest in it, and I thank her sincerely for that, and for making the time in a pressing schedule to read and comment on a draft manuscript. I am also grateful to Helen Morse who was has been so generous with her offers of childcare and general support over recent years.

A large number of people and institutions have helped me source archive records, illustrations, departmental histories, publications and contacts. I have tried to acknowledge everyone appropriately in the text, but would like to underscore their assistance here. Special thanks go to Margaret Wilkes, who helped me develop my 'map' of women geographers in 2005, and to Liz Baigent for getting up early and reading and commenting on two draft chapters. I am also grateful to the two anonymous RGS-IBG series readers for their encouraging and helpful comments, as well as to the series editors and Blackwell's commissioning editor, Jacqueline Scott, and copyeditor Tessa Hanford, for their patient support throughout the preparation of the manuscript.

I am grateful to Sarah Strong and Colleagues at the RGS Archive, the RSGS staff, Frances Soar at the GA and Sue Bird at the University of Oxford School of Geography and Environment Library, all of whom have assisted my work over the last decade and more. More recently I have been grateful to archivists at the British Library, National Library of Wales and National Library of Scotland, and numerous universities. Several have helped me trace sources and have extracted requested information for me from a distance, for which I am particularly grateful: Maria Castrillo (National Library of Scotland), Rachel Hart (St. Andrews), Chris Joy (Manchester High School for Girls), Ursula Mitchell (Queen's University, Belfast), Kate Mooney (London Institute of Education), James Peters (University of Manchester), Anna Petre (University of Oxford), Ian Salmon (Aberystwyth, University of Wales), Lorraine Screene (QMUL) and James Webley (University of Bristol). My thanks to the QMUL and Southampton geography departments for giving me copies of their departmental histories, and to the Sheffield department, the RSGS, RHUL Archives, LSE Archives, RGS-IBG Picture Library and National Library of Scotland for providing illustrations, as well as to Anne Buttimer, Richard Hickman, Dick Grove and Cuchlaine King for providing photographs from their own collections. My thanks to Paul Ravell at UWE, Jamie Owen at the RGS-IBG and Andy Morrison at Oxford Brookes for technical assistance with photographs. I am also grateful to colleagues who welcomed me to their own institutions and archive collections, including Mike Bradford, Felix Driver, Mike Hefferenan, David Matless, Chris Philo and Charlie Withers; and to Hugh Clout and Jan Monk who answered my queries and sent me information of interest.

I am particularly grateful to those who shared their own experiences of the geographical world and memories of others: Jay Appleton, the late William Balchin, Brain Blouet Rosemary Bromley, Sue Buckingham-Hatfield, Nicola and Sandy Crosbie, Jack Davies, Dick Grove, David Herbert, Linda McDowell, Doreen Massey, David McEvoy, Bill Mead, Geoffrey North, Bruce Proudfoot, Derek Spooner, Mike Tanner, Roy Ward and Michael Wise. I owe a special debt of gratitude to those who allowed me to conduct autobiographical interviews with them, or sent me their auto-biographical accounts: Anne Buttimer, Alice Coleman, Elizabeth Clutton, Nicola Crosbie, Gwyneth Davies, Catherine Delano-Smith, Gladys Hickman, Barbara Kennedy, Cuchlaine King, Sheila Jones, Kay MacIver, Janet Momsen, Christine McCulloch, Margaret Storrie, Jackie Tivers, and Margaret Wilkes.

My ultimate thanks go to Bill, Samuel and Breesha, who have lived with this project almost as much as I have, not least in my 'absence' when I have been away at archives, interrupted family holidays for archive visits or interviews, been preoccupied with the book and/or closeted in my study during evenings, Saturdays and swathes of school holidays. For their patience, understanding and continuous love and support I thank them, and dedicate this book to them.

Picture Acknowledgements

I am grateful to the following for granting permission to reproduce pictures: the RSGS for pictures of Marion Newbigin and Isobel Wylie Hutchison; the *SGM* for Figure 7.2; the RGS-IBG for pictures of Eva Taylor, Isabella Bird, Freya Stark, Gertrude Bell, Kate Marsden; *GJ* for Figures 4.5 and 6.4; *TIBG* for the pictures of Marjorie Sweeting and Monica Cole; the National Library of Scotland for a picture of Isobel Wylie Hutchison; the National Library of Wales for the picture of the Le Play Society field group in Guernsey; the School of Geography and Environment University of Oxford for Figure 5.1; the Principal and Fellows of Somerville College, Oxford for the portrait of Mary Somerville; the University of Sheffield Geography Department for the picture of Alice Garnett; Royal Holloway University of London for the photographs of Blanche Hosgood and Monica Cole; the London School of Economics for the photograph of Hilda Ormsby; Manchester High School for Girls for Figures 5.2 and 5.3; the Libertarian Alliance for the photograph of Alice Coleman; Cuchlaine King, Anne Buttimer, Dick Grove and Richard Hickman for photographs from their own collections.

Chapter One

Putting Women in their Place: Women in the Historiography of Geography

The kind of knowledge that emerges from a discipline depends very much on who produces that knowledge, what methods are used to procure knowledge, and what purposes knowledge is acquired for

Monk & Hanson 1982: 12

Introduction

This book is intended to offer a new perspective on the history of British geography by focusing on the geographical work of women from 1850 to 1970. In broad terms, historical studies allow us to trace the development of geographical ideas and can shed light on the nature and practice of geography today. As Holloway has argued, 'The study of history is important if we are to understand why society is organized the way it is and *how we can use our understanding of the past to become agents of change in the present*' (2005: 2, my emphasis). Understanding the social construction of a discipline's history also allows us to engage with that history epistemologically, to examine what is and is not accepted as 'knowledge' and how this defines membership of and practice within the academy. Women have been omitted largely from histories of geography (Domosh 1991a); and histories of geography that fail to consider what has been 'left out' – 'what has been constructed as not-geography' – tell only a partial story (Rose 1995).

Recent histories of British geography have stressed the role of enlightenment thought (Livingstone 1992; Livingstone & Withers 1999; Mayhew 2000) and the role of imperialism (Bell *et al.* 1995; Driver 2001). Others have traced shifts in theoretical and methodological schools of thought principally in the twentieth century (Cloke, Philo & Sadler 1991; Johnston & Sidaway 2004) or 'key thinkers' (Hubbard *et al.* 2004). Most of these have been consciously written in contrast to an earlier institutional approach to the discipline's history (e.g. Mill 1930; Freeman 1960; Brown 1980). All of these approaches have brought new insight to understanding the ways in which geographical knowledge has been shaped, and are to be welcomed, not least in bringing a more 'critical' approach to understanding the history of geography, typically grounded in the contextual history approach blended with theoretical underpinnings ranging from Kuhn to Foucault to Marx. Whilst these studies have addressed feminism as a post-1980s' school of thought, which has been significant in drawing attention to the underrepresentation of women and gender as an analytical concept, feminist approaches to the historiography of geography have been given little space. Feminist historiography has been articulated by Domosh (1991a,b), Rose (1993, 1995) Blunt and Rose (1994), Bell and McEwan (1996), the Women and Geography Study Group (1997), McEwan (1998a,b), Monk (2004, 2007), Maddrell (1997, 2004a, 2006, 2007, 2008) and others,[1] but there has been no sustained work to explicate the issues raised by these shorter engagements. As has been argued recently of political geography, 'The marginalisation – and even exclusion – of gender and of feminist perspectives has yielded a field that is partial in the understandings and knowledges produced within it' (Peake, Staeheli & Koffman 2004: 1). There remains a need for 'documenting and explaining the gendering of knowledge production in geography in general, and how this is reflected in different places' (Blumen & Bar-Gal 2006: 350).

This introduction will include five elements: (i) it will address the current place of women within the historiography of geography; (ii) it will discuss a 'more-than-contextual' approach to blending contextual and feminist approaches to history; (iii) the relationship between theoretical framing and methodologies will be explained, e.g. processes of selection, biographical approaches, oral history and reading texts such as obituaries and reviews; (iv) key contextual factors 1850–1970 will be indicated; and (v) central themes which emerge in women's geographical work will be outlined.

Women *qua* Women

Is it desirable or possible to discuss 'women' as a group, as women per se? Whilst the prevailing feminist discourse of the 1970s represented an image of a universal sisterhood which needed only to recognise itself and unite in order to counter discrimination, by the mid-1990s feminist theory and practice increasingly recognised the diversity amongst and between women. This was partly as a product of feminism being caught in a tension between its modernist roots and critiques of modernism, and partly resulting from the awareness of the differences or 'horizontal hostilities' (Pratt & Hanson

1994) between women – largely resulting from postmodern and postcolonial feminist critique which highlighted differences between women according to socioeconomic class, race, sexuality etc. (see Liu 1991; Mills 1991; Nicholson & Fraser 1990).The salience of gender as an analytical category and basis for common interests has been fiercely debated within and beyond geography undermining earlier confidence in *the* feminist project and necessitating the recognition of a number of *feminisms*, which in turn stress diversity and difference (see McDowell 1993a; Women and Geography Study Group 1997). However, the celebration of difference can obscure relations of power (Bondi 1990) including the hierarchy of white male privilege that has informed the creation of western intellectual tradition (Bordo 1990). These theoretical and political negotiations have led feminists to raise a number of questions, such as how to combine postmodern critiques of meta-narratives with the social-critical power of feminism/s? (Fraser & Nicholson 1990); how to refuse separation, but insist on non-identity? (McDowell 1993b). It is argued here that it is possible within a feminist historiography to blend strategic gendered subjectivity in methodology: i.e. to focus on women, within an analytical framework that acknowledges difference in its complexity. The different women geographers studied in this volume occupy different positions in time and space, in social class, education and politics. They have complex locations in relation to one another and to the institutions and discourses of geographical thought and practice, and this is what will be 'mapped out' in the following chapters.The complexity of the positionality and subjectivity of women travellers such as Mary Kingsley has been well documented by authors such as Mills (1991), Blunt (1994) McEwan (1998a) and Kearns (1998) (see Chapter Four). The same is true of women producing geographical work within the geographical establishment, for example Marion Newbigin (see ChapterThree) was both at the heart of a geographical institution and a producer of geographical knowledge, whilst simultaneously relatively marginalised from the growing university sector and the geographical establishment of the Royal Geographical Society (Maddrell 1997).

Rather than ascribing to an essentialised notion of gender, what is needed is a theoretical and methodological approach which recognises the *discursive* construction of 'feminine' in relation to 'masculine' and the common gendered social processes and strictures experienced by women in particular times and places through the cipher of 'feminine'. It is important to recognise that women in different places at different times know and experience the world, including their gender, in different ways. This last point is crucial to an account of women geographers which ranges from the mid-nineteenth to mid-twentieth century. For example it would be easy to argue that the current geographical discourse, including the 'cultural turn' and social, political and economic geographies of a wide variety of spheres including work, home, leisure and identity renders a focus on the gendered construction of knowledge unnecessary. However, to take such a position would, at least, neglect the contextual experience of geographers working in the nineteenth and early-to-mid twentieth century, for whom gender was one of the most significant categories in terms of their access to education and employment, as well as being contextual to the production and reception of their work. It is important to remember the particular gendered social mores which combined with those of class and race in the nineteenth and the first half of the twentieth-century, resulting in institutional and symbolic discrimination (see the section on contextual approaches below). Historians struggling with the representation of *women* in different histories have accepted the 'ontological experience of women as shaky' (Alberti 2002: 104) but recognise the need to continue the project of 'writing women into history'. Within the historiography of geography there is a methodological need to focus on women as subjects at this point, but a theoretical need to recognise their different experiences. Generalisations may be drawn from their individual experiences, but not universalisations.

Women's Place and Placing Women in the Historiography of Geography

> *Part of a good conceptual history is the recovery of forgotten ideas and personalities*
>
> Godlewska 1999: 9

A survey of the literature addressing women's absence from the history of geography and historical women 'geographers' will give a foundation to answering the question of why such a history as this is needed, a theme which will be returned to in the conclusion of this book. Domosh's (1991a) ground-breaking paper made several key points. She argued that attempts to contextualise the history of geography had ignored the gendered construction of much of that history; that something of women's contribution to the formation of geographical knowledge can be seen in the work of Victorian women travellers; that all women and some men were excluded from the class of 'geographer' because their views and activities did not conform to standards of what was acceptable as 'scientific' geography. Domosh further argued that gender relations and representations are integral to the social construction of knowledge and demonstrated how the social fabric influences the history of geography through the practices, discourse and the legitimation of knowledge within the subject. Looking particularly at Victorian women travellers, Domosh asked whether the social conditions in which women operated and conducted their exploration

constituted a 'woman's way of knowing', which has undertones of essential-
ism, but the most significant question Domosh posed was to ask what implications the recovery of women's geographical knowledge would have for the
reconstruction of geography?

Stoddart, in response to Domosh's case for a feminist historiography of
geography, argued that whilst there were women geographers who merited
the attention of historians of the subject, a feminist perspective was divisive
and unnecessary because these women had 'looked after themselves, their
careers and their scholarship perfectly well without such [feminist] assistance' (1991: 485). The fact that the women geographers Stoddart himself
identifies as meriting attention (Semple, Somerville, Newbigin, Ormsby
and Taylor), with the exception of the first two, are all missing or reduced
to fleeting references in existing histories of geography, seems to suggest
otherwise. Furthermore, whilst Domosh argued that Victorian women
travellers such as Isabella Bird contributed to geographical knowledge,
Stoddart argued that they could not be seen as geographers because they
took no measurements, failing to see that defining geography epistemologically as a science of measurement (and the exclusion of those who did
not take measurements) was precisely what Domosh was challenging. As
will be seen in Chapter Two, Isabella Bird and numerous other women
took courses in surveying when these became open to them at the Royal
Geographical Society and subsequently took field measurements in the
course of their travels, but this was far from the sum of their geographical
observations. Equally there were women (and men) who gathered information, recorded their observations and experiences of places, but took no
measurements as such. The notion of a singular 'woman's way of knowing'
would be questioned today in the light of developments in feminist and
gender theory in the 1990s, but the question of how women's gendered
status and socialisation influenced their access to geographical knowledge
and institutions, and in turn influenced how they saw the world geographically is a theme which will recur throughout succeeding chapters.
Ultimately, Domosh's argument that the inclusion of women's work in the
history of geography could make the subject more *inclusionary*, more
'human', lies at the heart of this book.

In *Feminism and Geography*, Rose (1993) argued that women were historically marginalised as producers and subjects of geographical knowledge
and that subsequent histories focusing on 'great men' – 'geography's paternal lines of descent' – produces a disciplinary territory from which the feminine is excluded. Rose noted that the erasure of 'outsiders' in a given
disciplinary tradition 'also works to erase the practice of exclusion itself.
Their complete invisibility makes the practice of their exclusion vanish'
(1995: 414), exclusion being achieved through accepted power relations
which legitimise the work of some and de-legitimise the work of others.

Rose concluded, 'it seems that, even if we can no longer be certain exactly what geography was in the past, in virtually all histories of geographical knowledges one apparently incontrovertible fact remains: geography, whatever it was, was almost always done by men' (Rose 1995: 414). This view was echoed by the Women and Geography Study Group's (1997) analysis of the visual representation at the Royal Geographical Society through portraits of past presidents and key Fellows; although Kobayashi (1995: 194), while critiquing Livingstone's (1992) lack of attention to women's geographical work, acknowledged that 'it is difficult to find works by women in a discipline that has been so male dominated'.

However, there can be a difference between how a discipline, its history, and institutions are *represented* in retrospect and the empirical detail of a given time, place and body of work. Rose argued for a multiple space for the history of geographical thought: 'We need an analytic space which can articulate boundaries, distinctions and disjuncture instead of erasing them, a space which can acknowledge exclusion as intrinsic to the processes of inclusion, a space through which the difference that gender makes to the production of geographical knowledges can be recognised' (1995: 416). This book may not fully address these criteria, but in excavating and explicating women's geographical work it represents a liminal space in the history of geographical thought, a bridge to more gender-inclusive historiographies of geography.

Where Have All the Women Gone?

A cursory glance at the histories of geographical institutions shows that the *Scottish Geographical Magazine* was edited by women from 1902 to 1939; women were members of the Royal Scottish and Manchester geographical societies from their foundation in 1884 and the Royal Geographical Society belatedly from 1913;[2] and approximately 15% of the original members of the Institute of British Geographers founded by university lecturers in 1933 were women (Maddrell 2004a). Women incontrovertibly were 'doers of geography' before 1970, which leaves one to ask two key questions: *what* were they doing? And *why* does so little of it feature in our disciplinary histories?

The women who have frequented the pages of existing histories are Mary Somerville (Mill 1930; Freeman 1960, 1980; Gregory 1988; Livingstone 1992); Marion Newbigin (Freeman 1960, 1976b; Dickinson 1969, 1976; Livingstone 1992) and Eva Taylor (Taylor 1957; Dickinson 1976; Freeman 1976b; Livingstone 1992; Heffernan 2003); the American geographer Ellen Churchill Semple also has a similar pattern of appearances (Maddrell 2004a). This shows the importance of this small group of women but equally demonstrates the reiterative nature of disciplinary histories, whereby key characters are re-inscribed in successive accounts. It also raises the issue of whether an author's work is mentioned or engaged with within histories; for example, of the three British women listed above, Livingstone (1992) discusses only Somerville's work in any detail. Whether by 'omission or comission' (Domosh 1991a), women have been shown to be absent from the histories of geography. Although there have been a number of studies characterised as having a 'critical' approach to the history of geographical knowledge, which is 'sensitive to the ways in which geographical knowledge has been implicated in relationships of power' (Driver 1992: 23) and although studies of individual women or events have provided windows on women's 'place' in and contribution to geography as a discipline (e.g. Blunt 1994; Bell & McEwan 1996; Maddrell 1997; Guelke & Morin 2001; Matless & Cameron 2006), there is little sense of an overview of the work of women geographers in Britain. Critical feminist approaches to histories of geography may be well rehearsed (McEwan 1998a), but a great deal more work is needed to substantiate women's part within the subject, and an understanding of the different work of women travellers, academics, educationalists and authors is absent from our understanding of the development of geography. Jan Monk (1998, 2003, 2004, 2007) has shown there are rich archives to be mined in the case of American women geographers and that these illustrate the varied experiences of different groups of women within specific geographical institutions; the same is true for British women geographers as a whole, even if the quality and quantity of sources for individuals might vary (see Bell & McEwan 1996; Maddrell 1997, 2004a–h).

Historians of women have been engaged with the subject of women 'entering male professional terrain' since the 1970s (Morantz-Sanchez 1995: 201), chronicling women's exclusion and inclusion and their negotiation of patriarchal institutions and masculinist cultures (Witz 1992; Wallach Scott 1992; Woollacott 1998) – but there is no such chronicle in the history of geography. This is not an argument for hagiographic recovery of nineteenth- and twentieth-century geographical 'heroines', but rather for an awareness of a largely invisible group and the nuances their work brings to our understanding of 'geography', the workings of the geographical community, and its discourses. The stories here are of some of the women working within geography in Britain 1850–1970, and whilst a single volume could not include all female producers and communicators of geographical knowledge, it is hoped that this beginning will encourage further engagement with women's place in the history and epistemology of geography.

Historiographies are always theoretically fraught (what Livingstone (1992) described in his own case as 'situated messiness'), they are complex and each 'reflects the partialities of its author' (Heffernan 2003: 4); considering women as subjects and objects in geography is no exception. Given their relative invisibility in received histories of the subject, there is a strategic need to assert gendered subjectivity in order to focus on women as a group in an attempt to address their absence from the historiography of geography. However, in doing so one must be conscious of the tension between this approach and recognising the socially constructed nature of gender and subjectivity. My approach adopted here combines feminist and contextual readings of the historiography of geography, in order to consider the 'place' or complex location/s of the women working in geography in the nineteenth and early twentieth century. It is important not simply to look at what work women undertook at different times in British history but also to look at 'the ways in which dominant ideas concerning femininity, women's roles, gender, class and, to some extent, race, have dictated the types of work deemed suitable for women; the value placed on women's work, the status of women workers and the strategies that women have employed to challenge these dominant ideas' (Holloway 2005: 3). In many ways these are the evolving social relations experienced by 'women geographers' (whatever their form of work), which I am trying to trace here.

Complex Locations and Embodied Genealogy: More-than-Contextual History

The contextual view has normalised within historical geographical writing as a result of the work of scholars such as Stoddart (1981), Driver (1988, 1992, 2001), Livingstone (1979, 1990, 1991, 1992) and Bell (1993, 1995b). By placing the changing practices of geographical thought in relation to ideas and practices external to the discipline, contextual approaches to history identified the relation between texts and context, i.e. the social construction of knowledge and sought to avoid the assumptions of Whiggish accounts of the development of the subject. Recognising the inevitable influences of present interests and selectivity of historical sources, Livingstone (1992) called for geographers to give heed to social context, metaphysical assumptions, professional aspirations and ideological allegiances, and goes on to suggest asking: what role did geography play in past society? Was geography used by particular groups for political, religious or economic purposes? Who benefited and who lost out by the introduction of new theory? Why were particular theories generated, welcomed, outlawed? Godlewska (1999: 9) similarly argues that conceptual history rests on asking basic questions about individuals' lives and work, such as 'What were their key ideas? Which were the decisive influences shaping not only their ideas but the method and presentational form of their work? ... What was, and what has been, the impact of their ideas and approaches to problems?'. She also points out that answering these questions requires extensive research and that this has been done for too few past geographers. It is hoped that this volume will contribute to answering these and related questions for the women geographers studied here.

However important, the contextual approach has limitations which are recognised by proponents (e.g. Driver 1992; Livingstone 1990, 1992) and it should not be seen as a theoretical and methodological panacea (ibid.). Contextual historians can fall into the trap of presentism, e.g. judging preceding scholars for not using the contextual approach; they can be guilty of overdetermining context, or be so wary of these traps that they fail to mention any reference to continuity or development of ideas. Using a contextual approach also raises the question of 'which context?'(Skinner 1969): it is crucial that the appropriate context is chosen for a given study. Neglect of issues relating to the social construction of gender within the historiography of geography is a function of selectivity of context, which has resulted in an intentionally or unintentionally masculinist account of the subject's development: both Stoddart (1991) and Livingstone (1995) acknowledge that women's work did not fit with their interests/framing of their particular histories of the discipline. Driver (1992: 36) has suggested that 'It might be argued that the ultimate fiction of "contextual" history consists less in its separation of "texts" and "contexts", than in its continual silence on the mediating role of the historian'.

Contextual material is vital when considering the place of women in the historiography of geography, but it is theoretically insufficient for interrogating the complexities of their gendered place/s. Rather what is needed is a context-sensitive feminist approach to interrogate the individual lives and work of women geographers. The numerous women considered in this book had different backgrounds, positions, strategies and achievements, but share in common that we know too little of them and their work. It is necessary to focus on these women as women, in order to constitute an inclusionary historiography, but also to combine feminist with materialist, postmodern and postcolonial forms of analysis in order to begin to understand the complexity of their lives and the character of the work they produced. This inevitably means combining theoretical approaches in a pragmatic discourse, with tailored methods and multiple categories, i.e. starting with them as women but going on to recognise in their differences the specificity of the 'politics of [their] location' (Rich, cited by Blunt & Rose 1994: 7) and multiple, fragmentary locations (Mohanty 1987), which result in 'less essentialist and more critical readings of the geography they produced' (McEwan 1998a).

Feminist History

A brief survey of feminist historical studies of other disciplines gives some sense of what feminist approaches to historiography can offer. A focus on women in history emerged from the 1970s as part of a more socially inclusionary approach to history, but had important (if neglected) precursors in the work of Alice Clark's (1919) study of seventeenth-century working women and Mary Beard's (1946) account of women in the discipline of history (Smith 1986, Alberti 2002). The 1970s' approaches to women's history tended to fall under the headings of 'women worthies', 'women's contribution' and 'victimology studies' (Harding 1986). Whilst inroads could be made into a predominantly masculine representation of history, these approaches all accepted given categories of subject matter and epistemology: women worthies tended to be thin on the ground, privileged and/or atypical, recording women's contribution requires accepting categories defined by male values and practices, and victimology studies can obscure the agency of women (ibid.). Evidence suggests that attempting to write women into existing categories is naive (Wallach Scott 1988), e.g. the periodisation of history organised around men's (public) activities makes many women's activities invisible (Harding 1986) and women's absence from history needs to be seen in relation to disciplinary power structures (Alberti 2002). The underrepresentation of women was a particular 'scholarly concern' for the editors of the 2004 *Oxford Dictionary of National Biography* *(ODNB)* given the 1882–1900 volumes' focus on 'the male, the metropolitan and the celebrated' (Baigent 2004): only 4.3% of entries in the *Concise ODNB* and 12% in the 1980s *Supplement* addressed the biographies of women. Even then too many of those few entries for women in the nineteenth century for example, related to women who were considered eccentric or notorious while the likes of important educationalists were underrepresented (Mitchell 1995). Aside from the issue of often scant sources, the key issue was conceptual: 'It is crucial to ensure that the criteria for selection are not being set in such a way to limit the recognition of women' (Garnett 1995). It is only by theorising gender as a category of analysis that new perspectives will be gained on old questions, new questions will arise and women will become visible as active participants in history (Wallach Scott 1996). Part of this reconceptualising in the *ODNB* related to acknowledging the 'complexity of the relation between public and private lives, and of establishing a proper balance between multiple roles of a subject', e.g. women silversmiths whose private role in their family company became public only after the death of their spouse (Garnett 1995).

Numerous disciplines have begun to address the question of gendering of their histories and unearthing/resurrecting formerly prominent but since obscured women is characteristic of a feminist reworking of any branch of history (Darling & Whitworth 2007). Part of this has been a critique of the ways in which western epistemologies or accepted 'ways of knowing' are gendered. Francis Bacon's seventeenth-century rhetorical description of scientific knowledge as the figurative domination of the female body of nature has been widely criticised on two key counts. Firstly, for its explicit oppositional dualism in which male, rationality and culture is placed in relation to female, irrationality and nature. Secondly for its implicit metaphor of domination and exploitation of nature/the female (see Fox Keller 1984; Jacobus *et al*. 1990; Rose 1993). Feminist work on the history and practice of science in the 1980s and 1990s was effective in contributing to debates about the social construction of knowledge and pushing the debates to question not only the privileging of discourses of 'objectivity' in science, but also the gendered character of those discourses (see Fox Keller 1984; Harding 1986, 1991; Haraway 1989, 1991). This led Harding and Haraway to argue for the recognition of the positionality of the researcher and that knowledge production is situated within individual, professional, disciplinary, social, economic and political contexts. It was further argued that acknowledgement of a researcher's positionality (e.g. the Feminist Standpoint) results in a more honest position than those who purport to have no positionality. Whilst Butler's (1990) notion of performative gender in *Gender Trouble* challenged the simplistic gender dualism implicit in the Feminist Standpoint position, the political necessity to speak as and for women persists (Alberti 2002). Fox Keller's statement about science can still be applied to geography: 'we cannot properly understand the development of modern science [geography] without attending to the role played by metaphors of gender in the formulation of the particular set of values, aims, and goals embodied in the scientific [geographical] enterprise' (1984: 43).

Recent historiographies of science have highlighted individual, symbolic and institutional mechanisms for excluding or marginalising women. The Royal Society, founded in 1662, did not admit women until 1945 when Marjory Stephenson and Kathleen Lonsdale were elected; Hertha Ayrton had been allowed to give her own lecture in 1904 but was rejected for fellowship on the grounds that she was married! (Fara 2004). Fara's study of women and science in the enlightenment has shown that women were active in scientific work in the eighteenth century, but have been excluded from traditional historical records. Women had access to science while it was conducted primarily in the domestic arena, but were barred from metropolitan institutional bodies because of gender. Women typically collaborated with male family members; those roles were often supportive and sometimes an extension of domestic roles, often time-demanding processes such as logging, calculating and filtering (Fara 2004). Fara's detailed work has both expanded the arena of scientific knowledge production to include the domestic, and undermined the trope of single-minded heroic endeavour in science.

Within the history of art, counterbalancing the emphasis on fine art and certain genres within that heading go some way towards redressing gender imbalances (Garnett 1995). Similarly within historical studies of business, medicine and science, wider arenas of action have been recognised beyond male-dominated professional and institutional organisations (ibid.). Other studies have gone on to suggest that existing ways of thinking about historical significance need rethinking. In *Women Medievalists and the Academy*, Chance (2005) has articulated ways in which the academy resisted female excellence, but also shows how numerous women medievalists in the nineteenth and early to mid-twentieth centuries used creative alliances and strategies in order to maintain their work. Some were supported by extraordinary male mentors; for others their work was subordinate to their husband's career and/or family life, but they participated in joint work and/or had a 'late flowering' of their individual work; some women worked as librarians on the fringe of research; others who were rejected by academia worked outside it, using private means for supporting their academic work through mundane day jobs; women who had homosexual relationships were usually free of the childcare responsibilities of their married counterparts, but risked double marginalisation within the academy (Chance 2005). In the case of architecture, it was found that there were relatively few 'great women architects' who would meet the 'women worthies' category and this led to reconceptualising what it meant to 'make' a building, i.e. that an architect did not work in a vacuum but was influenced by many actors including planners, social reformers, lobbyists, writers etc. (Darling & Whitworth 2007).

In the same way, the production of geographical knowledge is not limited to the academic researcher in the twentieth century any more than it was to the explorer in the nineteenth; there are a whole range of ways in which geographical knowledge was 'produced' and 'received', with 'producers' ranging from school teachers to travellers, popular writers, educational legislators, planners, conservationists, national and local geographical society members, and academics. The boundaries of who counts as a geographer is a political decision within the discipline (as Domosh (1991a,b) and Stoddart (1991) demonstrated), and although these boundaries have been stretched in relation to individuals in the history of geography, applying this approach to this study of women's geographical work will help to cast a wide net and catch female contributors who might otherwise be omitted. Those who held office in geographical societies, those who taught at universities and those who wrote influential texts will be represented here, but so too will the work of school teachers, teacher trainers, non-academic authors and those who might be called public servants, both 'major' and 'minor' figures. Perhaps what is most shocking about the omission of women in the historiography of geography is not simply that there were so many women engaged with geographical work, but that so many of them *were within the academy*, were university lecturers, society members and officers, and authors. These women need no stretching of traditional disciplinary boundaries to include them, yet they have been largely expunged from histories. Gender as an analytical concept is vital to understanding this process of marginalisation and exclusion. As Smith noted within the discipline of history: 'Studies of one or two great historians per generation often serve to make up historiography, but while we examine "objectivity", we rarely consider the shape of historiography itself and what it has meant to the profession to have its achievements exemplified in the biographies of a handful of great authors' (1996: 547).

In her study of women in science Fox Keller (1982) argued that women should gain access to what has been denied them (including a place in disciplinary histories), but at the same time legitimate areas of scientific culture previously rejected as 'feminine' and therefore 'unscientific'. Feminist work challenges masculine categories and values, as well as identifying ways in which space has been central to both masculinist power and feminist resistance (Blunt & Rose 1994). This includes the spaces of geographical institutions, educational establishments, textual space in the world of publishing and critical review, but also what Rose (1995) has described as the space or 'territory' encompassed by our disciplinary tradition as seen in histories of geography. Geographical knowledge is not merely the data and theory contained under the heading of 'geography', but also a discursive formation '… a specific way of knowing the world' that is a product of a constellation of concepts, practices and institutions (Driver & Rose 1992).

A critical study of the representation of women poets in the 1930s shows an interesting parallel with the relative absence of women geographers. Jane Dowson (1995: 296) has demonstrated that 'women were as involved in the process of producing poetry as women today seem to be, and that the poets are not obscure, but have been obscured by literary histories'. These were not closet but public and paid writers; however, whilst socially accepted, their work was not the subject of *critical engagement* and as a consequence, not written up in histories. In the chapters that follow there will be an emphasis on the *reception* of women's geographical work as well as the context of its production.

Placing Texts in Context

The production and reception of geographical and other texts is crucial to understanding the development of a discipline. Debates within contextual history and literary theory concerning the relationship between texts and contexts, knowledge and power; authorial intention and alternative readings; make surveys of geographical literature complex – and in many ways problematic – but nonetheless rewarding. The opposition of traditional Marxist and poststructuralist theory has resulted in the undesirable entrenched dichotomy of texts being identified as respectively either fulfilling a narrow range of determinate functions transparently dictated by the workings of capitalism or as some pure space of discourse beyond the world (Driver 1992). Similarly, the frequent characterisation of histories as either 'internal, cognitive history' or 'external, contextual history' (Glick 1984: 280), has been challenged and an acknowledgement of the creation of geographical knowledge through discourse allows an alternative to this conventional dichotomy, thereby avoiding a caricature of knowledge as 'pure' or 'corrupt', when all knowledge has been subject to external and internal influences. If knowledge is discursively constructed then an understanding of institutions and practices is vital to the historical project (Driver & Rose 1992; Wallach Scott 1988). Discourses work in social contexts with material consequences: this is both the site and content of 'new' self-critical discursive historiography of geography (Driver & Rose 1992). An attempt made here to draw on both perspectives in conjunction with feminist theory to allow for the intersection of socioeconomic, cultural and political practices with gender as well as recognising the active role of texts in producing as well as reproducing ideologies and power relations. See Table 1.1 for a selection of key dates that indicate some of the social and political events and legislation which impacted on women's rights and constructions of gender.[3]

Foucault's discussion of power and knowledge has been central to much of the debate on the significance of texts, particularly his key argument about the reciprocal nature of these two: power produces knowledge and knowledge presupposes and constitutes the relations of power in a transformative process (what Said (1978: 32) describes as the 'increasingly profitable dialectic of information and control'). Largely influenced by this Foucaultian view it is widely accepted that texts are not mere reflectors of the material world but are relations of power in themselves (e.g. Said 1978; Mills 1991; Driver 1992; Matless 1992). Understanding the context of the production and reception of geographical texts in the form of publications is central to this project. Foucault argues the question (and therefore the strategy) is to consider 'what it means for them [statements] to have appeared when and where they did ... (Foucault 1972: 109). Pearce (1991) describes this as a text's location in the cultural complexity of the specific moment in its historical production, arguing that 'It is not the contradictions within a text that reveal its ideological complexity, but rather the historical discourse by which that text is inscribed (1991: 24). However, a postmodern suspicion of transcendental themes and an awareness of knowledge as a situated social practice raises questions regarding the mechanistic interpretation of the relationship between text and context (Driver 1992; Skinner 1969): cultural concepts lack transparent and shared meanings (Wallach Scott 1988). The recognition of the instability and contestation of meanings facilitates and demands an examination of the politics behind the

Table 1.1   Key dates 1850–1971

Date	Key event, legislation etc.
1832	First Parliamentary Reform Bill
1848–9	Queens College and Bedford College founded for women
1853–6	Crimean War
1857	Matrimonial Clauses Act
1864–9	Contagious Diseases Acts
1867	Second Parliamentary Reform Bill
1869	Endowed Schools Act extended benefits to independent girls' schools
1870	Elementary Education Act
1870 (England and Wales)	Married Women's Property Act
1881 (Scotland)	Married Women's Property Act
1874	Women's Trade Union League (WTUL)
1876	Enabling Act allowed universities to award degrees to women
1878	Women permitted to take degrees at the University of London
1884	1884 Third Parliamentary Reform Bill
1886	Infants Act
1892	Four Scottish universities admitted women to degrees
1893	University of Wales admitted women to degrees
1897	National Union of Women's Suffrage Society (NUWSS)
1900–2	First and Second South African Wars
1903	Women's Social and Political Union (WSPU)
1909	League for Opposing Women's Suffrage
1914–18	First World War
1918	Representation of the People Bill entitled women over 30 years of age and home owners to vote
1919	Sex Disqualification (Removal) Act
1926	General Strike
1928	Universal Suffrage
1939–45	Second World War
1944	Education Act
1958	Removal of marriage bar in civil service and teaching
1969	First polytechnics designated
1970	Equal Opportunities Act
1971	Open University courses start

Source: After Blunt & Willis, 2

conflictual processes that establish those meanings, including the politics of gender, all of which is pertinent to the analysis of the reception of women's geographical work. The use of biography is part of a wider contextual approach to situating historical subjects and is particularly important here as it helps to situate individual careers and publications and other geographical work in relation to the specificity of an individual's education, upbringing, politics or life history. Biography will be discussed in detail under the heading of methodology below.

Methodology: A Note on Selection, Sources, Representation and Ethics

Following an explanation of the biographical approach, four categories of sources which merit particular note as both useful and problematic are discussed: the use of archives, oral histories, obituaries and reviews of publications.

Using a contextual (auto)biographical approach

Autobiography, whether in textual or oral form, represents a conscious form of *self*-representation within that frame. Anne Buttimer pioneered the use of autobiography in studying the intersection of individual lives and geographical thought and practice through the Dialogue Project conducted with Torsten Hägerstrand in Lund in the 1970s. Since then it has been used by the American Association of Geographers filmed interview series, Blunt's (2005) exploration of the 'hidden histories' of Anglo-Indians, and much other qualitative work. Biographical studies within geography and geographical thought have also been developed by a number of geographers including Daniels and Nash's (2004) discussion of the relationship between life histories and life geographies and Thomas' (2004) study of Lady Curzon, where she stresses the synergy achieved between understanding a biographical subject in relation to friendship and family networks *and* wider social, economic and political networks or contexts. Barnes (2001) used a related contextual biographical approach to interrogate individuals' roles in shifts in disciplinary practice around the quantitative revolution in relation to their specific life trajectories as well as wider intellectual and socioeconomic and political contexts. Among other things, the autobiographical approach helps the researcher to identify and understand the 'diversity of keys' at play within the discipline, and how the use of stories facilitates understanding of other 'worlds' as well as critically reflecting on one's own 'story' (Buttimer 1983).

Autobiographies can complement formal archive-based histories addressing the history of geography and the social construction of its thought and practice. Furthermore, an understanding of an individual's schooling, home and other formative influences can shed a different light on her/his own position in relation to the practice of geography (Buttimer 1983). As Chance (2005: xxx) noted on autobiographical sources from women medievalists: 'rare glimpses of the woman scholar herself offer unusual insights into her own perceptions of her life and career' which cannot be found in published work. If our sources are confined to the formal output of geographers, we have a limited perspective: 'What he or she may write in books and journals yields an image of discrete knowledge products, but may yield little understanding of the intellectual processes unfolding within that person's life' (Buttimer 1983: 3). Biographies are increasingly recognised as significant in the history of ideas and biographical excerpts or sketches are being incorporated into individual studies in the history of geography, as well as other texts (e.g. Cloke, Philo & Sadler 1991; Women and Geography Study Group 1997, 2004; Hubbard *et al.* 2004), echoing Freeman's (1960) *A Hundred Years of Geography* which included an appendix of biographical sketches.

In this study, where possible, autobiographical interviews were undertaken with geographers who were working prior to 1970, but inevitably the majority of women discussed in this book have long since died. Of those known/thought to be still living, a few have proved untraceable and a few have not responded to invitations to participate. Where autobiographical interviews were not possible, biographies have been pieced together, from the oral histories of others where relevant, from an individual's papers where these exist, from public records such as birth and death certificates, obituaries, employment records, geographical and institutional archives, publications and secondary sources. These sources have also been used to give these women 'voice', principally through the reproduction of their own words, but this requires recognition of both the fragmentary and multiple character of those 'voices' found in different textual forms of self-representation (Woollacott 1998).

The (auto)biographical approach allows one to see 'women using agency, not as some abstract or undefined expression of autonomy, but in specific instances of creative resistance, self-promoting complicity and wilful discursive self-formulation' (Woollacott 1998: 338). The majority of the women studied in this volume were white and middle class, but few left significant personal archives of correspondence or diaries, leaving their subjectivity as something to be excavated and teased out through the institutional archives of geographical societies, university records, publications and obituaries. This involves taking these texts (often professional texts) and interrogating them for clues 'as to meanings for the subjects' sense of themselves, and looking for patterns of language and points of reference' (Woollacott 1998: 333). Written subjective constructions vary according to the form of writing or speaking – that is, a different slant on an individual's subjectivity can be gleaned from reports, academic papers, interviews, speeches and policy documents (Woollacott 1998). Whilst studying individuals through their self-representation facilitates the reading of the personal in the professional, inevitably it must be seen (as indeed must all biography) as partial and fragmentary (Wollacott 1998). However, it does at the same time accommodate the historically specific, the varied and even contradictory subjectivities on the part of individual women, and women collectively, as they negotiated their complex position/s within geography as travellers, academics, authors and educationalists.

With these detailed biographies, an image of what Braidotti (1994) calls the 'embodied genealogy' of each individual, begins to emerge – the specificity of the lived, female bodily experience, within masculine modes of thought, practice and values. But as Braidotti has suggested, any sense of unity is based on recognition of the complexity of individual positionality, not a universalised image of sisterhood. Where there are 'rhizomatic connections' (Deleuze & Guattari 1988: 7) between women, hidden or otherwise, their own differences and connections to practices and discourses will be contested from other subject positions (Gedalof 1996), not least in placing their work in the context of their 'invisible' markers of white western thought. All this gives a hint of new and challenging perspectives on women in geography and geography as a whole. As Baigent (2004: 545) noted, using a biographical approach 'puts geography firmly in its place. That place is not just embedded in an economic and political context but in a personal one …' These biographical studies raise difficult theoretical questions about the politics of the women's location and where we place them in more inclusionary histories of the discipline (see Rose 1995). The presence of these women problematises representation of geography as a masculinist endeavour. If they were accepted/incorporated into the discipline, do we deny their agency or accept the constraints on their agency? It also complicates our perception of some of the so-called 'founding fathers' of modern geography, men like Mackinder, Herbertson, Roxby and Fleure who appointed the first-generation women to university posts (see Chapters Four and Five). These questions will be returned to in the conclusion.

Inevitably the material evidence of women's geographical work has varied in quantity and quality and this will be apparent in the range of sources used when discussing and analysing a particular individual's work. I have tried to reach a balance between covering a range of individual women and employing as wide and detailed sources as are available. The result is that there is a huge discrepancy in terms of sources and material on different individuals, but I have retained the commitment to a contextual biographical approach because I believe it to be the most effective in giving a sense of an individual's work in personal and professional context. However, no matter how detailed the sources, it is not possible to reconstruct an 'authentic' inner experience of the lives of women from the past (Alberti 2002), there always has to be at least a note of qualification and speculation. I have also persisted in including 'minor' figures to keep the study representative of the breadth of geographical work and practice, reflecting a view that there is as much to be learned from so-called 'minor' as 'major' figures in the discipline (Livingstone 1992; Guelke & Morin 2001; Lorimer 2003; Maddrell 2004a, 2006). I cannot pretend that there has been equal information available on the subjects discussed here, but hope that a balance is achieved between those with large archive sources and those who have given detailed autobiographical accounts, and between 'major' and 'minor' figures, indeed that this crude dichotomy is at least blurred, if not eradicated. It is also my hope that the reader will appreciate the nuance which is brought to the bigger picture by the numerous shorter biographies as much as the nuances found within more detailed biographies.

Archives

... history is not merely a project of fact retrieval ... but also a set of complex processes of selection, interpretation, and even creative invention – processes set in motion by, among other things, one's personal encounter with the archive, the history of the archive itself, and the pressure of the contemporary moment on one's reading of what is to be found there

Burton 2005: 8

Archives represent the field for historical research (Burton 2005) and this study is no different. Widespread archive sources have been used including public records (birth, marriage, death and probate records); personal archives (e.g. the Taylor and Campbell collections at the British Library, the Smee papers at the Northamptonshire Record Office, and the Sylvester and Davies papers at the National Library of Wales); numerous departmental minutes and employment records (especially from universities); and geographical society records (from the Royal Geographical Society-Institute of British Geographers, the Royal Scottish Geographical Society, the Geographical Association and the Liverpool, Manchester and Tyneside geographical societies). Archives can offer wonderful insight to the processes of an appointment or publication, an individual or institution's character or reputation, but they have to be used in light of their limitations. These limitations have been well articulated within and beyond the discipline of geography, but I would like to highlight a few key points here. Drawing on Derrida's *Archive Fever*, Withers (2002) demonstrates how the etymology and associated status of archives varies across languages/countries, indicative of the ways in which the contents of archives, and the interpretation of those contents, are always *mediated* by their contexts. This does not only apply to the archives of the establishment, as Mayall's (2005) account of a British suffragette movement demonstrates, the archives of counter-hegemonic groups can be equally susceptible to the imperative of a particular historical narrative.

In recent years canonical notions of what archives are and how they should be accessed and used have been challenged by new forms of archives and overtly political use of their contents (Burton 2005). Furthermore, the identification of formal archive material as 'fiction', and fiction as archive has challenged the ontological status of the archive in history (ibid.). While feminist scholars have played a part in questioning the status of the archive as unmediated 'truth teller', they have also contributed to extending the boundaries of what constitutes as 'archive', e.g. Burton (2003) on the domestic house as archive, and materials previously considered peripheral have been prioritised in a 're-ordering the archive' (Thomas 2004) in order to access marginalised material on female subjects.

As much as archives can reveal they can also obscure: they can represent past (and present) power relations and can reiterate exclusion from that power, they are often fragmented and need to be placed in the context of their production (see Barnett 1998; Withers 2002; Pohlandt-McCormick 2005 for example), not least in their most recent incarnation as online resources (Burton 2005). Many archive records have been found for the women discussed here, but the uneven availability of sources has represented a challenge in both the over- and underrepresentation of individuals. Accessing records for geographers who have died since 1970 has also been unpredictable and dealt with in different ways by different institutions ranging from refusal to answering selected questions, to full access. Wherever archive materials have been found, I have taken this as an *opportunity* to engage with that individual's personal/geographical biography. Just as imperial archives have been used to support recent indigenous land claims (Perry 2005) and other emancipatory endeavours, primary archive sources are recognised and relevant materials in the contemporary as well as historical feminist project. In the case of this particular project, archives have contributed much to the substantiation of individual stories and women's collective status as long-standing, varied and productive tillers of geographical territory rather than as recently arrived stakeholders.

Oral histories

Oral histories are part of the biographical approach adopted here and are used to give autobiographical 'voice' and biographical commentary from those who studied or worked with the women geographers discussed.

As Withers (2004: 317) has argued, although far from unproblematic, 'Attention to memory and to the different forms taken by memory's representation can ... offer insight into questions of geography's reception: what geographical knowledge meant, and for whom it meant, in historical context'. Oral histories, as part of that memory process and representation, can demonstrate the limitations of preceding historiographies (Roque Ramirez 2005) and have been a popular methodology in feminist work across disciplines because they allow the subject to speak in their own words, they can reveal 'hidden' aspects to history not 'visible' in textual forms, especially the personal memories of everyday experience (Blunt 2005). Despite this, there is a persistent view that oral histories are undermined by their subjectivity, but this implies: (i) that written sources are not subjective and (ii) that 'subjective' sources are invalid (Burton 2005). Portelli has argued that it is impossible to separate occurrences from how they are remembered: 'subjectivity is as much the business of history as the more visible 'facts' – what the informant believes is indeed a historical fact (that is, the fact that he or she believes it) just as much as what 'really happened' (Portelli 1981: 100, cited by Kirk 2003: 130). I have used oral history accounts to bring an alternative perspective to that gleaned from written sources (see Buttimer 1983). These accounts have to be recognised as based on a particular positionality, but when triangulated with other sources (which must also be recognised as 'positional'), they provide invaluable insight and detail to an individual's biography, personality and experience of the arena of geography and geographical knowledge. The use of multiple 'lenses' allows a rounded, 'three-dimensional' perspective (Kanner et al. 1997: lvi). I undertook 12 oral history interviews, most of which were taped and transcribed; they were shaped with a series of biographical and career questions, including areas previously identified as significant (e.g. teaching and research balance, pastoral responsibilities, mentoring and fieldwork) and a final section explicitly addressed gendered experience. These interviews were forms of professional self-representation and self-storying, which were novel to most of those interviewed, and several expressed concern that they were not telling me what I wanted, presupposing a particular agenda on my part. Despite reassurances that I was interested in *their* story, nonetheless, knowledge of the wider story I was bringing together and the questions I used to frame interviews inevitably influenced the telling of interviewees' stories (Sidaway (1997) echoes this experience). These formal interviews were supplemented by many informal oral history accounts and personal communications from colleagues or students of the women geographers discussed here.

Using oral histories raises ethical as well as methodological questions. Occasionally I was told things 'off the record' which I respected, although rather like a court of law, comments may be struck from the official record, it is harder to obliterate them from one's mind and remove all influence from the verdict. Very occasionally I felt an interviewee may have told me personal details which they may have been happy to tell me, but not necessarily intended for public consumption, and I have either double-checked with them or made an editorial choice to anonymise these responses in general rather than individual accounts. I also decided that while I would cite comments from others on women who were deceased, I would not cite those on living women, relying on published and public sources alongside their own interview material. Conscious that by writing about them I was (re)placing these women in the public arena in an unforeseen and/or unpredictable way, both of these decisions were made in order to respect the privacy of those who had contributed so much to this study. These ethical dilemmas are everyday occurrences for qualitative researchers, but are less familiar in the setting of historiography, suggesting this is the point at which historiography blurs with the ethnography of geographical practice.

Obituaries and reviews

Obituaries are a form of textual memorial, sometimes referred to as 'the first draft of history'. They are a very compact form of biography, which within the geographical community are usually written by someone who knew the deceased well (often a colleague and/or friend), although earlier shorter notices were often anonymous and attribution policy varies with the publication. *The Times* has always had a policy of anonymous authorship for obituaries and Brunskill (2005) argues the main advantage of this is that the obituary will be written and read about the deceased's life rather than their relation to the author. However, this is frustrating to a contextual discursive approach when the relationship between author and subject is central to interpreting the text.

Obituaries usually conform to the dictum 'don't speak ill of the dead' and are generally celebratory accounts of someone's life, which can be explicitly or implicitly hagiographic. However, there is an art to reading between the lines in obituaries and there are particular words and phrases which are heavy with subtext, such as 'a determined character', 'not always easy', 'could be difficult', which can be used to signal a more complex subject position of the deceased and this will be returned to in the conclusion. Length of obituaries tends to reflect a combination of the fame of the person both during their career and at their time of death (and the difference between these two can be significant for women who may have a long retirement), the complexity of their professional lives and detailed knowledge of this, and – of growing significance in academic journals – the often limited space available within publications for obituaries. Obituaries for the same person may vary in style according to the publication, for example one written for *The Times* or *Transactions* may be quite formal, stressing the chronological facts of life and career, others in local newspapers or departmental newsletters may stress the more personal characteristics. However, there can be a reiterative element in multiple obituaries, sometimes because they have been written by the same author, or one obituarist has sourced details from another and not quite escaped from the original's structure, omissions, turn of phrase or evaluation. Although I have been aware of these issues, it does not guarantee that I have escaped these reiterative traps myself. Whilst multiple obituaries have been used wherever possible, in many cases they have had limited comparative value.

Reviews

Reviews of publications are vital references for understanding the critical reception of books and occasionally papers, although the latter more commonly elicit responses in short commentaries or more typically prior to 1950, letters to the journal of publication. Reviews represent a critical response to a given text at a particular time and place. However, recent work has shown how reviews can vary enormously according to context even within relatively bounded pockets of space and time (Livingstone 2005; Keighren 2006). This should not be surprising given the nature of reviews, but it underscores the importance of being aware of the ways in which reviews can take on a sort of 'textually fixed authority' (Blunt 1994) and the dangers of attributing too much to a given review. Readers' responses to literature are 'constrained by their ability to perceive, read and interpret as discursively constructed subjects' (Blunt 1994: 117) and this includes reviewers. Reviews are often relatively unmediated (they are not peer reviewed and often only receive minimal editorial intervention) and although the reviewer has to justify his/her evaluation of a given text, s/he is not free from existing predilections and interests. Indeed the review process can represent the veiled politics of disciplinary and institutional loyalties and rivalries, which can range from wishing to support or diminish the career or reputation of the author of the text, as well as reflecting the disciplinary status of the reviewers themselves. Reviews need to be read critically to evaluate whether they constitute a 'justified' position, compared with other reviews and a sense of sales or reprints of the text. Sales figures are often unattainable as publishing houses close or merge and records are lost or simply not kept in the long term, but numbers and dates of reprints and new editions give some sense of the ongoing demand for a particular book and citations also give some sense of the extent and ways in which others engaged with its content. Berg (2001: 511) has argued that the 'objective' system of 'blind' peer reviewing papers for publication privileges a masculinist view of objectivity as 'disembodied, impartial, and unlocated'. Where sources are available, the evidence of the review process is discussed in the light of this claim.

Published reviews are the best source of critical response, but inevitably given the wide scope of this book it has not been possible to excavate the responses of every review for every publication and map the influence of each text through the discipline, as has been admirably demonstrated recently in the case of Semple's *Environmental Influence* (Keighren 2006). Even allowing for the ways in which the reception of a text can vary within one city (Livingstone 2005), reviews, although partial, provide a sense of a text's perceived strengths and weaknesses and are valuable as such.

It should also be noted that I am not taking 'geographical work' to refer only to textual output. For many geographers (male and female) working within the emerging academic discipline of the early twentieth century, undertaking writing a research paper or monograph was an impossible luxury: wide syllabuses had to be taught, lecture materials had to be researched and gathered, fieldwork organised, practical classes delivered and school texts written. Many inter-war geography lecturers considered university teaching both all important and all consuming. I have sought to respect the geographical work of those who published little or nothing by discussing their teaching and impact on their departments and students, again derived where possible from departmental records, published departmental reports, university archives (such as employment records), obituaries and oral histories.

Implications for the Historiography of Geography

This book is offered as a platform to explore further the historiography of geography armed with a knowledge of the range and character of geographical work produced by women between 1850 and 1970. These dates have been chosen to (roughly) encompass Mary Somerville's *Physical Geography* (1848) and Alice Garnett's tenures as president/vice president of the Institute of British Geographers, the Geographical Association and the Royal Geographical Society (1968–1970). It was also chosen to address the commonly held misapprehension of many contemporary geographers that women were not significant producers of knowledge prior to the 1970s. Whilst a single volume could not hope to include or address all female producers and communicators of geographical knowledge over a 120-year period, it is hoped that this beginning will encourage further exploration and engagement with a wider gender-sensitive history of geographical thought.

This work feminises the history of geography by repopulating it with women and providing a substantiation of their work that has been largely absent from disciplinary histories. It also raises questions about the practice of how geographical knowledge has been defined and how the history of the discipline has been constructed and reproduced. Gender inclusivity raises epistemological and ontological questions as well, which make it necessary to rethink the nature of our historiography. As Blunt and Rose have argued: 'It is crucial to locate women within the historiography of geography, but this act should question the very basis of that historiography rather than reproduce it, albeit in a revised form' (1994: 9). It is hoped that with the body of women's geographical work recovered/excavated here, there will never again be any justification for omitting women's geographical work from histories of British geographical thought and practice. It is also hoped that this study might exemplify the benefits of 'hidden' histories (Blunt 2005) and 'minor' figures (Livingstone 1992; Lorimer 2003) as well as demonstrating the processes by which work becomes marginalised in the course of making histories, not least by gendered discourses. The ways in which the historiography of geography might be framed differently will be returned to in the concluding chapter.

The following chapters are made up of more than 50 biographical studies, which are organised broadly according to type of geographical work (travel writing, educational, academic, etc.) and chronology. A brief biographical outline is followed by an analysis of the production and reception of each individual's geographical work. The key themes that thread through these biographies are: the number of women geographers and their presence in geographical societies and higher education; the subject and methodological groupings of those women within the discipline; their experience of war work; the significance of fieldwork to their geographical work; and issues of access, recognition and promotion within geographical institutions. A number of discursive constructions will also be examined: definitions of geography and geographical practice; negotiations of gender in women's writing; the role of gender in the production and reception of women's geographical work; and discourses of representing and memorialising women geographers. Each of these themes will be revisited in the Conclusion.

Fieldwork merits a particular note here for two key reasons; first, because it has been central to many debates about the epistemology and practice of geography; and secondly, because it has been associated with the masculinisation of the discipline. As Bracken and Mawdsley (2004: 280) note, 'Fieldwork has always been central to the enterprise and imaginary of geography', but it has become a contested space. The gendering and Eurocentric character of fieldwork has been debated within the discipline since the early 1990s (e.g. Domosh 1991a,b; Stoddart 1991; Driver 1992; Rose 1993;

Bee, Madge & Wellens 1998; McEwan 1998a, Maguire 1998). A question at the heart of these debates is whether fieldwork, especially in physical geography, is a masculinist domain? It has been argued that fieldwork is masculine by dint of a combination of epistemological grounding in scientific methods and an overemphasis on an 'initiation rites' element' (Rose 1993; Sparke 1996), which relies on personal physical attributes such as strength and fitness, which in turn are frequently associated with masculine cultural norms and competitiveness (as well as an assumption of able-bodiedness). Rose (1993) added to this the notion of 'aesthetic masculinity' seen in the privileging of the visual in both physical and human geography, the 'masculine gaze' being just as likely to feminise the landscape in human geography studies as the methods of the 'scientific method'. Constructions of fieldwork as masculinist have been helpful, not least in evaluating current field discourses and practices, but have also been criticised for being too simplistic (Powell 2002; Bracken & Mawdsley 2004). Broad arguments have been made about the exclusion of women from scientific geography in the nineteenth century as a result of the professionalisation of science and exclusion from the Royal Geographical Society (McEwan 1998a); empirical studies have demonstrated women's participation in contemporary physical geography (Dumayne-Peaty & Wellens 1998); and qualitative studies (e.g. Maguire 1998) have explored ways in which discursive constructs intersect with the practices of fieldwork. However, with the exceptions of Sack's (2004) and Monk's (2004, 2007) largely US studies, there has been little data collection on British women's pre-1970 fieldwork, or comparative analysis between the nineteenth century and the present geographical practices, as Powell's (2002) discussion of the historiography of fieldwork demonstrates. Evidence of extensive and varied field study in the following chapters makes a significant contribution to these ongoing debates concerning British women geographers and fieldwork.

Chapter Two

Women and British Geographical Societies: Medals, Membership, Inclusion and Exclusion

This chapter explores the various stages of and attitudes towards women's membership of geographical societies in the UK. Geographical societies were particularly significant as the institutional face of geography prior to the subject's professionalisation as a university discipline with concomitant qualifications in the early twentieth century (see Chapter Six). They provided accreditation through membership, prizes, publication and invitations to address the given society. Evidence for institutional openness and closure to women is examined, with reference to how this varied over time and space, and degrees to which it was contested and by whom. Women were admitted to the non-metropolitan geographical societies from their foundation in the 1880s onwards: the Royal Scottish (1884), Manchester (1884), Tyneside (1887), Liverpool (1891), Southampton (1897) and Hull (1910) geographical societies; and a brief history of the foundation of a selection of these societies is provided with reference to women's membership. Particular attention is given to an analysis of the Royal Geographical Society's (RGS) women's membership debate 1892–3 and its implications for the epistemology of geography. The biographies of Jane Franklin, Mary Somerville and Isabella Bird demonstrate different engagements with geographical societies, including the varying responses to their work by different societies, how attitudes varied within geographical societies, and how the RGS in particular changed its institutional response to women as producers of geographical knowledge over time.

Women's Membership of British Geographical Societies

The RGS was founded in London in 1830, and was part of a nationwide flowering of philosophical, scientific and arts societies, as well as a

European-wide diffusion of geographical societies (see Stoddart 1986; Morris 1990). John Barrow of the Admiralty chaired the first meeting and argued that the success of the Society (which had already gained royal patronage and the support of some leading political figures) would depend not on the Council alone, but on the 'many individuals eminent in the Arts, Sciences, and Literature, and from the distinguished officers of the Army and Navy, whose names appear on the list of members'.[2] In many ways this reflects the Society's early membership: in 1830 almost all of the 460 members were men of high social standing (Stoddart 1986; Maddrell 2007). As the nineteenth century progressed the RGS continued as 'a club for travellers and explorers, supported by gentlemen and made intellectually respectable by the scientists' such as Hooker, Everest, Murchison, Sedgwick, Darwin and Wallace, and the armed services represented a consistently high percentage of RGS membership (17–19 percent, 1830–1900) (Stoddart 1986). Whilst some members sought to combine geographical knowledge with other scientific or cultural knowledge, or to promote the subject within the universities, these were not necessarily distinct from more applied motives, and the majority of members were men who wanted to belong to a club which focused on travel, 'discovery' and the interests of Empire. In his presidential address to the RGS in 1885 Lord Aberdare identified the 1880s as a period of unprecedented geographical activity, including the 'Scramble for Africa', which was embodied in the formalisation of European territorial annexation at the 1884–5 Berlin Conference. Inevitably the politics and organisation of Empire affected the prevailing ethos within the RGS and this strong link between Empire and British geography and geographical institutions such as the RGS has been widely acknowledged (e.g. see Hudson 1977; Godlewska & Smith 1994; Bell et al. 1995; Driver 2001).

At the time of the debate (1892–3) about women's membership of the RGS, Queen Victoria, as monarch, was the patron of the Society. The RGS had also previously awarded medals to Lady Jane Franklin (1860) and Mary Somerville (1869) for their contributions to geographical knowledge; indeed the announcement of Somerville's medal had been met by loud cheers at the RGS (The Times, 25 May 1869), but neither women were ever proposed as fellows of the Society. Other British geographical societies were founded in the 1880s in Scotland, Manchester and Tyneside; the Royal Scottish Geographical Society (RSGS) also opened a London branch in 1892 (see MacKenzie 1992, 1995). All of these non-metropolitan societies admitted women from the outset,[3] in keeping with a number of other learned societies which extended their membership to women in the second half of the nineteenth century. Women were admitted to the Zoological, Botanical, Statistical, Asiatic, Hellenic and Anthropological societies and 'every other geographical society in the empire' (The Times, 29 May 1893; Stoddart 1986) – but not yet to the Royal, Linnaean or Geological societies. Debates around women's intellectual abilities and public roles need to be placed in the context of the patriarchal structure of British society in the nineteenth century. Women's educational and employment opportunities improved in the second half of the century (Holloway 2005), as did their legal status, but this was not a smooth or uncontested transition. That women had only had access to universities since 1876 and widows had only been able to be legal guardians of their own children since the Infants Act of 1886, gives some sense of how basic and new these rights were (see Table 1.1). Women's enfranchisement was fiercely contested in the last quarter of the nineteenth and first quarter of the twentieth century and this issue coloured the motivation for, and responses to, many other campaigns for gender equality.

The question of women's membership of the RGS had been raised as early as 1847 and was later urged on the Council by Roderick Impey Murchison in 1853, after women had been admitted to RGS meetings under his presidency (1852–3) and to the British Association for the Advancement of Science (BAAS) in 1853 (Stafford 1989). Murchison may have been prompted by his intellectually active wife, Charlotte, and Stafford (1989) credits him with a genuine desire to see women admitted as full members of London's scientific societies. Murchison later articulated the benefits of women's membership in terms of gendered citizenship, arguing that women could educate their sons: geographical knowledge 'to be by them communicated to the sons of England' (Murchsion 1897). He was not successful, but the question reappeared periodically and in 1887 the RGS council had 'given its approval in principle and agreed to reconsider the matter when there was evidence of a demand' (Bell & McEwan 1996: 296).

Whilst women could attend RGS lectures as guests of male fellows, access to the facilities of the Society required a male fellow as intercessory, e.g. world traveller Isabella Bird had to borrow books under the name of her publisher. Similarly, Alexandrine Tinné's paper on her journeys in the Nile region to the Society in 1889 was read by her son; and the request for an honorary RGS fellowship for Tinné, forwarded by respected explorer J.H. Speke, was unsuccessful. These examples point to the fact that women were actively engaged with the production of geographical knowledge, but that this was obscured by the institutionalised gendered regulations of the RGS. While women's relation to geographical knowledge remained hidden it did not threaten the male hegemony of travel or the Society itself (Birkett 1989) at a time when the epistemological status of geography as a 'manly science' was questioned in academic circles (Keltie 1885); the homosocialility of the space of the RGS was also protected.

These gendered structures were challenged in a number of ways in 1891 and 1892. First, three women had addressed the geographical section at the annual BAAS meeting held in Cardiff in 1891, when Section E under the presidency of E.G. Ravenstein invited Miss E.M. Clerke, Isabella Bird

(Mrs Bishop) and May French Sheldon to speak. Secondly, the RSGS suggested reciprocal membership between the two societies, but this was rejected, in part because of women members in the RSGS. This caused the RSGS to set up a London branch, similar to its other regional branches in Scotland and Isabella Bird spoke at the second London branch meeting in 1892. The Marquis of Lothian, chairing the first London RSGS meeting stressed that the scope of the RSGS was wider than that of the RGS, not least in its acknowledgement of women's ability to produce geographical knowledge (Bell 1995a; Bell & McEwan 1996). In response, RGS secretary, Douglas Freshfield, invited Isabella Bird to speak at a Society meeting. In her reply Bird, unable to speak due to ill health, highlighted the hypocrisy of the RGS which would invite her to speak but not allow her to be a member. This response, much cited by Freshfield, has been credited as the trigger of the 1892–3 controversy around women's membership of the RGS, but was far from the single flashpoint as the Society had already received several requests from women seeking membership. Bird herself claimed that her involvement with the issue was unintentional and she felt herself to be misrepresented in what became a very public debate which featured in the national press. In a letter to her publisher and friend, John Murray, she wrote: 'I am annoyed to see that Mr Freshfield both in a circular and in a letter to Saturday's *Times* has referred to my declining to read a paper for the RGS in an inaccurate way which makes me look ridiculous. My health was breaking down at the time, and I could not prepare a paper and I added in declining in a friendly note these words as nearly as possible: "it scarcely seems consistent in a society which does not claim to recognise the work of women to read a paper". I have never made any claim to be a geographer, and hope that none of my friends have ever made it for me. As a traveller and observer I have done a good deal of hard and honest work and may yet do more but I never put forward any claim to have that recognised by the RGS. If I had thought any use would have been made of my note I should not even have written the above remark'.[4] The letter to Murray is rather defensive, perhaps fearing criticism from her publisher or more generally wishing to distance herself from being the centre of what was to become a fervent, highly gendered and sometimes ridiculous public debate. That Bird did not see herself as geographer per se was a reflection of the dominant definition of that term: she did not see herself as matching the criteria set by the RGS of a surveying explorer, imperial agent or natural scientist.

In July of 1892 the RGS Council made an executive decision to open the membership of the society to women on the basis that: 'The increasing number of ladies, eminent as travellers and contributors to the stock of geographical knowledge, and the number of women now interested as students, or teachers, in our branch of science, coupled with the evidence brought forward of a desire among both classes to enjoy the practical privileges conferred by our Fellowship, were, in the opinion of the Council, sufficient reason for at once making the proposed extension, which will it is believed, be to the advantage of the Society' (*Proceedings of the Royal Geographical Society* 1892: 553; also published in *The Times*, 6 July 1892).

This announcement mentions the expectation of approval in the membership, which could be seen as naive, misled or hopeful of forestalling criticism. The furore that erupted was to spread over two years and continue in a lower key for two decades (see Bell & McEwan 1996). According to Mill (1930) the optimism on the part of the Council seems to have been well founded and only undermined by a small group of influential reactionaries which centred on George Curzon and Admirals Cave and Inglefield (Birkett 1989). In the event, 22 women were elected to membership by early 1893, but their presence was immediately challenged by opponents at the Annual General Meeting of the Society. Opponents to women's admission to the RGS maintained they were not necessarily against the entrance of women, but rather concerned with the unconstitutional act of the Council in admitting women.[5] Admiral Cave's claims to disinterested procedural concern are somewhat undermined by his comments expressing fear that women would undeservedly gain certificates of competence by adopting the letters FRGS after their names and would interfere with the running of the RGS (Maddrell 2007). Drawing on discourses of tradition and masculine exclusivity and authority, Cave declared – at a time when the RGS was barely 60 years old – that he 'should be very sorry to see this ancient society run by ladies' (ibid.). Admiral Sir Leopold McClintock, who had captained the *Fox* for Lady Franklin in 1857, reiterated the spirit of the Society's founding aims when he argued: 'The people we wish to attract are explorers, geographers, men employed in trade and commerce and missionaries who seek for geographical knowledge, and I do not think that the admission of a large number of ladies will add to our utility in these respects'. Those opposing women's fellowship of the RGS thereby discursively placed geographical knowledge as something to be found abroad rather than at home; placed geography squarely in the domain of masculine endeavour and employment; and notably excluded teachers from their constituency, a group which would include women and the lower middle classes. The reference to a 'large number of ladies' was also emotive, suggesting the Society would be flooded by women who would feminise the institution, which had to date in effect run as a gentleman's club. That this spectre of a large number of women could be called up so easily also suggests women would have little difficulty meeting the existing membership requirements, which were based on personal recommendation and a general interest in things geographical (Maddrell 2007). These arguments are clearly based on the desire of a particular group within the RGS to maintain the Society as an exclusively masculine physical space and discursive arena, based on the ethos of a male social club which depended on the organisational/structural exclusion of women.

When George Curzon returned from his travels in Persia (where he had crossed paths with Isabella Bird) he had great influence on the debate as a powerful social figure and member of the RGS Council, not least in undermining the Council's previous unanimous support for women fellows. Curzon considered the growing number of women travellers as undesirable 'female globetrotteresses' whose membership of the RGS would reduce its already falling market value (Birkett 1989). Whilst he held a guarded respect for Isabella Bird, Curzon wrote in *The Times* (1893): 'We contest *in toto* the general capability of women to contribute to scientific knowledge ... their sex and their training render them equally unfitted for exploration'. Thus the epistemology of geography was limited to exploration and the application of the scientific method, of which women were considered biologically and educationally incapable – despite the RGS medal being awarded to scientist Mary Somerville, and the growing number of women attaining science degrees at this time (see Newbigin Chapter Three). This represents epistemological exclusion as well as embodied biological exclusion through gender-defined structures. However, this view was not limited to men. Just as there was a strong group of women anti-suffragists centred around Augusta Humphrey Ward, some women opposed women's access to scientific and learned societies. Mary Kingsley was of this view and refused to support Marian Farquharson's attempts to gain access to the Linnaean Society and repudiated any desire to become a member of the RGS herself (see Chapter Four).[6]

The ensuing debate over women's membership spilled over from the RGS to the media and society at large and was satirised in a ditty in *Punch* magazine (10 June 1893), which underscored insecurities surrounding prospective women travellers' sexuality and domesticity (note the equation of travel with geography as well):

> A lady explorer? A traveller in skirts?
> The notions just a trifle seraphic:
> Let them stay and mind the babies, or hem our ragged shirts;
> But they mustn't, can't and shan't be geographic.

Douglas Freshfield responded with a couplet drawing on discursive structures of masculine gentility:

> The question our dissentients bellow
> Is 'Can a lady be a fellow?'
> That, Sirs, will be no question when
> Our fellows are all gentlemen.[7]

Various compromise positions were proposed, largely aimed at limiting the number of women and/or their status and power within the Society. The final resolution put to the Special General Meeting in April 1893 proposed that women should be elected as ordinary fellows; that regulations should refer to both the masculine and feminine gender, but that women would not be qualified to serve on Council or hold any other offices. Some of the women recently elected attended the meeting, but none are recorded as contributing to the debate (Maddrell 2007).

Among others, Robert Needham Cust (vice president of the Royal Anthropological Institute) and journalist traveller Henry Morton Stanley argued the case for women's membership – both having proposed women for election that year (Maria Cust and May French Sheldon, respectively). It was argued that women had been admitted to a number of comparable learned societies without detriment to those societies and the contradictory status of the Queen as Society patron was discussed, as were the benefits of a body of women trained to contribute to filling in the blanks on maps.

In the event, RGS fellows present voted not to allow the admission of women by 147 votes to 105. A later postal vote of members, which overwhelmingly supported women's membership (by 1165 votes to 465 against), was not allowed because it was not sanctioned by the Society (Mill 1930). The 22 women fellows already admitted by Council were allowed to stay but no others would be admitted (see Table 2.1 also Blunt 1994). Douglas Freshfield famously resigned his position as honorary secretary to the Society in protest at the outcome, and others such as W.H.D. Rouse chastised the Society in its journal (Bell & McEwan 1996). The annual subscription of the RGS had to be increased in 1893[8] suggesting the Council may have been motivated as much by financial imperatives as by the politics of equal access in proposing the admission of women as Fellows.

Isabella Bird was the first woman to present a formal paper at the RGS in 1897 (see below) and was followed by other women, both elected and unelected who spoke to the Society and published in its journals, and 'participated 'on the fringes' (Bell & McEwan 1996). Over the years pressure grew from outside the Society from women seeking membership and from supporters within the Society who petitioned Council. Women who assumed that because they had published in the Society's journals they would be eligible for membership met refusal with disbelief (Bell & McEwan 1996). These women reasonably asked what better credential was there for RGS membership than to have their work published by the Society? It appears that there was not an issue about the quality of their geographical work and the route to the public production of that work was open to them. However, there was a tension between Council members and individual office holders who acted as gatekeepers (notably the secretary and journal editor) and the

Table 2.1   Women admitted to the Royal Geographical Society 1892–3

Mrs Isabella Bird Bishop	Mrs Edward Maberly
Mrs Zelie Isabelle Colville	Mrs Juliet Mylne
Miss Maria Eleanor Vere Cust	Mrs Julia MacLennan
Lady Cotterell Dormer	Mrs Elizabeth Prentis Mortimer
Miss Agnes Darbishire	Mrs Nicholas Roderick O'Connor
Mrs Lily Grove	Mrs Mary Louisa O'Donoghue
Miss E. Grey	Mrs Emmeline Porcher
Mrs Edward Patten Jackson	Miss Christina Maria Rivington
Mrs Beatrice Hope Johnstone	Mrs May French Sheldon
Miss Julia Lindley	Miss Florence M. Small
Miss Kate Marsden	Lady Fox Young

Source: Women's Membership, RGS-IBG Archives

unchanged regulations and views of some fellows who objected to the bodily presence of women in the RGS.

Women and their male supporters continued to petition the RGS,[9] including B. Pullen Burry who wrote to the RGS asking that a notice appear in the journal stating that membership of the Lyceum Club now required 'original geographical work', listing the work and publications of some members.[10] Somewhat ironically, it was Curzon, then president of the RGS, who oversaw the admission of women as fellows of the RGS in 1913 (see Bell & McEwan 1996). Whilst the Council had been ahead of mainstream public opinion in 1892 in acknowledging women's abilities to produce geographical work, by 1913 the proposal to admit women merely reflected women's widening participation in society, especially in education, including significant numbers on university geography courses (see Chapters Five and Six). However, women's admission was still contested in the Society's chambers and in the public press. For example, in 1912 the *Times Literary Supplement* argued that 'it cannot be said that any piece of actual exploration of the first importance has been accomplished by a woman' (cited by Birkett 1989: 194). Curzon wrote to RGS members and to *The Times* justifying the change in policy by using all the arguments supporting women's membership put forward in 1892–3 – including the reassurance that a flood of women's applications were not expected and adding that the introduction of a standard of qualification for membership was being revived: 'It will always be within the power of Council, as the electing body, to keep within reasonable limits the feminine addition to our ranks',[11] thereby stressing (male) Council control and hinting at the informal ways in which a female presence could be regulated despite apparent open membership. In this approach he emphasised continuity of practice rather than radical change and made a distinction between membership of the RGS and broader arguments about the political enfranchisement of women fiercely debated at the time (Bell & McEwan 1996). This debate indicates that the issue of women's membership was still being contested in 1913 and that although Curzon had changed his position on women's ability to work within the geographical episteme, others held deeply entrenched and largely social fears about women's membership. Ultimately social change and women's accreditation through education made it difficult for the RGS reactionary faction to exclude women. Gertrude Bell was awarded the RGS Gill Memorial Award in 1913 and could be seen as representing all that was desirable in a woman fellow: she was highly educated, socially and economically privileged, and her work included surveying unmapped areas; it might be added that as a result of her work she was also frequently abroad.

The 163 women admitted as fellows of the RGS in 1913 fell into three broad categories: about half were explorers/travellers; about a quarter were scientists and/or teachers; and the remaining quarter were made up of various professional women, including an emerging class of public servants. Many of these women had a high public profile, through their publications, memberships of other learned societies, occupation and social connections; whilst others relied on academic credentials and networks (Bell & McEwan 1996). Those who travelled with husbands or family on imperial duty and those who worked for government (such as Gertrude Bell and Violet Markham) represent a continuation of the strong link between the RGS and state institutions and interests. The first cohort of women admitted in 1913 relied largely upon a small network of male supporters to propose them, but once fellows themselves, they were quick to use their entitlement to propose other women (ibid.).

In Britain's other geographical societies women were admitted from the outset, these societies thereby avoided the splits and controversies about women's membership experienced by the RGS (Nigel & Brown 1971). In addition to being full members, women were regular speakers and in some cases served on the societies' councils. Perhaps reflecting the strong local nonconformist community which promoted female education and the strong lobby for female suffrage in the town, Manchester Geographical Society (MGS) was the first to elect women to its governing body when Mrs A.H. Wood and Miss Fanny Rutherford were elected to its committee in 1885. However, it should be noted that the president and all 12 vice presidents were men; its large committee of 46 also included two bishops, seven MPs, 11 JPs, two Lords and the Lord Mayor, as well as seven academics, so its social composition was still firmly linked to establishment and state, as well as the commercial interests the Society wished to promote (see Nigel & Brown 1971 for committee list). Miss M.K. Sturgeon was the first woman to give a paper at the MGS, when she spoke on 'The teacher of elementary geography' in January 1887. May French Sheldon's 1891 visit aside, women travellers or educationalists only occasionally addressed the Society, despite there being a typical programme of around 20 speakers a year. It is only after 1904 that at least an annual appearance of a woman speaker became the norm, including regulars such as L. Edna Walter recorded as BSc and Her Majesty's Inspector (HMI) and Kate Qualtrough, noted as FRGS when she spoke in 1914.[12] Women represented a relatively high proportion of the Liverpool Geographical Society membership, but only a small proportion of speakers. In 1901, 10 years after the Society's foundation, women constituted approximately 18 percent of the 730 membership, but only 42 percent of these favoured full membership, the majority being associate members; most of the latter were unmarried, with some identified as trainee or practising teachers, others were part of a family membership. In the previous year only one of the 14 speakers was a woman (Mrs Rickmers, speaking on travel), while eight were fellows of the RGS, as were 13 of the first 16 speakers to the Tyneside Geographical Society,[13] illustrating the wide-reaching influence of the RGS in the English regional geographical societies. Travel writer Ella Christie and *SGM* editor Marion Newbigin were the first women to be appointed to the Council of the RSGS in 1923, some five years before universal suffrage was attained in 1928. The first woman to serve on the RGS Council in 1932 was Elizabeth (Wilhemina) Ness, a rich socialite who travelled widely and was a generous benefactor of the Society; she was soon followed by Eva Taylor. Hilda Ormsby served on the Council of the Institute of British Geographers (IBG; founded 1933) in 1936 and 1937, but was not followed by another woman until 1949 when women became a permanent presence on the Council (see Chapter Eight and Figures 10.1–10.6). These patterns of institutional involvement and recognition will be returned to in the Conclusion. The second part of this chapter addresses the biographies and geographical work of three women: Jane Franklin, Mary Somerville and Isabella Bird. Franklin and Bird as travellers could just as well appear in Chapter Four, but they are discussed here as pertinent to the debate around women's geographical abilities and membership of the RGS.

Lady Jane Franklin

Jane Griffin was born in 1792, the middle of three daughters of John Griffin and his wife Mary Guillemard (who died when Jane was four). As a silk merchant her father regularly went abroad and consequently Jane had travelled widely in western Europe, Scandinavia and Russia before her marriage in 1828 (Middleton 2004a). She married family friend, naval commander and polar explorer, John Franklin on 5 November 1828, after the death of his first wife Eleanor. Franklin was knighted in 1829 and was posted to a ship in the eastern Mediterranean, during which time Jane travelled in Asia Minor with Franklin's niece Sophy Cracroft and her flea-proof iron bedstead which she took with her on all her travels (Middleton 2004a, Robinson 1990). Jane then accompanied her husband to Van Dieman's Land when he was appointed to the post of Governor in 1837. She took her role as Governor's wife very seriously (Birkett 2004a), campaigned for the rights of female convicts (Middlelton 2004a; Flemming 1998), and with her husband established a school and a scientific society, which later became the Royal Society of Tasmania (Middleton 2004a). However, in time the Franklins' progressive views fell out of favour and Franklin was recalled, departing under something of a cloud (Birkett 2004a; Flemming 1998).

When Sir John set sail on his fourth Arctic journey in the prestigious and alluring search for the Northwest Passage in 1845, he left with three years' provisions and Lady Franklin pursued her own travels in Europe and North America. Franklin and his 129 crew perished in the Arctic, causing shock nationwide (Stoddart 1986). Jane persistently lobbied for and privately financed some of the protracted series of searches for Franklin and his crew, which took place over more than a decade until their deaths were confirmed. Subsequently Jane Franklin was the first woman to be awarded a medal from the RGS, when she was awarded the Patron's medal in 1859/60. However, this was not in recognition of her own extensive travels, but for her role in expanding geographical knowledge through the search parties which she galvanised in the hope of finding her husband. The medal was accepted on her behalf by her supporter (and indeed medal nominator) Sir Roderick Murchison, twice RGS president and friend of Franklin (Stafford 1989). At a time when independent travel and female intervention in the public domain were not deemed appropriate (including for women of the middle and upper classes), Jane Franklin was 'presented as a woman who was spurred on to duty by a ghost' (Birkett 1989: 188). In so far as the medal was vicarious recognition of her husband's career, the award in 1860 was in keeping with the RGS tradition of recognising explorers and surveyors rather than scholars in the first 50 years of the Society (Stoddart 1986), and numerous men associated with the search for both Franklin and the Northwest Passage received RGS awards, often on Murchison's recommendation. Although not appearing on Samuel Pearce's 1851 portrait of the Arctic Council, Jane Franklin became 'one of the most influential figures on the Council', which often met in her home, and she continued to be consulted on Arctic matters in later years (Fleming 1998: 382). After her death on 18 July 1875 at 83 years of age, Jane Franklin's funeral was conducted by the first Bishop of Tasmania and her pall bearers and mourners included numerous officers from the Arctic expeditions, including Leopold McClintock. Buried at Kensal Green Cemetery, her wifely devotion was noted as a footnote to John Franklin's memorial at Westminster Abbey. Jane Franklin's family left her papers to the Scott Polar Research Institute, with instructions to extract material relating to polar expeditions and to 'burn the rest' (Middleton 2004a).

Jane Franklin was recognised by both the RGS and wider society, principally in terms of her role in promoting and funding the searches for her husband and his crew. However, the extent and range of her travels would have exceeded those of many fellows of the RGS at this time, which suggests that had fellowship been open to women during her lifetime she would certainly have met any entry requirements. While stopped over in Cape Town en route to Van Dieman's Land Jane and a small party made a 'typically spirited' ascent of Table Mountain setting off at three o'clock in the morning, all against the advice of Franklin and Alexander Machonochie (Clay 2001). Later, she would be the first European woman to climb Mount Wellington's 4000-feet peak and to travel overland from Melbourne to Sydney (Robinson 1990; Russell 1994). After the death of her husband was finally confirmed in 1857 Jane Franklin and Sophie Cracroft undertook 'frenetic' travels in Japan, India, Brazil and North America (Robinson 1990). She also played a role in forging political links between Hawaii and Britain in the face of growing American and French influence, and hosted Queen Emma of Hawaii during her visit to Britain in 1865 (Stafford 1989; Birkett 2004a). In common with many educated women and men of her class, Jane Franklin kept a meticulous diary from a young age, as well as being a copious correspondent, cataloguing her travels and experiences. Had Jane Franklin lived in the late 1800s, when women's travel writing emerged as an oeuvre, she would almost certainly have sought to publish her travel accounts; as it was, an edited selection of her diaries and correspondence were published by Willingham Franklin Rawnsley in 1923. The writings have been edited, so her 'voice' has been tailored by another, but they remain a useful account of both her life and the places she visited.

Mary Somerville

Mary Somerville (Figure 2.1) is widely credited with writing the first book on physical geography in the English language and has been designated 'the first woman scientist in English history' and 'the first English geographer' (Sanderson 1974). The merits of these claims will be discussed after a brief

Figure 2.1   Mary Somerville self-portrait.
Source: courtesy of the Principal and Fellows of Somerville College, Oxford

overview of her life and analysis of her key contribution to geographical literature.

Biographical overview

Mary Somerville (née Fairfax) was born in 1780 in Jedburgh, daughter of Margaret Charters Fairfax and Lieutenant (later Admiral Sir) William George Fairfax. She was the fifth of seven children, three of whom died in early years (Creese 2004a).

Her childhood was shared between the small seaport of Burntisland and Edinburgh on either side of the Firth of the Forth. At about the age of eight or nine years she was unable to write or do simple sums and about a year later was sent to Miss Primrose's Academy, a boarding school for girls at Musselburgh, Midlothian. Mary was very unhappy during her year there and learned little except the basics of handwriting, English and French grammar. Nonetheless, this was to be the total of her formal education. With the exception of a few classes in painting, music, handwriting and arithmetic, which were considered appropriate for a young lady's education at this time, Somerville was largely self-taught (Patterson 1969; Phillips 1990). In addition to French, she learned Greek and Latin in order to read Euclid and Newton's *Principia*, respectively. These were no mean achievements, particularly in the light of family opposition; they considered her more serious study both socially inappropriate for a young lady and potentially injurious to her health (Phillips 1990; Neeley 2001). What her parents did allow was occasional lessons with the village schoolmaster, who taught her the use of terrestrial and celestial globes, knowledge which undoubtedly underpinned her later interest in large-scale systems, astronomy and geography.

In 1804 Mary married Samuel Greig, a cousin serving in the Russian navy. Greig was appointed as Russian consul to London and the couple lived there until his death in 1807. Greig had not been supportive of Mary's studies and it was only when she took their two sons (born 1805 and 1806) to Edinburgh that she found, for the first time, that she had both the social and financial independence to pursue her mathematical studies seriously. In 1811 she won a competition medal from the journal *Mathematical Repository* and was advised by its editor William Wallace (later professor in mathematics at Edinburgh University) on further mathematical study. In old age Mary reflected on the long route from her first encounter with an algebraic puzzle seen in a ladies' magazine (a feature of such magazines which was to disappear as the nineteenth century progressed), through years of secretive and limited study to a period when she could study openly (Somerville 1873).

As a young widow Mary Greig moved in Edinburgh's vibrant intellectual society, with Jefferey, Brougham, Sir Walter Scott and Playfair, and she was treated as something of a protégée as a result of her interest and talent (and no doubt her gender and charm played a part too). Her second marriage to her first cousin, 41-year-old widower Dr William Somerville, took place in 1812. In contrast to her first husband, William Somerville approved of the education of women and supported Mary in her studies and research. The new family grew, with daughters Margaret, Martha and Mary Charlotte being born in 1813, 1815 and 1817, respectively. However, reflecting high mortality rates amongst children in the nineteenth century, the family also suffered the deaths of three children: Mary's second son died in 1814, aged nine, another son died in infancy in early 1815, and daughter Margaret died in 1823 aged 10 (Creese 2004a).

In 1816 the Somervilles moved to London and lived first on Hanover Square, then the Chelsea Hospital from 1819 when William was appointed as physician and surgeon. The move to central London, along with William Somerville's membership of the Royal Society, gave them access into a new literary and scientific circle including Roderick and Charlotte Murchison, Sir John Herschel, Michael Faraday, Harriet Martineau, the Lyells, Davys, Carpenters and Babbages. Many among this network of early nineteenth-century scientists and writers became firm friends, who accepted Mary as an intellectual equal and included her in their scientific discussions.[14] This network was to be crucial to Somerville's success (Neeley 2001) and appears to have fostered and/or given encouragement for her more radical political beliefs. She attended Lyell's lectures at King's College, London in 1832 with Charlotte Murchison and her daughters in order to make a point about women's access to the university and later gave the support of her now famous name to formal campaigns for women's access to university and the vote (her name headed John Stuart Mill's petition for women's suffrage in 1867) (Patterson 1969). However, Somerville tended to remain silent when personally cited in controversies, such as the debate around women's entrance to the BAAS in 1832, and this has allowed for varied interpretations of her position to be made by others (Neeley 2001). Somerville also made many European intellectual contacts during tours of Europe undertaken 1817–18, 1824 and 1832–3 (for example the Somervilles were entertained in France by François Arago, Marquis Pierre de La Place, Alexander von Humboldt and Augustin de Candolle).

Somerville's first publication in the scientific field appeared in 1826 when a paper entitled 'On the magnetizing power of the more refrangible rays' was published in the *Philosophical Transactions of the Royal Society at London*. A paper on a similar theme based on correspondence with John Herschel was published by the Royal Society in 1845. Somerville's key publications were: *The Mechanism of the Heavens* (1831), *On the Connexion of the Physical Sciences* (1834), *Physical Geography* (1848) and *On Molecular and Microscopic Science* (1869) (published in her 89th year). The focus here will naturally be on her geographical work, but both the character of that work and its reception need to be placed in the wider context of her scientific writing and reputation, not least because by the time *Physical Geography* was published in 1848, Mary Somerville was already known as a scientific author.

Over her lifetime Somerville received many national and international accolades and honours, including Somerville Island in the Canadian Arctic being named in her honour by friend Edward Parry (see Patterson 1969; Creese 2004a). The Royal Society, in a near immediate response to the publication of *The Mechanism*, commissioned a bust of Somerville in 1832, which was sculpted by Chantrey and paid for by public subscription. Honorary membership of the Royal Astronomical Society (with Caroline Herschel), the London Mechanics Institute and the Bristol Institute followed in 1835. After receiving a complimentary copy of *Connexion*, A. McFarlane wrote on behalf of the London Mechanic's Institute in appreciation of the honour 'and to assure you they attach a high value to this mark of your favour towards an establishment having for its object the extension of useful information to the working classes of the community'.[15] This suggests Somerville supported universal education some 35 years ahead of the Elementary Education Act (1870). Somerville was awarded a diploma by the recently formed American Geographical and Statistical Society in 1857 and the first gold medal awarded by the Geographical Society of Florence. One of the last awards Somerville was to receive was the Victoria or Patron's medal from the RGS (instigated by Murchison) in 1869. In 1835 Sir Robert Peel placed Somerville on the civil list when she was granted a King's Warrant of £200 (later raised to £300) per annum for her 'eminence in science and literature', as part of a new state initiative to use civil pensions to encourage scientific work.[16] Somerville's award of a state pension was a personal triumph and provided much needed financial support for her family during William's illness as well as the remainder of her long life (Neeley 2001).

The Somerville household, including daughters Martha and Mary, moved to Italy in 1840. This was partly due to William's ill health since 1838, but it was also a financial necessity, as William had also incurred debts in 1835 after guaranteeing loans for a cousin. The cheaper cost of living in Italy allowed the household to live on Mary Somerville's income from her book royalties and civil pension. William died in Florence on 26 June 1860 and Mary's eldest son Woronzow Greig died in 1865. Mary Somerville herself died on 29 November 1872 in Naples. Her scientific library was bequeathed to Hitchin Ladies' College (later Girton College, Cambridge) and all of her estate was left to her daughters.

The production of *Physical Geography*

Somerville's popular reputation was established with the publication of her first book *The Mechanism of the Heavens* (1831), when she not only translated Laplace's work, but also added explanatory notes and diagrams, making the ideas more accessible in her clear and elegant style. A major undertaking, this significant piece of work took Somerville three years and covered four of Laplace's volumes. Somerville's text proved too expensive for its intended place in Lord Brougham's Society of the Diffusion of Useful Knowledge series and it was published in 1831 by John Murray. It was 'generally well received' (Creese 2004a) and the related short work *Preliminary Dissertation on the Mechanism of the Heavens* (1832) was published the following year. This work prompted her next book *On the Connexion of the Physical Sciences* (1834), a widely acclaimed exposition of the main branches of science of the time, which went to 10 editions, making it a commercial as well as a critical success. *On the Connexion* was dedicated to the young Queen Victoria and in the preface Somerville states her intention 'to make the laws by which the material world is governed, more familiar to my countrywomen'. Whilst not apologising for her work, Somerville limits her claims to readership and status by stating that her authorial intention is to educate women. There are two points to be made here. First, whilst at one level writing for women may have been the rationale by which she negotiated her ongoing publications, on the basis of past experience Somerville would have been disappointed not to have received critical approval from the leading male scientists of her day. Secondly, female education was a cause which she espoused, so this might be read as a political point in relation to ongoing debates about women's suitability for higher and scientific education.

Regularly updated with emerging research findings *On the Connexion* became a best seller 'which functioned for a time as an annual progress report for physical science' (Creese 2004a). Somerville developed the idea of the interrelationships and interdependencies between natural phenomena, a theme dating from her preface to *The Mechanism*, which was to figure continually in her work. Several editions of *On the Connexion* included a quote from Francis Bacon on the frontispiece: 'No natural phenomenon can be adequately studied by itself alone – but, to be understood, it must be considered as it stands connected with all nature' (cited by Livingstone 1992: 173).

Somerville went on to publish *Physical Geography* in 1848 with John Murray, but it had been nine years in the writing, completion of the text having been delayed by William Somerville's illness and the need to revise *Connexions* for its 1841 second edition (Baker 1948). There was a clear demand for a book of this type, as the president of the RGS bemoaned the lack of a satisfactory book on the subject in English in his 1846 address (ibid.). *Physical Geography* was a development of geographical themes in *On the Connexion*, an element of which can also be traced back to Somerville's early notebooks which show what was a popular fascination with latitude and longitude as well as mathematics, physics, geometry and astronomy.[17] *Physical Geography* demonstrated the breadth of Somerville's knowledge of the sciences (*Our Obituary Record*, 14 February 1872) and applies her usual format of analogy, comparison and generalisation, stressing the dynamism of the earth, the role of Creator and Providence (Neeley 2001). *Physical Geography* went into seven editions, each one incorporating new findings and ideas.

Somerville defined physical geography as: 'a description of the earth, the sea and the air, with their inhabitants animal and vegetable, of the distribution of organised beings, and the causes of their distribution' (1848: 1), and showed how physical processes and interrelationships account for topography. She also combined physical geography with the consideration of geographical history and natural phenomena and made her first incursion to discussing human/social phenomena. Her integrative approach to physical geography as a subject was in keeping with the work of Ritter and Humboldt, but she brought the perspective of mid-nineteenth-century science to her analysis. Somerville's explanations of the causes of floods in India and the distribution of vegetation are two examples of her analysis of causal relations (Baker 1948). As Livingstone has noted, 'To her, geography reached beyond integrative description; it was a causal science of distribution' (1992: 173).

The book relied on published sources and its format was a series of systematic chapters with examples organised by regions, ending with reference to animal life, including humans. In this, Somerville placed people in the environment and stressed the interdependence of both, albeit allowing people the right to use and dominate nature. Somerville credited the breadth of her approach to Humboldt's model (Neeley 2001), but a similar layout can also be found in Maria Zornlin's *Recreations in Physical Geography* published in the UK and USA in 1840. Zornlin, who had previously published *Recreations in Geology* (1839), likened geology to ancient history and physical geography to modern history. *Recreations in Physical Geography* was organised into systematic chapters, including the geographical distribution of plants, animals and people, and Zornlin, like Somerville after her, collated her material from the most reliable and up-to-date sources available, including the publications of the RGS, travellers' reports to the BAAS, Murray's *Encyclopaedia of Geography* and scientific texts by authors such as Whewell. Zornlin went on to write a third book in her series, entitled *Recreations in Hydrology* (1842) and her books undermine claims for the novelty of Somerville's *Physical Geography*.

The seven editions of Somerville's *Physical Geography* provide a window on changes in geographical knowledge 1848–77. For example, her short account of Africa and its inhabitants emphasised what was not known about the continent. She included incorrect deductions on the basis of her secondary data, such as arguing against claims of snow-capped mountains on African tablelands because of the absence of large permanent rivers, but was far from alone in making that deduction. By 1877 a more detailed empirical account of key topographical features was given, but this was still largely confined to central and southern Africa – reflecting the main areas of British exploration. Somerville's discussion of different races and social practices is interesting in that it both reflects and challenges the prevailing views of British society (Maddrell 1998, 2004c). She writes of humans as a single species rather than following the model of polygenesis, which was a widely held view at the time of her writing. She wrote of civilised mercantile nations in central Africa (1848: 112) and implied that even if colonisation was inevitable and largely beneficial through the spread of science, technology and Christianity, nonetheless, indigenous peoples had rights (Neeley 2001). Somerville appears to be unique in describing tropical fever (the 'Angel of Death') as guarding the interior of West Africa from the aggressions of Europeans (Maddrell 1998). Whilst having something positive to say about every race (Neeley 2001), Somerville nonetheless reinforces many of the racial stereotypes, hierarchies and associated value judgements circulating in British thought at the time, for example, her description of the Caucasian race as including 'the handsomest and most intelligent portion of mankind' (1862: 247). She also seems to accept the decimation of indigenous peoples as a sad but unavoidable fate (1862: 539), which may reflect what Neeley (2001: 134) describes as 'a striking similarity in the models she [Somerville] used for thinking about the natural and the social worlds'. Despite these views Somerville avoids more extreme deterministic and reductionist explanations through her belief in multi-causality and interaction of factors, including, ultimately that of individuals (Neeley 2001).

Somerville's work has been characterised as pre-Darwinian (Livingstone 1992), and it certainly pre-dates the publication of *The Origin of Species* (1859), but this label potentially obscures Somerville's subtle navigation of her own Christian beliefs and emerging scientific data and theory concerning evolution. Her implicit support for the geologists' view of the age and development of the earth in *Physical Geography*, rather than Mosaic accounts of creation, is indicative of the fact that she was prepared to be scientifically courageous at the time (Sanderson 1974), and indeed this resulted in questions being raised in parliament and the discrediting of her work from the pulpit of York Minster (Patterson 1969). Best described as 'mapping the geological theories of Charles Lyell onto the Biblical account', her rhetoric demonstrated the reciprocal nature of theistic beliefs and scientific findings (Neeley 2001: 138–40) and the hybrid position taken by many scientists which have been obscured by the representation of an oppositional relationship between science and religion at this time.

Traveller and assistant secretary to the RGS, H.W. Bates, edited the sixth edition of *Physical Geography* in Somerville's old age (and the seventh edition posthumously), on the understanding from Somerville that it should not include any explicit evolutionary theory. Somerville knew and admired Darwin and refers to his travel observations, as well as his work as a naturalist; she had used Lady Lyell as an intermediary to request use of diagrams from Darwin's book on orchids and sent him a copy of her subsequent book.[18] Patterson (1969) attributes Somerville's reticence to a combination of reluctance to court controversy and criticism as well as awareness of the imperfections in the theory. Whilst Somerville's books tended to represent the scientific consensus of the day, she was also capable of including dissenting views (Neeley 2001); intentionally or not, her implicit uniformitarianism (adopted from Lyell's geology) helped pave the way for Darwinian influence in geography and she is credited with demonstrating the cognitive claims of geography as a science, looking at causal relations as well as descriptions.

The reception of *Physical Geography*

Mary Somerville quickly came to be widely acknowledged in her lifetime as the 'Queen of Science' – indeed the term 'scientist' was first used by Whewell in his favourable 1834 review of Somerville's *Connexion* (Neeley 2001). Despite Somerville's own fears of its inadequacies on seeing the translation of Humboldt's (1845) *Cosmos*, *Physical Geography* was widely acclaimed, including by Humboldt himself, by Herschel (former president of the Royal Astronomical Society and the BAAS) and by Hamilton (president of the RGS) – their praise being sufficient to 'guarantee its general merit' (Baker 1948: 208). Humboldt wrote: 'I know of no work in physical geography in any language that one could compare to yours' (*Personal Recollections*: 287–8). Herschel drew on Somerville's book for his articles on meteorology and physical geography in the *Encyclopaedia Britannica*, the entry on physical geography being published in book form in 1861 (Baker 1948). Contemporary reviewer Henry Holland asserted that Somerville's ideas in *Connexion* preceded and influenced Humboldt's *Cosmos* (Neeley 2001). In the twentieth century Hartshorne made claims that Somerville influenced the work of Elisée Reclus, but this has yet to be substantiated (ibid.). Reviewers recognised the vitality and clarity of Somerville's writing style and the breadth of knowledge required to write such a book. One reviewer captured the impact of the book for the recently emerging discipline of geography: 'so recent in truth that Mrs. Somerville's volumes come before us as the first English work bearing that title, and distinctly comprehending what belongs to this great subject'.[19] Reviewers were not uncritical, suggesting where less detail and more explanation or diagrams were needed, but overall they stress the value of her expertise and innovative approach in her synthesising text.

The president of the RGS praised her work: 'The importance of this subject renders it all the more satisfactory that it should have been taken up and treated by a writer already so well and so favourably known' (Hamilton 1848). Such approbation from the RGS was no doubt a crucial endorsement of Somerville's contribution to geographical literature and knowledge; but reference to her established reputation causes one to ask whether the book would have been accepted by the Society on its own merits had Somerville not been 'so well and favourably known' in both social and intellectual circles. We can only speculate whether Hamilton's comments were also a dig at the less well-known Maria Zornlin's earlier 1840 text on the same subject. Hamilton continued: 'The graphic style and manner in which the different branches of the subject, not in themselves new, have been arranged and discussed by Mrs. Somerville will, I trust, make her unpretending volumes universal favourites' (ibid.), thereby combining praise and recommendation with a clear steer that this work was not original but was useful and 'unpretending' – a term which might be a compliment or could be flagging that it did not 'pretend' to more than was acceptable from the pen of a woman author, no matter how highly esteemed elsewhere.[20] Henry Holland used similar terminology in his praise for *Connexion*: 'unassuming in form and pretensions, but so original in design and perfect in execution' (cited by Neeley 2001: 157). Holland, reviewing *Physical Geography* in the light of Somerville's wider writing, status and known piety, went on to claim moral authority for her writing on the human race in her final chapter, which was identified as a welcome guide through taxing questions as more of the world became known to and claimed by Europeans (ibid.).

Somerville's work was used widely in education: *The Mechanism* was a required text at Cambridge, and *Physical Geography* was long found on university reading lists, and was adopted as a set text in government colleges in India. More generally *Physical Geography* 'contributed, with other aids, to rectify the barren and tedious method of teaching geography by nomenclature merely, which has long prevailed in England, and is still not wholly discarded' (*The Times*, 4 December 1872). Henry Holland also maintained that many aspects of Somerville's text held their own when compared with Humboldt's formidable *Cosmos*. In retrospect *Physical Geography* has been claimed as the only academic advance in geography in a period when the RGS was preoccupied with exploration and cartography (Freeman 1980). Stylistically it was pioneering with its lack of a political/ state framework, and it has been suggested that it prefigured the regional approach to geography to emerge at the turn of the century, and that her synthetic approach 'shaped the definition and aims of a new discipline' (Neeley 2001; Creese 2004a). Somerville's original wide definition of the subject reflected an admirable and realistic holistic approach when she first wrote the book (Livingstone 1992), but this was ultimately its undoing in the long term. A hundred years after its publication Baker (1948) described *Physical Geography* as 'failing' in the light of unforeseen increased specialisation of scientific and geographical knowledge in the second half of the nineteenth century and succeeded by the work of Arthur Guyot. Whilst Baker allows that Somerville could not have predicted these developments, the term 'failed' is a rather harsh judgement for a book which went to seven editions, was in print for over 30 years, was widely read, and was Somerville's best-selling text (see Neeley 2001; Creese 2004a). Perhaps it is better to say that it had had its day and was succeeded by the work of specialists in both geology and geography. However, there is evidence that *Physical Geography* continued to be read alongside emerging literature such as

Guyot and Ritter, and that it influenced American polymath and conservationist George Perkins Marsh in his environmental concerns and Japanese scholar Uchimura's *On Geography* (1903) which drew on Somerville, Guyot, Marsh and others (Martin 2005). This suggests a wider international impact than previously acknowledged in British accounts of Somerville's *Physical Geography*, as well as links to conservation discourses through the work of Marsh.

Reference to Mary Somerville's *Physical Geography* can be found in a wide range of geographical texts: from anniversary accounts of the RGS (Markham 1881, Mill 1930), to surveys of geographical thought and practice (Freeman 1980, Livingstone 1992), and physical geography texts (Gregory 1985). However, Livingstone apart, these references are largely that, references, typically a mention of her text and/or medal, or a quotation of her definition of physical geography, otherwise her work is absent from other recent surveys of geographical thought and history.[21]

Gender, memory and historiography

Somerville personified many of the desirable attributes of a woman in Victorian society and this undoubtedly made her more acceptable to her peers and the general public, often being credited for combining scientific acumen with 'feminine accomplishments'. This combination put her above social criticism for engaging with science as well as lending weight to campaigns for women's education. Somerville used her position strategically, rarely initiating debate but supporting causes close to her heart, particularly in later years. This was seen notably in the petition for women's access to London University (1862) and subsequent petitions for women's suffrage. Somerville became a member of the General Committee for Woman Suffrage in London (albeit from a distance), but was more outspoken in her support for female education than suffrage, although Neeley (2001) suggests Somerville's values should be read from her actions as much as her words.

However valorised, Somerville had to negotiate her career through the norms of gendered social etiquette in her time, place and class. It is notable that her publishing career was not established until after her childbearing years, and her book royalties, as well as much her scientific correspondence, were directed to her through her husband. As a woman, she was unable to be a member of many learned societies including the Royal Society and the RGS, despite being honoured by both. The same was also true of Cambridge University, where she and her husband were official and honoured guests of the university in 1834 when they stayed at Trinity College where Whewell was master (Neeley 2001). It was only at the age of 46 that her first work was published, and her own longevity and the support of her daughters

Martha and Mary were vital to her more than 40-year writing career (Maddrell 2004c).

Somerville College, Oxford was named for her in 1879 as was a mathematics scholarship for women students at the University of Oxford and the Somerville library at Girton College, Cambridge. Her obituaries record the high regard of men of science and nation (*The Times*, 2 December 1872) and her unquestionable status as 'Queen of Science' (*The Morning Post*, 2 December 1872). One was more feminist in tone, noting her 'most extraordinary intellectual capacity', not to stress the uniqueness of her abilities but rather that 'her life and works constitute an unanswerable refutation of the fallacy that a woman's brain is of necessity unfitted for abstract studies; that it is impossible for a woman to be successful in scientific research, or to attain proficiency in mathematics' (*Our Obituary Record*, 14 February 1872). She is nonetheless credited for retaining her 'feminine' skills and characteristics (ibid.);[22] indeed her 'womanliness' 'made her more acceptable as a "scientific lady" to both the public at large and her scientific peers' (Patterson 1969: 319). Frances Power Cobbe (1872), Somerville's most politically active feminist friend, believed that despite the catalogue of awards Somerville collected, she did not receive the recognition in financial or other terms, which she would have benefited from had she been a man.

Although Somerville's ideas and writings were overtaken by new developments and theories in the explosion of scientific knowledge in the second half of the nineteenth century, her books in particular, with their clarity of style and visualisation of natural phenomena, remained influential for decades. Her most significant legacy might be seen in her articulation of the interconnected nature of science in general and geographical knowledge in particular. Thomas Kuhn has suggested that Somerville's work on interlinkages represents a vital intermediate stage in scientific knowledge, which presaged other more specific scientific developments such as energy conservation (Neeley 2001). Somerville's work was never entirely forgotten within histories of science and geography, but has frequently been marginalised or minimised. She has been described as a brilliant scientist 'treated badly' by history (H. Rose 1994), which is particularly surprising given Somerville's impact whilst alive, as Creese (2004a) noted, 'Perhaps no other woman of science until Marie Curie was as widely recognised in her own time'. Neeley (2001) describes Somerville's marginalisation in histories of science as a conscious 'forgetting', which would have seemed implausible at the time of her death, and that this forgetting can be attributed principally to three factors. First, that Somerville 'fell outside the patterns of history', because she had no major discovery or technique to attach to her name. This 'heroic discoverer' model, which dominated narratives of the history of science, has clear parallels with the main criteria for fame in geographical circles in the mid- to late nineteenth century. This model was established in the history of science by Whewell's 1837 *History of the Inductive Sciences*. Whewell was an admirer of Somerville's work but defended himself against contemporary criticism for his omission of her from his text on the basis that she had not made an original discovery; he courted the Somervilles at Cambridge, but he 'was not willing to create a space for her in history' (Neeley 2001: 219). Secondly, as science was increasingly discursively separated from questions of faith, work such as Somerville's, which was written from a teleological point of view was marginalised. Thirdly, the process of specialisation in science was accompanied not only by a marginalisation of more integrative work, but also by a clear process of masculinisation as the scientific community sought cultural status (see Fara 2004). Debates about women's political rights and place in society also intensified in the late nineteenth and early twentieth century (especially as suffragettes became more militant), and these political questions caused social conservatives to reject Somerville's work as indicative of wider claims for women's abilities and entitlements. Neeley also suggests that Somerville's ability to conform to feminine norms of her time and class whilst ensuring her acceptance whilst alive, may have contributed to the marginalisation of her work later. Somerville's autobiography, written late in life and edited posthumously, represents her as a woman of science but seems to stress her faith and traditional feminine characteristics, including her self-doubt as a scientific author (which was not found in her publications) and it is this, coupled with her diplomatic silence on some controversial issues, which accommodated those who wished to minimise her achievements or dissociate her from controversial issues. For Neeley it is *Personal Recollections* that is the lynchpin on which Somerville's reputation transforms from great to good, not least in her own supposition that women might not have the 'spark of genius' (Patterson 1969; Sanderson 1974). Always careful not to claim too much for herself, Somerville and her editor-daughter seem to focus on her mission to contribute to the ongoing debate concerning women's ability to study the sciences at the highest level. This seems to have caused them to tailor the discursive structure of Somerville's life's account above all in terms of scientific attainment without loss of feminine characteristics (Neeley 2001).

In terms of the geographical canon, Somerville has had early and regular reference (e.g. Markham 1881; Mill 1930; Baker 1948; Sanderson 1974; Gregory 1985; Livingstone 1992). Baker (1948) embedded Somerville very much in the subject's history, by linking her through Bates to Mackinder (it was Bates who later invited Mackinder to make his pivotal address to the RGS in 1887): 'In this strange way Bates, the editor of Mary Somerville's book was the link between the geographer of 1848 and the maker of modern geography in Britain' (Baker 1948: 221). Baker's observation, based on a simplistic whiggish notion of the progressive lineage of ideas, is interesting as Mackinder implicitly gains legitimacy from the link with Somerville.

The number of editions and the length of time *Physical Geography* was used as a text supports this bridging analogy and Mackinder repeatedly stressed physical geography as the foundation of all geographical knowledge, but this has to be seen as a reflection of a wider geographical discourse of which Somerville's book was a part, albeit an important one. There was a lack of continuity between the mid-nineteenth-century publications of geographical authors such as Somerville and Huxley and the turn-of-the-century establishment of professional geographical scholars in British universities (Martin 2005). This hiatus represented something of a vacuum, which may have allowed repeated editions to be published but did not provide an *academic* arena for developing and building on their geographical work. Sanderson (1974) overstates the case when suggesting that Somerville should have the title of first English geographer as well as first English woman scientist, but undoubtedly Somerville had an important impact on both the scientific status and wide readership of things geographical in the nineteenth century. Whether it is her attempt to frame a reconciliation of nineteenth-century scientific findings about the earth and its processes with theistic beliefs, or her discussion of steam power's 'annihilation of time and space' or her holistic approach to physical processes and the relationship of peoples to their environment that moved beyond state boundaries, Somerville's *Physical Geography* represents a personal perspective which combined a wide spread of contemporary beliefs and intellectual frameworks and as such offers us an important window on geographical knowledge and its developments in the nineteenth century. Recent scientific biographies and the republication of Somerville's works by Thoemmes Press are indicative of both the quality of her writing and renewed interest in Somerville's work within the history of science; equally they represent a significant resource and opportunity for insight to the history of geography, not least in her definition of physical geography and expositions on the relationship between people and environment, and physical processes such as the heat budget.

Isabella Bird

Isabella Bird (Figure 2.2) was born on 15 October 1831 at her mother's family home of Boroughbridge Hall, Yorkshire. Her mother Dora (née Lawson) was second wife to Edward Bird, a middle-class mature entry to the clergy who was related to the Wilberforce family and shared many of their values. The family moved with Edward's postings to Tattenhall in Cheshire, Birmingham and Wyton, Huntingdonshire (Middleton 2004a). She had no formal schooling but was recognised as an intelligent child and studied with her cultured parents, gaining an education far wider than most Victorian girls. Isabella and sister Hennie studied French, literature, history,

Figure 2.2   Isabella Bird.
Source: Courtesy of the RGS-IBG Archives

drawing and the Scriptures with their mother, and Latin and botany with their father; Isabella went on to independent study of metaphysical poetry, chemistry and biology. The Bird family also encouraged charitable and Christian missionary work (Barr 1985; Birkett 1989; Middleton 2004b) and included the role models of two strong-minded aunts who were missionaries in India and Persia (Gergits 1996).

From her youth Bird suffered from a debilitating spinal complaint and related depression. At 18 years of age she had an operation to remove a benign spinal tumour, but this was only partially successful and poor health was a permanent feature of her life. A post-operative holiday is credited with initiating her love of Scotland and better health motivated her wider travels. Bird travelled to the USA in 1854 and 1857, and achieved a modicum of success with the anonymous publication of her account of her first journey, *An Englishwoman in America* (1856). This was perhaps most significant in initiating a lifelong friendship with her reputable publisher, John Murray. Her father died in 1860 soon after her second trip to America, which necessitated the family move to the more economical Castle Terrace, Edinburgh.

During this time Bird submitted to semi-invalidity and devoted herself to 'worthy' causes and writings (such as for the magazine *Leisure Hour* and *Family Treasury*), but also joined Edinburgh's intellectual and philanthropic community through her friendship with Ella Blackie, wife of Professor John Blackie. Her publication of 'Notes on Old Edinburgh' provided 'a dramatic and powerful description of slum conditions' (Gergits 1996: 33) and is indicative of her engagement with social issues in the city. It was only at over 40 years of age and after her mother died in 1866 that Bird undertook more adventurous travel, albeit on medical advice, starting with Europe and an unhappy trip to Australia (Barr 1985).

On assurances of the curative properties of the air on the west coast of America, Bird set sail from Australia, only to find herself disembarking at the Sandwich Islands (Hawaii) at the request of a fellow passenger whose son was ill on board. This uncalculated decision was to alter Bird's life. It was here that Bird was to discover the twofold liberty of travelling in a foreign country and the use of a Mexican saddle (as opposed to the conventional side saddle) and the freedom of what she called the Hawaiian riding dress (Barr 1985), This combination allowed her to ride astride, whilst retaining feminine propriety. Although later she had to defend the latter as ladylike in the face of media comments to the contrary (Birkett 1989). These two innovations were to ease her back pain in travelling and liberate her from her socialised concern about decency of attire, the value of which cannot be overstated in terms of enabling her future travels. She became renowned for her resilience when travelling compared to her semi-invalidity when at home (Birkett 1989).

During her seven-month stay in the Sandwich Islands Bird spent time with the European missionary community, relished travelling with Hawaiian guides, learnt about the local topography and customs and ascended the 14000 feet of Mauna Loa with William Green (British Consul and amateur volcanologist). She also met and received an award from the king. Disregarding the temptation to stay and settle after a proposal of marriage, Bird pressed on to America, to her original aim – The Rockies – where she climbed Long's Peak (the 'American Matterhorn') with notorious local trapper Jim Nugent, 'Mountain Jim'. Bird was tempted by romantic entanglement with Nugent, but his alcohol dependency and wild character were too great an obstacle. She travelled on to Colorado and other better known places in America, where her hatred of 'tourist traps' was to be her making. It was her book *Six Months in the Sandwich Islands* (1875) that brought her fame and acclaim. *The Spectator* described the book as 'remarkable, fascinating and beautifully written' and *Nature* acclaimed the accuracy and breadth of her botanical knowledge (Barr 1985; Birkett 1989). The account of her American adventure *A Lady's Life in the Rocky Mountains* (1879) ran to eight editions, staying in print until 1912. Bird travelled to Japan and

Malaya 1878–9, returning via Cairo and Mount Sinai where she contracted rheumatism of the chest muscles. In Japan she travelled to places and met people largely unknown to Europeans, writing about the interior and minority groups such as the Ainu. *Unbeaten Tracks in Japan* (1880) proved to be very popular (Middleton 2004b), and there is still a strong academic and popular interest in Bird in Japan where three monuments to her mark places she visited (Cornwell 2004).

The death of Bird's sister Hennie from typhoid in June 1880 had a devastating effect on Bird and was to be a major factor in changing the style of her writing (Barr 1985). Bird published *The Golden Chersonese* in 1883, the last of her books to draw on her letters to Hennie, which had shaped the lively style of her texts. It was this book which was to gain her critical acclaim and recognition as an accomplished travel writer (if not much commercial success). The book included her grim account of the prison and execution grounds at Canton, which Murray wanted to excise, but Bird insisted on keeping on the grounds that people should know what happened in 'the dark places of the earth' (Middleton 1965: 36). More recently her book has been critiqued as full of western assumptions (Gergits 1996). Bird, still in mourning, married her long-time suitor and family doctor, John Bishop, on 8 March 1881 and didn't travel during their five-year marriage when both partners were plagued with ill-health. John Bishop died on 6 March 1886, two days before their fifth wedding anniversary. After three months' training in nursing, a spell of journalistic work for Murray in Ireland and a short-lived attempt to run a nursing home in Maida Vale, Bird set off in 1889 for Kashmir and Tibet (Middleton 2004b). Finding consolation for bereavement in her faith, Bird was baptised by Dr Spurgeon as a symbol of her commitment to missionary work and she went on to establish the John Bishop Memorial Hospital in Srinigar and financed several other hospitals and orphanages under the auspices of the Church Missionary Society (Barr 1985).

Travelling through the Punjab to Simla, Bird met Major Herbert Sawyer, who persuaded Bird to accompany him, acting as a cover and assistant for his clandestine surveying of politically sensitive areas in western Persia for the Indian Army. Consequently Bird had to have her manuscript of *Journeys in Persia and Kurdistan* (1891) checked by the Internal Department of the India War Office before publication. While Sawyer appeared as 'M' in Bird's account, the only mention he made of her in his official report was to describe how she had to draw a pistol when they were threatened by a riotous group (Middleton 1965). Bird's travels overlapped with Nathaniel Curzon, who they met on a steamer on the Tigris. Curzon's own book on Persia appeared after Bird's and cited her work 20 times; he also wrote a complementary review of Bird's own book. On return home Bird addressed Section E of the BAAS in 1891 on the 'The Bakhtiari Country' (see *Proceedings of the Royal Geographical Society* 1891: 6333–5). She also used her experience to contribute to political debates in Britain concerning the status of the Kurds, Nestorians and Armenians, briefing Prime Minister Lloyd George over dinner at the Murrays' home and addressing a House of Commons committee (Barr 1985). Her planned book based on her ongoing travels in Kashmir and Tibet (*A Lady's Ride through the Western Himalayas*) was never written, but she published two short religious treatise on Tibet (1894) and Japan (1898).

Aged 63 and in poor health Bird set off for Japan in January 1894, spending three years moving between Japan, Korea and China, at a time of significant regional unrest including the outbreak of the Sino–Japanese war. The Korean monarch, caught between the power of the Japanese and the country's historical antipathy to them, used Bird to send a secret letter to the British Foreign Secretary (Middleton 2004b). Her subsequent book, *Korea and Her Neighbours* (1898), was well received and went into a second imprint after only 10 days.

In early 1896 (in her 65th year) Bird began a journey from Shanghai up the river Yangtze, chartering a local barge to take her from Ichang, braving the rapids. When travelling cross-country in open chair, she aroused hostility and suspicion from locals in some areas, which culminated in a murderous riot as she entered the city of Liang-shen Hsien when she was knocked unconscious by stones thrown at her (Chisholm 1898). She spent some time recuperating with missionaries before pressing on to the west, despite the barriers of officialdom, and ultimately received a warmer reception in the Buddhist Mantzu area (ibid.). On her return she presented a paper on her travels to the RSGS and the RGS. The first woman to present a formal paper to the latter, she was welcomed by the Society's president as an 'old friend' and author of national repute. Although she (perhaps pre-emptively) apologised for her lack of scientific knowledge at the beginning of her address, the president stressed the arduous and risky nature of the journey and the benefits accrued to geographical knowledge: 'We must remember that Mrs Bishop considered any risk worth running in order to advance geographical discovery. I therefore think that the thanks of geographers are due to her for having undertaken this perilous and very important journey' (*The Geographical Journal* 1898: 50). *The Yangtze Valley and Beyond* (1899), richly illustrated with her own photographs, was the product of her final major journey. After failing to settle in her new house, Hartford Hurst on the Ouse, and forced to abandon plans for another trip to China, she spent six months of 1901 in Morocco, travelling some 1000 miles (*SGM* 1904). For Bird 'life was intolerable for her without the prospect of a journey ahead' (Trollope 1983: 145) and when she died at the age of 73, her trunks were packed ready for her next journey to China. Isabella Bird Bishop died on 7 October 1904, having returned to Edinburgh in her last years to be near friends.

Publications

The personal note in Bird's writing, which had been an important element of her commercial success, had been achieved through editing her letters written en route to her sister. Inevitably this sense of familiar engagement was reduced after Hennie's death, resulting in a more detailed and 'professional' character to Bird's later books. This stylistic shift could also be attributed to her membership of the RGS and an attempt to meet the requirements of the geographical discourse. However, this shift was not universally welcomed and Middleton (1965: 40) described her later work as on the one hand including some of her finest descriptive writing, but on the other 'over-weighted ... by trade statistics and moral reflections'.

Having kept a low profile in the RGS women's membership debate, she was nonetheless identified as the tinder point of the debate by others within the RGS. This makes it difficult to separate the reception of her work from reaction to her perceived role in bringing the issue of women's admission to the fore with its unsatisfactory and undignified conclusion. Bird went on to use her fellowship to her benefit, undertaking training in geographical skills. Whilst she felt her work was not necessarily distinctly *geographical*, she did feel she had made a contribution to the general knowledge of the countries she visited (Birkett 1989). She also felt that her contribution offered something distinctive (as Mary Kingsley also hinted of her own work), having travelled as a woman, unaccompanied by other Europeans as a rule, and as a traveller who made the acquaintance of the local people (Domosh 1991a). *Korea and Her Neighbours* (1898) was widely and favourably reviewed, being applauded in *Nation* (Barr 1985). In *The Geographical Journal* it was reviewed by G.G. Chisholm (1898) and his rather reserved commentary, which mortified Bird,[23] says as much about his own interests and his perception of the interests (geographical and otherwise) of the RGS, as it does about Bird's book. The review opens: 'A work by Mrs Bishop is sure to be full of interest, of novelties, of instruction, and food for reflection' (p. 288), echoing Curzon's earlier reference to her 'novel and interesting information' (1892: 300), an example of reiterative opinion and the process whereby once a text is identified as female authored, critics choose appropriate (limited and limiting) adjectives and phrases (Moi 1985; Mills 1991). Chisholm 'corrects' her on a number of points including the spelling of place names and concentrates his discussion on what he sees as the distinctively geographical content of the book. He comments on Bird's population figures as well as the physical geography in a display of his own text-based knowledge of the area, but does recognise that she added some important information about the Han river bed and valley. Reflecting the Society's and his own interests he advised British statesmen and merchants to pay attention to the sections on trade.

Bird also felt that her work would have received more recognition had she been a man. Reiterating the claims made for Mary Somerville by Frances Power Cobbe, Bird argued '... that I had been a man I should undoubtedly have received some recognition from the RGS. I consider myself to deserve it at least as much as Mr. Warrington Smyth [fellow traveller in China]!'[24] Bird's work compares favourably with that of other fellows of the RGS, for example the account of Wellby and Malcolm's journey across Tibet (*The Geographical Journal* 1897) differs little from Bird's account of her travels in Tibet: both used surveying methods, recorded temperature and noted changes at altitude and took photographs. Wellby and Malcolm collected 80 species of plants and accidentally 'discovered' the source of the Yangtze (their original aim), when a merchant they travelled with told them they had met at the Chumar river – the source of the Yangtze. The difference between the achievements of the two expeditions is that the geographical discourse of the time epistemologically privileged a serendipitous 'discovery' over the social comment and observations of Bird. The socio-intellectual climate that prioritised river sources over cultural observation was in part a gendered one (Maddrell 2004d). After the furore of the RGS debate, Bird made a telling gesture by asking her publisher to send a copy of *The Yangtze Valley and Beyond* to the RSGS in appreciation of its acknowledgement of her work by granting her honorary membership of the Society. Bird did not conform to the masculinist criteria of geographical exploration in the same way as Mary Kinglsey (see Chapter Four), but she contributed to an understanding of the places she visited and increasingly complied with the geographical discourse in providing enmeshed empirical and commercial information on those places. This included a degree of sensitivity to women's lives, seen for example in her observations about the servitude of Korean women to their men folk's white clothing (Bird 1898) (in contrast to Curzon's (1892) comments on the same subject), and the gender equality in the Mantzu communities she visited (Middleton 1965).

Bird's writing was that of a travel writer rather than a scientist (according to contemporary definitions), but her photographic and RGS training had an impact on her writing (Maddrell 2004d). Whilst on her journey up the Yangtze Bird used her new skills of photography to great effect, photographing the landscape, local people and practices, rinsing her prints in the Yangtze itself for a lack of clean water on board the boat. When travelling up the River Han in Korea Bird talked with the locals and occupied herself 'taking geographical notes, temperatures, altitudes, barometric readings and measurements of the river (nearly all unfortunately lost in a rapid on the downward journey), collecting and drying plants, photographing and developing the negatives under difficulties ...' (cited by Barr 1985: 277). She also appears to have tried to escape from being defined exclusively as a female subject and author: 'I am specially pleased,' she told John Murray who had sent her a batch of favourable reviews 'that the reviewers have not made any puerile remarks on the feminine authorship of the book or awarded praise or blame on that score' (Barr 1985: 188). Contrary to this hope Bird is principally remembered as a *woman* traveller and has been celebrated as such, through the republication of her work by Virago in the 1980s, numerous biographies and press coverage on the centenary of her death (Cornwell 2004). Her work was central to both the debates around women's membership of the RGS and more recently, albeit fleetingly, in Domosh and Stoddart's 1991 exchanges about feminist historiography. There is much more to her work than Stoddart (1991) credited (based on accounts of her early writing), but there has been relatively little engagement with her publications within geography compared to literature. Her work merits much more attention within the history of geographical thought and historical geographical studies than I can offer here.

Conclusion

At its foundation in 1830, the RGS shaped the geographical discourse as a subject principally concerned with gathering field measurements and commercial data from abroad. The elite social status of the founders, with close links to the personnel and institutions of the state, as well as other learned societies, defined its ideal fellowship explicitly in terms of 'high-ranking' male membership. The anxieties expressed in the women's fellowship debate reflected fears of intellectual and social dilution, but more particularly the feminisation of the physical space of the Society and the epistemological space of a 'manly' discipline. In many ways the furore that emerged over women's fellowship reflected the full range of views concerning women's abilities and roles in public life in wider society. Details of the debate show that many of the existing male RGS fellows supported women's membership, as did all the other non-metropolitan geographical societies. This complicates the image of nineteenth-century British geographical societies as representing a wholesale embodiment of patriarchal closure. Patriarchal closure of the RGS was achieved in 1893, not due to the consensus of members, but to the ability of a faction to mobilise sufficient support in London at the Special Meeting when women's membership was considered, as well as making legal challenges to any procedural errors. Not only was there a geography to women's access to geographical societies prior to 1913, but also the early foundation of the RGS at a time when women had less access to formal education and intellectual arenas meant its gendered assumptions about membership were enshrined in its constitution. The institutional biographies of the non-metropolitan societies and individual biographies outlined here demonstrate the varied ways in which women's travel and geographical writing did contribute to geographical knowledge, but also how success was contingent on appropriate avenues for publication and careful negotiations of gendered roles in the nineteenth and early twentieth century. The period 1830 to 1913 was a time of great social, economic and political change as could be seen in the very different experiences of Mary Somerville and Marion Newbigin (see Chapter Three), who gained initial credibility through their social connections and university qualifications, respectively. Questions of gendered institutional and epistemological inclusion and exclusion will recur throughout following chapters and will be returned to in the Conclusion. The next chapter addresses the role of women as editors of geographical society publications.

Chapter Three

Marion Newbigin and the Liminal Role of the Geographical Editor: Hired Help or Disciplinary Gatekeeper?

Editors are by their nature boundary and gatekeepers, not least in academia: gatekeepers in their ability to allow individuals into or exclude them from print in a given journal, with all that this implies about disciplinary recognition; boundary keepers in terms of implicitly marking the limits of acceptable knowledge through their acceptance of papers. They manage the pace and nature of changes to those epistemological limits through encouraging or limiting emerging work, notably from the next generation of writers. In an era dominated by print communication, editors mediated between the society council and its membership, largely through the pages of the society's journal, which was despatched on behalf of the society to homes, workplaces and libraries. Yet depending on the journal and the autonomy given to a particular editor, the role can be seen as one of academic administration, employee rather than academic figurehead. The overwhelming editorial presence in British geography 1850–1970 was male, but there were women editors, whose roles have had varying degrees of impact on the discipline, notably at the *Scottish Geographical Magazine* (*SGM*). The most important of these was Marion Newbigin, who was the first and longest-serving female geographical editor, one of those who bridged the institutionalisation of geography in private societies and the professionalisation of the discipline in public universities. The decision to devote this chapter principally to the work of Newbigin is not to attribute to her greater significance than other figures within the history of women geographers, but is a reflection of both the importance of the editorial role and the difficulty of placing her in other chapters because of her hybrid liminality. As will become apparent from her biography below, Newbigin was completing a science degree at the same time that women's membership of the RGS was being so fiercely debated (1892–3), making her more academically qualified than many of the male fellows. She was drawn into geographical circles in

Scotland through her zoological work and was appointed to the paid post of assistant editor and later editor of the *SGM*. She taught and examined in British universities on a part-time basis, often seeming to be a 'flying geographer', travelling backwards and forwards for meetings and lectures. She published extensively and her output is analysed below, but it is widely held that it was her editorial post which was the locus of her most significant influence within early twentieth-century geography. Newbigin negotiated these ambiguities, in no small part through her own publication record as she established herself as a geographical author, and she moved from assistant editor to a position of disciplinary leadership. However, this was not always a matter of smooth progression or one of financial security.

Biographical Background

Marion Newbigin was born on 23 September 1869, the same year Mary Somerville received her medal from the RGS. By the time Newbigin was an adult, many of the changes in women's education that Somerville had hoped and campaigned for had been achieved. There was free universal elementary education and suitably qualified (and monied) women had access to university courses from 1876, even if access to different universities and actual qualifications was patchy. At the time of her birth, Newbigin's parents, Emma (née France) and James Leslie Newbigin, lived at Greenbatt Cottage, Alnwick, in Northumbria.[1] Her father was a chemist and druggist, and, perhaps influenced by his interests, Newbigin chose to study natural sciences at University College of Wales, Aberystwyth and Edinburgh's Extra-Mural School of Medicine for Women. James Newbigin was certainly unusual in believing that daughters should be educated to the full, whereas sons should support themselves from their teens, but this was a regime both Newbigin daughters and sons seemed to thrive under (Boog Watson 1967–8). After studying at Heriot-Watt College in Edinburgh she passed her London University matriculation examination in June 1890 and entered the University College of Wales, Aberystwyth in September 1891, aged 22. Whilst there she studied chemistry, physics, biology and mathematics, passing her London Intermediate examination in 1892 and was awarded a BSc (Hons) in zoology (second division) in 1893. She went on to gain a DSc in 1898, both degrees awarded by London University's external degree conferment. The first woman to take a degree at Aberystwyth had done so only in 1888, so Newbigin was among the early cohorts of women graduates; she was only the second woman at Aberystwyth to be awarded a DSc.[2]

Facilitated by the admission policy of the new higher education college at Aberystwyth and the conferment of external degrees by London University, Newbigin was among the first women in Britain to study and gain qualifications in the emerging natural sciences. In turn she was lecturing and examining these subjects in higher education from 1897 when she succeeded her tutor J. Arthur Thompson as lecturer in biology and zoology at the Edinburgh Extra-Mural School of Medicine for Women. This training in natural sciences formed the basis for Newbigin's involvement with the *Challenger* project in Edinburgh, which laid the foundation of modern oceanography and contributed to zoology and meteorology. Newbigin was part of a huge team of specialists analysing the data collected by the *Challenger* (1872–6). She undertook early original research on coloration in plants and animals, working at the Millport Marine Biological Station and in the laboratories at the Royal College of Physicians. Newbigin produced several papers, including joint publications with N.D. Paton and others, her research culminating in the publication of *Colour in Nature* (1898) and *Life by the Seashore* (1901) (Maddrell 1997, 2004b; Creese 2004b).

It may have been Newbigin's involvement with the *Challenger* project which led to her subsequent interest in geography (Lochhead 1984). The project was certainly influential on several of the new generation of professional geographers such as H.R. Dickson, H.R. Mill, A.J. Herbertson and J.Y. Buchanan. However, Newbigin's sister, Hilda, attributed the move to geography largely to the influence of James Geikie, whose geological lectures Newbigin attended 1890–1.[3] Newbigin never officially explained her career move to geography, but in correspondence with RGS secretary Edward Heawood, she implied that financial necessity was a driving force in her career switch: 'I had once a certain limited reputation as a biologist – before I abandoned biology for bread and butter …'.[4] Newbigin was a single woman, and appears to have been dependent upon her own income generation, living in a shared house with her sisters Hilda and Alice (later joined by Maude on her retirement). Hilda reported that Marion undertook long hours of marking Civil Service examination scripts – up to a thousand at a time – on top of her editorial work, writing, lecturing and external examining, in order to 'keep the pot boiling'.[5] Marion and her sisters obviously shared broad interests and skills bases as two of her sisters also worked in higher education: Alice held a post at the Edinburgh College of Agriculture and Maude was lecturer in history and deputy principal of the Day Training College at Portsmouth (Creese 2004b). Florence Newbigin seems to have played the role of academic helpmeet, being Marion's travel companion at home and abroad as well as producing the line drawings, sketch maps and indices for several of her sister's later books.

In terms of the history of geography, it was Newbigin's 1902 appointment as assistant editor of the *SGM*, on the recommendation of James Geikie, honorary editor of the journal, which was a pivotal moment in her career as a 'geographer' because it brought her into the RSGS and wider geographical establishment. Newbigin continued as editor of the *SGM* until her death in 1934, wielding a growing and ultimately tremendous influence on the journal and its authors over three decades. She demanded high standards (Freeman 1976a), but also commissioned and supported new authors, which could be very time consuming (Taylor 1934a). Newbigin also worked as assistant editor of the Scottish Forestry Society's journal. Few papers relating to Newbigin's work have been kept at the RSGS, but letters in the RGS correspondence files show her tact in dealing with leading figures in the RGS establishment as well as her confidence in her own views.

Marion Newbigin never held a permanent full-time university post, but made a living from what we would now describe as a portfolio career. Having found an academic post at the Edinburgh Extra-Mural School of Medicine for Women, she lost it when women were admitted to the mainstream university school in 1916. Newbigin, along with all her sisters, were 'strong supporters of the feminist cause' (Boog Watson 1967–8; Creese 2004b) and would have welcomed this move to equal access in higher education, but as an obiturarist wrote: 'when a new victory for feminism closed the Women's School of Medicine the life of at least one distinguished woman teacher was further complicated' (*SGM* 1934: 331). Whilst women may have gained access to higher education and qualifications at this time, academic employment opportunities for women were few and far between, largely limited to women's and teacher training colleges. Geography itself was only just emerging as a university discipline in the UK and G.G. Chisholm filled the first modern geography lectureship in Scotland when he was appointed at Edinburgh University in 1908. Newbigin lectured on plant and animal geography at Bedford College, London as a visiting lecturer for 15 years and at Patrick Geddes' summer schools at the Edinburgh Outlook Tower, as well as being an occasional lecturer at Edinburgh University. She was external examiner at Aberdeen (during the three years Thomson held the chair of zoology), at Aberystwyth geography department circa 1919–21,[6] at Glasgow and Manchester universities and a long-standing examiner on the Board of Geographical Studies of the University of London. She is reported as disliking lecturing in her later years, but in terms of both content and style she was remembered as a 'lecturer of that conspicuous ability which seems to become rarer every day' who attracted 'unusually large audiences' (*SGM* 1934).

Newbigin was elected a fellow of the RGS on 20 November 1916, three years after the general admission of women. Clearly no one from the RGS had offered to put her name forward for fellowship (although she had been asked to review books) and Newbigin found it necessary to write to the RGS secretary: 'I am very anxious to become a fellow of the RGS, do you suppose there would be any objection?'[7] She was duly proposed by RGS secretary Heawood himself and J.E. Bartholomew, a leading figure in the RSGS. Newbigin was also active in the Geographical Section of the BAAS, which provided an important forum for academic geographers from 1913 onwards (Beaver 1982). She gave papers in 1916, 1922 and 1924 and was appointed president of the Section in 1922; she also served as vice president in 1921 and 1923–4.[8] Newbigin was the first woman president of Section E and whilst her presidential address on 'Human geography: first principles and some applications' was a mainstream topic of interest to the emerging academic discipline (Beaver 1982), the morning's programme was unique, with three out of the four speakers being women, Ellen Semple[9] and Hilda Ormsby having preceded Newbigin herself. Within five years of joining the RGS Newbigin was honoured by the Society when she received its Back award in 1921.[10] Replying in diffident terms, Newbigin wrote formally to the RGS: 'My grateful thanks for their kind appreciation of my modest contribution to geography as shown by the award of the Back Grant. Such appreciation is a great incentive for further work.' Newbigin reported that she would use the award to purchase maps and atlases, but was gently chastised by Hinks for not attending the award dinner in person.[11] In 1923 Newbigin was awarded the RSGS Livingstone Gold Medal and she and Ella Christie were the first women to be appointed to the RSGS Council in the same year. It is no coincidence that these honours and signs of incorporation followed hot on the heels of an initial recognition within the BAAS: this represents a common model for both women and men which might be described as the 'accumulator effect' of awards, as seen in the case of Mary Somerville in Chapter Two.

Newbigin travelled widely in Europe, drawing on her observations in her writing. Correspondence shows she was in Palermo, Sicily when she received news of her award from the RGS and Freeman (1980) comments that her (1907) paper on the Swiss Valais was undoubtedly based on fieldwork. Newbigin's presence in the landscapes described in her writing is often depersonalised and implicit, along the lines of 'the traveller will note ...' etc., but her views are explicitly expressed in her travel book *Frequented Ways* (1922). The long subtitle of this text is indicative of its purpose: 'A general survey of the landforms, climates and vegetation of Western Europe, considered in their relation to the life of man: including a detailed study of some typical regions' – no light travel account this. Starting with what Newbigin describes as a 'panegyric' to Europe, she extols the virtues of Europe as a breeding ground of civilisation and goes on to list the merits of the geographical variety to be found within the relatively compact continent of Europe, flying in the face of those who insisted that true travel could only occur in the New World. For Newbigin, travel was the natural occupation of the educated classes: 'we all travel as much as we can' (1922: 1), but it was also a personal passion and she hoped that despite the detail associated with offering geographical insight to the places visited, 'the joy of intensive travel shines through' (*op. cit.*, v).

Figure 3.1   Marion Newbigin.
Source: courtesy of the RSGS

Marion Newbigin died suddenly on 20 July 1934 at her home on Chamberlain Road, Edinburgh. Her sister attributed Newbigin's death at 64 years of age to the heart trouble she had experienced since a young girl and hinted at the long arduous examining work she undertook (and had just returned from). The only portrait of Newbigin in circulation (see Figure 3.1) shows a mature and confident but not necessarily emotionally open woman. However, we don't know for what purpose the photograph was taken, and serious faces were the convention for late-nineteenth- and early-twentieth-century photographic portraits. She had worked hard to establish a reputation as a serious scholar and this is what is represented in the photograph. Whilst remembered for her exacting standards and lively mind, obituarists also commented on her fresh outlook, sense of humour, the time she gave others and her love of her garden and flowers, making the distinction between the professional and private person (see *SGM* 1934). *Plant and Animal Geography* (1936) was published posthumously, completed with editorial help from H.J. Fleure and the addition of two chapters on soils by Margaret Dunlop. That others could pick up and complete her book is indicative of the level of organisation in Newbigin's work, as was the fact that after her death everything was found ready for the 50-year jubilee edition of the *SGM* (see Freeman 1976a). Whilst the quality and influence of the *SGM* was considered Newbigin's 'proper eulogium' (*SGM* 1934), she wrote 12 books and over 20 papers, which was a large textual legacy in her day and this will be discussed and evaluated prior to considering her place within the historiography of geography.

The Production of Newbigin's Work

Newbigin's geographical writings can be grouped under two broad headings: (a) physical/scientific geography (and biogeography in particular); and (b) regional/political; with epistemology/pedagogy making up a smaller third theme in her work. However, whilst individual texts might reflect a particular theme in the main, the different themes can be found interwoven in many cases. The significant contribution of Newbigin's training, travels, fieldwork and photography to her writing have been noted above, but to this list we need to add her wide knowledge of literature and her use of local knowledge as vital elements of her work. Newbigin's grounding in relevant literature, especially French texts, is evidenced in occasional references and short bibliographies (as well as the ways in which she conceptualised her subject). Her comment that 'one cannot be said to know any part of the earth's surface until one realises how it appears in the eyes of its inhabitants' (1922: v) suggests a deliberate engagement with locals to gain an embedded view rather than the limited perspective of an outside observer.

Physical/Scientific geography

Newbigin's original training was to have a permanent bearing on her vision of the nature and practice of geography, as well as having an influence on the tenet of Scottish geography in the early twentieth century. Although earlier more generalist writings, such as Zornlin's and Somerville's physical geographies, had included chapters on plants, Newbigin, as a biologist in the first instance, gave greater attention to flora and fauna than was generally the case in geographical work of her time. In her introductory chapter to *Man and His Conquest of Nature* (1917 third edition) she repeatedly refers to the concepts and explanations of the botanist and similarly the zoologist in her later chapter on the sea. Using her specialised knowledge of biological influences within geographical discourses of the early twentieth century, Newbigin defined the subject in a way that has definite resonances with Somerville's (1848: 1) definition: 'The main object of geography may be said to be to bring out the relation between the life of organisms – of plants, animals and, in particular, of man – and the physical conditions which prevail on the surface of the globe' (Newbigin 1929: xv). This can be seen carried through in her biogeographies of *Animal Geography* (1913) and *Plant and Animal Geography* (1936). Whilst Hilda Newbigin described the latter as written 'for fun', Livingstone (1992) suggests the former was written with a view to keeping zoology within the geographical remit.

In a paper initially read to an audience largely made up of teachers at the Geographical Association in London, 'The geographical treatment of rivers' (1916) addresses the status and practice of physical geography in relation to geology and uses rivers as an example of how an inherently geographical approach can be taken. Describing the state of British physical geography as 'very chaotic' Newbigin argued that geography would never gain its proper place in British universities or public esteem unless geographers created their own disciplinary epistemology and ontology rather than merely reproducing the work of others (in this case geologists). In the case of rivers Newbigin argues that the geologist is principally concerned with the river as process, as historical modelling agent and often absorbed by the unusual. In contrast she suggested that geographers should primarily be concerned with the 'normal' rather than exceptional, with the lower rather than the upper reaches of the river, not least because this is where rivers have the most impact on the lives of people and it is this relationship between people and environment which should be the focus of geography. Thus a geographical treatment of rivers should address questions of accessibility and navigability, routes into and out of interiors, settlement, trade and agriculture. Without diminishing the value of what she defined as geological processes, Newbigin challenged her audience and readers to see these as a starting point, to take geographical study and understanding beyond the geologists' remit (as seen in the works of Lapparent and de Martonne), and in so doing make a clearer demarcation for geographical knowledge.

An important part of Newbigin's scientific approach to geographical work was the incorporation of Darwinian theory: she saw the publication of *The Origin of Species* in 1859 as 'marking the beginning of the distinctively modern period of geographical science' (1911a: 1) in which 'He [Darwin] showed that there is a delicately adjusted balance between organisms and their surroundings, taken in their widest sense' (*op. cit.*: 11). Given that *Origin* was published 10 years before her birth and that she grew up with a father who was scientifically trained, we can speculate that Newbigin may have learned Darwin at home, but regardless of that her natural science courses and *Challenger* work would have been imbued with Darwinian theory. *Origin* oriented Newbigin's research, as perhaps might be expected given her biological training, and her work is conceptually interwoven with references to succession, evolution and competition. These ideas are applied to flora and fauna, for example 'The tropical forest is an example of what the botanist calls a "closed association" – that is, one where the struggle for life is so keen that no new species can force its way among the others' (1917: 8), but this idea was equally applied to people. Her whole thesis in *Man and His Conquest of Nature* is compressed in the title of the book and its introductory chapter 'The battlefield'. Newbigin articulated cases where people as a species have been 'worsted' by extremes in nature, but otherwise had uniquely succeeded in colonising the whole earth. However, influenced by French geographers Bruhnes, de Martonne and Vidal de la Blache (see Newbigin 1917: v), she tempered the popular role accrued to struggle and competition, cited Darwin's own stress on the metaphorical character of struggle in his work and concluded 'From the geographer's standpoint, therefore, it is better to say that Darwin's work has added a new interest to the study of interrelations' (Newbigin 1911a: 9). Newbigin is also sensitive to the damage done by human domination over other species, particularly those from what she calls 'civilised' nations, noting for example the threat to vulnerable human and animal groups and the impact of industrialised hunting of whales and seals. She refers to the extinction of larger animals as 'one curious revenge which wild Nature has taken upon her conqueror …. That [some] plants and animals should prosper while others disappear is one of the curious features of the struggle' (1917: 14–15).

As a scientist in science-dominated Edinburgh, it was clearly important to Newbigin personally, and to the wider membership of the RSGS, that geography be deemed a science. She credited Darwin with having endowed geography with scientific status which enabled it to escape from its 'chaotic jungle' or gazetteer-like 'matrix of mere facts', like a Rodin sculpture from a block of marble (Newbigin 1911: 13). Her early geographical writing very much bears the mark of scientific methodology and Darwinian thought, but hers was an evolving intellectual position – her regional studies were always grounded in the relationship between people and the environment, but in later work she increasingly recognised the plasticity of that relationship.

Regional and political geography

Having already written a 1905 editorial on the value of geography, Newbigin's first research paper in the *SGM* was a regional study of the Kingussie District in southern Invernesshire, appearing in 1906. At the time of writing the regional approach to geography was relatively new but would become the standard discursive framework and this paper follows the standard structure for the regional study: general description and delineation of the region discussed; climate; natural resources; population; and means of communication. Her study is detailed and supplemented with graphs and photographs, presumably her own as they are not otherwise attributed. In this article Newbigin demonstrates her own ability to write within the regional discourse, and in so doing she establishes her geographical credentials and provides a template for others, as well as positioning herself as knowledgeable about Scotland.

Drawing on detailed empirical data for evidence, such as climate and census figures, Newbigin puts these to work to explain the cause of economic decline and possible routes to recovery in the area, noting in her introduction that case studies of prosperous economic regions have dominated the new geography, but that 'the reason why certain regions have had their economic development arrested, has attracted less attention, though geographically it is quite as interesting' (1906: 285) (a point echoed recently by economic geographers such as Samers 2001). In her analysis Newbigin notes the Anglicization of the parish with the decline of Gaelic speakers, as well as the push factors which have led to male emigration causing a gender imbalance in the region. It is in her conclusion that Newbigin makes her most interesting observations. Clearly drawing on personal observation, she compared the economy and opportunities of Kingussie with similar highland areas in Switzerland, drawing particular attention to the economic impact of the lack of road access in Kingussie. Her reference to the comparative lack of a 'vigorous peasantry' in Kingussie may seem patronising to readers in the twenty-first century, but in essence she is talking about signs of well being for the working classes. She concluded with a forceful argument for democratic access in the area as opposed to the constraints of elite grouse and deer parks, asserting that geographers should rightly be concerned with these questions, albeit justified by the interests of national efficiency rather than more labour-centred politics. Democracy is a recurrent thread in her work.

Newbigin came to see the region as the most appropriate unit of geographical study, not least because it allowed intensive survey of representative sections of the globe or nation whilst avoiding superficial generalisations that more detailed knowledge might disprove. Given her training in and continued commitment to the scientific method, this process of verification was a very important part of Newbigin's epistemology, but her appreciation for the intensive survey within the regional approach was also based on her understanding of its empirical and explanatory efficacy: this was the means which 'makes it possible to bring out specific instances of man's response to his physical environment, with a certainty not otherwise obtainable' (Newbigin 1929: xvii). The last two books completed by Newbigin were *A New Regional Geography of the World* (1929) and *Southern Europe* (1932).

Both her local and wider regional studies were firmly based on Newbigin's understanding of the physical environment and her evaluation of its impact on human ways of life. In her discussions of the Balkans (1915a, 1915b, 1920a), Newbigin illustrated the relationship between the mismatch of topography and political units and the persistent political problems experienced in the region (Maddrell 1997). Some of her analysis is framed by elements of environmental determinism, national and gendered stereotypes, for example, when she discusses Mediterranean agriculture or reproduces representations of Turks as economically parasitic (Maddrell 1997), and in her representations of the Albanian people (Lyde 1915; Freeman 1980: 95). In many ways the adoption of the scientific method within Newbigin's (and others') geographical practice gave geography the appearance of being rooted in scientific objectivity but could obscure what were 'profoundly ideological questions' (Bell 1995a).

Newbigin's earlier work relied upon an application of classification for peoples, reproducing a long-standing tradition within geography texts, taking a 'scientific' approach which is deeply embedded with cultural values and hierarchy (Maddrell 1997, 1998). Newbigin's later discussions of race and nationality (1917 and 1918) seemed to undergo a marked change; they continued to represent a strong relationship between place and human characteristics, but being more sensitive to the plasticity of those characteristics. Contrary to her 1915 papers, Newbigin emphasised the human adaptive response to new environments, exemplified by Greek settlers in Asia Minor who lost characteristics associated with seafaring people and became 'tillers of the earth' (Newbigin 1918). Although in correspondence she described anthropology and folklore as off her beat,[12] this shift in her analysis can be attributed to giving greater causality to human rather than biological functions. Acknowledging the influence of Marrett's (1917) volume on anthropology, Newbigin contends that geographers – and politicians – had been overly influenced by the disciplinary inheritance of biological theory which adopted the belief that the mechanism that produces and maintains diversity in other organisms operates in the same way for people (Newbigin 1918). In relation to race this meant that within scientific discourse it was taken to 'imply that (1) that racial characteristics in man are fixed and hereditary, and (2) that they show adaptation to a particular environment, in which they have arisen' (1917: 318). She went on to argue that greater mobility combined with 'mental, moral and social qualities' made it possible for humans to adapt to any environment, ultimately meaning that 'the physical differentiate of race ... are of little importance under modern conditions' (1917: 321). This causes her to conclude further that 'race' was a political tool and that nationality was something that could change – both were relevant to the pressing contemporary political issue in Europe, peace in the post-war Balkans.

Newbigin continued to unravel neat distinctions between racial–national characteristics in her secondary-/university-level texts (1920a, 1929). In her book on Scotland she problematised the popularly held view of distinction between Highlander and Lowlander, making the case for a more environmental cause, differences being: 'less the result of the racial composition of the population in the two areas, than the different natural surroundings, and, associated with these, a difference in social and historical relations' (Newbigin 1920a: 89–90). However, these clear elements of environmental determinism seem to contradict her earlier book *Man and His Conquest of*

Nature (1912). These apparent inconsistencies were part of her ongoing evolving perception of the people–environment relationship as well as an attempt to match causal factors to individual settings, that is, to recognise difference rather than overgeneralise. In the same way she counselled against using broad climatic zones or outstanding relief features alone to explain the earth's variety as these can mask the often complex 'real and profound differences' in human response to their surroundings (1929: xv). In this she was of like mind with H.J. Fleure, but contra A.J Herbertson, L.W. Lyde and followers, who used large-scale climatic zones.

It is in Newbigin's regional studies that we can identify traces of gender sensitivity. They are only traces, rather than dominant themes, but nonetheless raise pertinent issues in the specific regions studied. The impact of male economic migration was noted above in the case of Kingussie. She discussed the gendered division of labour in *Man and His Conquest of Nature*, particularly in relation to the fishing industry. Newbigin also used this opportunity to extol the benefits of work for women, both for their own mental and physical fitness and the benefit of society. She argued that as society becomes more sophisticated avenues of work for women become more limited, until only the least interesting and least remunerative remain. 'She then acts as a drag on the whole community, slackening its natural growth and development, till perhaps the whole social fabric totters and falls, or else a new distribution of labour is found, and the women once more become independent workers' (1917: 155). Newbigin repeatedly asserted that outsiders with their land-based perspective often misunderstand the fishing industry and its gendered division of labour. She challenged both representations of fishermen as lazy and women as passive dependents ('men must work and women must weep'). She stressed women's household and financial management as well as their work in processing and selling the fish and, making the link to Ibsen's strong female literary characters, she used Norwegian examples where the women manage most land-based socioeconomic functions as the men are away deep-sea fishing much of the time: they drive cars, run their small farms and 'keep the post-offices and do the work which elsewhere is man's work, and just because of their manifold activities, they are capable and managing' (1917: 156). These examples are supported by illustrations, one showing Norwegian fishermen on their boat with their catch, one of a Belgian fisherwoman, and another of a Newhaven fishwife. Thus she embeds quite a strong feminist discourse promoting the individual and social value of women's work within her case study of fishing, something which was pertinent to socio-political debates in Britain at the time of first publication (1912) and even more so by the 1917 edition when women were engaged in war work as well as ongoing suffrage campaigns.

Newbigin also showed awareness of women's gendered experiences and roles in her more political regional studies of the Balkans and nationality.

In her paper on 'Race and nationality' (1917) Newbigin highlighted the lack of attention given to the influence of women in the analysis of forced emigration and naturalisation policies. Whilst men as potential soldiers were forced to move, Newbigin illustrated how women, even if not potential combatants, could be subversive of forced national unity, albeit in a quiet and unnoticed manner. Although seeing these women principally as 'mother', she recognised the *political* nature of their actions as *mothers* and challenged the assumption of women as private and politically passive. Newbigin may have used the male pronoun as generic (making exceptions for 'Mother Earth'), as was typical of the social and literary conventions her day, but she threaded very case-specific gender analysis through her geographical studies where she thought it appropriate.

Epistemological and pedagogical themes

Newbigin's views on the philosophy of geography are implicit in all her writings, they can be found in her papers explicitly addressing geographical methodology and also in the prefaces and introductions to her many books. Reading even a selection of her varied output gives a sense of her evangelical call to the scientific method, but also her desire to fulfil the potential of what geography, based on that method, could be.

The stated aims in Newbigin's (1907) paper on studying weather are revealing of her views on the purpose of education and how this relates to scientific knowledge.

> We want in the first instance to train the powers of observation and develop the intelligence with a view not only of making better citizens, but also of increasing the happiness of life; and in the second place we want to give an insight into the methods of science. To show that the methods of science are everywhere in essence the same and to suggest that, owing to the fact that the further the scientific research is carried, the more obvious it is that nature is orderly and uniform, and that therefore there are few phenomena too trivial to be worth studying by someone – these in my opinion are points of great and increasing importance (1907: 627–8).

She went on to argue that scientific studies were now accessible to all rather than just the elite and that 'the stability of the modern community depends, literally and absolutely, upon a widespread sympathy with the aims of science, if not upon a widespread knowledge of its contents. We must "educate our masters" if we are even to hold our position as a nation' (ibid.). All of which reflected turn-of-the-century national anxieties about levels of intelligence, fitness of workers and conscripts to the armed forces as well as the expansion of the franchise (see Thane 1990), but points to scientific/

geographical knowledge as a means to negotiate these concerns and socio-political changes in much the same way that the government offered innovations to the elementary curriculum in the form of physical exercise and technical skills as a palliative for perceived weaknesses in the working classes (see Maddrell 1996). The paper was based on a talk to teachers at Aberdeen University and at the Outlook Tower in Edinburgh and the clear practical advice on how to make, keep and build on basic weather charts, as well as the importance of gaining the child's interest in the subject, shows Newbigin's own sensitivity to what constitutes good teaching as well as good scientific practice.

The Reception of Newbigin's Work and her Place in the History of Geography

Described as the 'doyenne of Edinburgh geography' (Clout 2003), Newbigin is best known for her 32-year editorship of the *SGM*, through which she had enormous – 'disproportionate' even (Stoddart 1991) – influence on the nature of geography as disseminated through the journal. As editor she became a much respected gatekeeper for Scottish geography who was known for expecting high literary and scientific standards and is credited with making the *SGM* a leading British geographical journal (Adams, Crosbie & Gordon 1984; Herries Davies 2004). Indeed, Newbigin has been held as not only the best role model of a geographical editor, but also as the only one in the British Isles to have had a real impact on the shape of the journal and practice of the discipline, not least in orienting geography away from concerns with Empire (Herries Davies 2004). Interestingly, Newbigin wrote in correspondence that she hated the word 'Empire' and considered it largely a product of chance,[13] which explains the relative absence of Empire as a theme in papers in the *SGM* during her editorship (Lochhead 1984), although she did herself write one small text on Empire in 1914. Newbigin was also known for her less tangible legacy of fostering the next generation of geographical authors (Freeman 1976a). As Eva Taylor recorded in Newbigin's obituary: 'It is difficult to over-estimate the part which Dr. Newbigin played in encouraging original work in what is still a new and therefore difficult subject, and one in which the canons of sound scholarship are not yet established' (Taylor 1934a: 367). Newbigin's role at the RSGS is seen as equivalent with that of Fleure at the Geographical Association and Markham, Keltie and Hinks at the RGS (Wright 1957), although Newbigin herself had cause to feel that her work at the RSGS was taken for granted.[14] Her contemporary at the RGS, H.R. Mill, is credited by Newbigin's sisters for having written a sympathetic and appreciative notice for *The Times* at the time of Marion's death and described her in his biography as the British counterpart of the American geographer, Ellen Churchill Semple. He thereby gives her recognition by association with an influential figure, but only in relation to another woman. Mill considered Newbigin had died before her merits were fully recognised (Mill 1951), but seems to set definite limits to his admiration of her. They served together on the Board of Geographical Studies of the University of London, during which he 'admired the soundness of her judgement and the substantial basis of knowledge on which her mind worked' (Mill 1951: 94); he also acknowledged her as having 'won a high place in the educational world', but not as someone who was a contributor of original knowledge. This seems strange given that Newbigin received her RSGS Livingstone gold medal for 'numerous contributions to geographical science, based largely on her own observations', while Mill who received his RSGS gold medal in the same year was nominated for 'his long-continued geographical work of a high order' (*SGM* 1924: 30). There is no evidence of ill-feeling *per se*, but Mill may have wished to make a distinction between his own achievements and those of Newbigin. The RGS archives hold a letter from Newbigin to Mill, in which she responds to his sense of affront at the *SGM*'s 'inadequate review' of his new edition of *Instrumental Geography*. Newbigin soothed him, assuring him that the limited notice of the new issue was merely a reflection of the assumed excellence of the book and its already established 'widespread supremacy among geographical text books';[15] but reviews are important to both critical esteem and sales and this perceived slight may have coloured Mill's view of Newbigin. Academic alliances can be made and lost on such things. Newbigin herself is wary of these traps and worried about reviewing Fleure's book in 1921 because they were on 'very friendly terms' and because it was so close in subject of her own *Aftermath*. Similarly, she declined to review Lyde's work on the grounds that it would seem 'a trifle awkward for me to return the compliment so speedily?' Her sense of humour shines through when she adds 'I am afraid also I should not be able to resist saying something about Albania', referencing Lyde's critique of her own work.[16]

At the time of her death Newbigin was credited with being the first geographer to thoroughly apply the scientific method consistently to her work (*SGM* 1934: 333): 'No geographer of our acquaintance has lavished on *his* work so much of the true scientific method as Marion Newbigin' [my emphasis]. She is also credited with laying the foundations of the scientific method on which Scottish geography was built (Adams *et al.* 1984). This singling out may be overstated, but what is certain is that Newbigin and others coming to geography with a training in natural sciences were schooled in the importance of primary research, fieldwork and study of the natural environment – very different attributes from those geographers with a background in the classics or arts (Lochhead 1984). Her long-lasting contribution is generally recognised in the field of biogeography and the definition of geography as an ecological subject, but as illustrated above, Newbigin also undertook significant work in regional and other aspects of geography. Fleure described her work as combining scientific and humanistic thought and she considered his approach most close to her own – not surprising given his role at Aberystwyth and their common move from biological sciences to geography. Newbigin's position in the RSGS was undoubtedly mediated in the first instance by the patronage of James Geikie, but she was soon established as someone of inherent professional ability and commitment. Whilst her work was recognised with awards from the RSGS and the RGS, her research was not always deemed as conforming to the geographical discourse of the latter, something she clearly resented. Placed at the heart of the Scottish geographical institution and playing a significant role in the Geographical Section of the BAAS, she nonetheless remained to an extent marginalised from the London geographical establishment. Despite this, Newbigin's qualifications, publications and research far exceeded those of most people who took the title of geographer or fellow of a geographical society and she made a major contribution to the construction of geography as a science. However, as the 'objective' scientific method is problematised by studies of the social construction of science, in turn, Newbigin's application of the scientific method to geographical studies is problematised (Maddrell 1997).

Modern Geography (1911a) was welcomed by *The Geographical Journal* (1911: 527) as a demonstration of geography's distinct identity, something needed for the emerging academic discipline: 'We find here geography set forth as possessing a definite outlook of its own, sharing material with various sciences, but not leaning upon any of them; it can hardly be denied after reading this volume, that the subject is its own support'. In keeping with other first-generation geographers, Newbigin gave the discipline status by placing geography within a genealogy of previous study. It was she who established the beginnings of modern geography in the pivotal year of 1859, when Humboldt and Ritter died and *The Origin of Species* was published (Newbigin 1911a: 7–8), an observation which was often repeated in subsequent mid- to late twentieth-century histories of geographical thought.

Although superseded by Civic's influential book, the merits of Newbigin's work on the Balkans was recognised at the time of publication, for example MacMunn and Coster (1922: ii), who described it as 'a regional essay of a type all too rare in the English language' and Lyde who commented in his *GJ* review: 'It modestly professes to be only a 'war' book, but it will take its place at once as a standard book on the geography of the Balkan peninsula. We expect from Miss Newbigin work that is always stimulating and artistic as well as sound and satisfying; this volume reaches a high-water mark in its admirable selection of essentials and in the deft and suggestive treatment of the material thus selected' (1915: 386). His only real criticism was of Newbigin's representation of the Albanian people: 'To describe them as "a comic opera people" (p. 231) is unworthy of all the rest of the book with its very marked and distinctive merits, and it shirks the most difficult problem in the whole peninsula' (1915: 387). *Geographical Aspects of Balkan Problems* was also singled out retrospectively as of particular merit in the literature on the First World War (Freeman 1980). T.W. Freeman thought Newbigin's post-war book *Aftermath* showed 'remarkable social concern for the building of a better Europe' and tells us a great deal about her as a person as well as a geographer (Freeman 1976a). Her regional work influenced Fleure (Livingstone 1992; Martin 2005) and her text on Southern Europe although having a little too much on physical processes for a regional and economic geography in the opinion of one reviewer (H.C.K.H. 1933) was welcomed as a much-needed 'comprehensive contribution' to British literature (ibid.); it was widely read and reprinted, used in the preparation of Naval Intelligence Handbooks during the Second World War and was still in use in the 1950s. In addition to this political–regional work, she also produced historical geographies, seen in the case of *Mediterranean Lands* (1924a) and a 1927 study of Canada. Whilst the former was roundly criticised by a classics scholar in *GJ* for numerous flaws, her book on Canada was described by H.P.B. in the same journal as 'probably the most brilliant book dealing with the history of Canada that has yet to be written' and is still acknowledged as a significant contribution to historical geography (Martin 2005). Fleure in his posthumous tribute to Newbigin in *Plant and Animal Geography* (1936: vii–viii) summarised her contribution to geography in glowing terms: 'it was indeed the combination of scientific and humanist thought, touched by genius and enriched by travel and study, that made Dr Marion Newbigin so valuable a contributor to the progress of knowledge and still more of understanding'. *Plant and Animal Geography*, based on lectures Newbigin gave at Bedford College, was reprinted seven times, and ultimately reprinted as a University Paperback. In the preface to the 1968 edition Monica Cole stated that the book was still the essential introduction to biogeography, on which more recent developments in quantitative and experimental biogeographies were built. Indeed, she went so far as to suggest that biogeography had faded from popularity in the years after Newbigin's death because the subject could not be pushed further without new research in biology and related fields, something only achieved in the 1950s and 1960s.

Newbigin was not above criticism in her day and her early work did not gain universal approval at the RGS. Hinks wrote in 1917 criticising her editorial judgement for printing what he thought was an erroneous review of *The Military Map* which he had dismissed in the *GJ*. Reviewing her paper to be addressed to the RGS, Douglas Freshfield (early proponent of women's membership of the RGS) judged 'The paper is original, but can hardly be called geographical ... It consists mainly of an argument in support of the view that the identity of race has been greatly overvalued as a constituent of nationality'.[17] Despite Freshfield's fears concerning the paper's length and suitability for an evening meeting, the paper was presented and published (after what Newbigin felt was unnecessary editorial intervention). In the ensuing discussion Newbigin was praised for addressing such a pertinent topic as aligning borders and people in the Balkans and for her suggestions on this matter. Military man and president of the society, Sir Thomas Holdich, argued that such matters should be increasingly addressed by the RGS and to government, and whilst challenging her on a couple of small points, Professor Lyde praised her for being 'our most stimulating and, I believe, also our most suggestive geographer. Tonight she has been provocative, which I think is a great thing' (*GJ* 1917: 330). The adjective 'suggestive' may not read well today (Maddrell 1997), but in the context of his other comments seems to refer to Newbigin's ability to encourage and provoke fresh thought on the subject. Freeman (1980) uses the same term referring to her *Mediterranean Lands*. More recently, aspects of Newbigin's thought and writing, which are now dated, have been criticised, for example, for her use of the term 'conqueror' (Freeman 1980); for the transfer of hierarchical forms of classification and explanation from natural science to social settings (Maddrell 1997); and for lacking the critical judgement of Geddes' work on the environment (Meller 1990).

Conclusion

In Scottish accounts Newbigin is credited with being one of the 'parents' of modern British geography for her contribution to geographical thought and to the infrastructure of the emerging discipline (*SGM* 1934; Adams *et al.* 1984). Walter Freeman (who may well have benefited from Newbigin's editorial advice himself as he took his first academic post as assistant lecturer in geography at Edinburgh University 1933–5) regarded her as the 'beau ideal' of editors (Herries Davies 2004). Freeman's historiographical work in the 1960s, 1970s and 1980s always referred to Newbigin (see Freeman 1960, 1961, 1976a, 1980, 1984) and in doing so played a vital part in keeping alive part of the esteem for Newbigin's role in geography, something which is frequently lost in more recent histories of geography unless they have a particular Scottish interest. In addition to Newbigin's influential role as editor of the *SGM*, her publications spanned the whole field of geography as it was recognised at the time. I have given an introduction to that variety but this huge body of work merits more attention within the history of the discipline than space can afford here.

Other Geographical Editors

Other women also undertook the role of editor of geographical publications. Newbigin was succeeded at the *SGM* by **Harriet Wanklyn** 1935–6, followed by **Lois Latham** 1937–40, her post being terminated as the society was unable to pay her salary (Freeman 1984). During the period 1944–53 **Isobel Wylie-Hutchison** held the position as honorary editor of the *SGM*, the only woman to hold this office. **Joy Tivy** was *SGM* editor 1954–63 while lecturing at Glasgow University. After early retirement from lecturing at the Edinburgh department of geography, **Catherine Snodgrass** became editor 1964–7. A political animal herself, she sympathised with the student political movement in the 1960s and thought the *SGM* could 'make its contribution in the present-day situation by having a more topical and Scottish approach', although this did not preclude non-Scottish work (cited by Freeman 1984: 60) (see Chapters Seven and Eight). Snodgrass had to relinquish her editorship for health reasons and was to be the last female editor of the Society's journal until 1999, but women constituted a major editorial presence at the RSGS in the period studied, with women editing the SGM for 49 of the 86 years from 1884 to 1970 (see Chapter Ten). **Mabel Wright** was briefly editor for two volumes of the *Journal of the Manchester Geographical Society* when it was revived by Fleure in 1934, but the readership complained its content was 'too academic' and Wright resigned (Nigel & Brown 1971; Freeman 1984). **Gladys Wrigley,** born in York in 1885, to a businessman father and teacher mother, studied geology at Aberystwyth, graduating in 1907, followed by several years teaching in secondary schools in Wales (Fairchild 1976) and attendance at one of the Oxford Summer Schools in Geography (Light 1950). She was a protégée of H.J. Fleure, who considered her one of the most able students he taught and she returned to Aberystwyth at his request to assist in teaching geography and geology 1908–9 (she had leave of absence from her school to be a student herself on the Diploma in Geography course!) (Bowen 1970). In 1911 Wrigley won a studentship to a US university of choice and she opted to study geography, physiography and anthropology at Yale with Isaiah Bowman (where she was also considered one of the brightest on the course (Light 1950). She went on to become a graduate student with Ellsworth Huntington and Isaiah Bowman; and when the latter was appointed director of the American Geographical Society (AGS) in 1915 he in turn appointed her as his assistant.[18] She subsequently became the editor of the Society's revamped publication *Geographical Review* in 1929, a post she was to hold for 30 years. As Monk notes, it is difficult to identify the influence of women who worked at the AGS, because their contributions to the journal were often unsigned and 'extensive editorial labours were expended on articles and books that came out under other people's names' (2003: 246). As editor Wrigley favoured empirical work addressing wide-ranging, innovative and timely geographical themes, collating papers into coherent editions of the journal. 'That she herself had substantial research talents and was a stylish writer is evident from the few signed articles of her early years (Wrigley 1916, 1917, 1918)' (Monk 2003: 247). She recorded her own perspective on her work as editor (Wrigley 1952), but in an echo of Newbigin's posthumous acclaim, Wrigley is credited with promoting geography in the English-speaking world at a crucial time in the discipline's development (Bowen 1970) and giving the *Geographical Review* international standing as a scholarly journal (Light 1950; Fairchild 1976) through her inspired and meticulous editorship. Clearly a gifted editor who contributed much to collective American and international geography through her editorship, however, given her abilities, her authorial and editorial self-effacement was regretted during her lifetime (Light 1950) and one can only speculate on her impact on the discipline had she written more. Gladys Wrigley received awards for her life's work from the Geographical Society of Chicago and AAG (1948, 1951); she retired in 1949 and died in 1975.

Chapter Four

Women Travellers: Inside or Outside the Canon?

Introduction

Women travellers are important to the discussion of the role of women in geography for several reasons. First because there have always been strong links between travel and geographical knowledge, as articulated in the founding aims of the nineteenth-century geographical societies; secondly because women travel writers' ability to contribute to geographical knowledge was at the heart of the RGS women's fellowship debate 1892–3; thirdly their work was often popular, so widely read and discussed; and fourthly there are stronger links between women travellers' geographical work and the interests of state than is often recognised.

Much work has been undertaken on women's travel writing since Middleton's (1965) *Victorian Lady Travellers*, both in literary and geographical studies. This can be seen in anthologies of women travel writers (e.g. Russell 1986; Birkett 1989, 2004a; Robinson 1990; Morris & O'Connor 1996); individual biographical studies of independent travellers such as Isabella Bird (Barr 1985), Lady Hester Stanhope (Gibb 2005), Mary Kingsley (Blunt 1994) and Isobel Wylie Hutchison (Hoyle 2001); accounts of women travelling with their husbands such as Lady Mary Wortley Montagu (Secor 1999), Katherine Routledge (Van Tilburg 2003) and Lady Caroline Brassey (Ryan 2006); and women working for or representing institutions such as missionary Mary Slessor (McEwan 1995) and emigration proponent Violet Markham (Bell 1995b). Victorian women travellers have attracted much attention and this chapter will discuss the work of some women from this period, but it will also attempt to redress the relative neglect of twentieth-century travellers within geographical debates through consideration of the work of Gertrude Bell, Violet Cressy-Marcks, Freya

Stark and others, in order to examine shifts in the relationships between women's travel work/writing and social and political change, geographical institutions and discourses.

Middleton distinguishes Victorian women who travelled for their own amusement and later, often university-educated, women who typically refined their travel experience by 'beginning to master the languages, record the geography and even to interest themselves in the politics of their chosen field' (Middleton 1994: 123). Whilst there is a discernible shift in the levels of pre-travel training, for example in surveying or languages, Middleton's distinction seems too clear cut. It does not do justice to those like Kate Marsden or Mary Slessor who certainly did not travel for 'amusement'; or Isabella Bird and Mary Kingsley who contributed to scientific and geographical knowledge as well as contemporary political debates. If a distinction is to be made, it is perhaps better to see those who had access to training at the RGS and other institutions and/or had university educations and networks behind them, relative to those for whom these things were not possible or not desirable.

The period 1850 to 1970 incorporates almost the whole span of the growth, formalisation, high noon, decline and demise of the British Empire and the relation of both authors and Britain to the Empire is a recurring theme in most of the travel writing selected here. Although Empire is often represented as an entirely masculine colonial domain, women travellers were frequently agents of Empire (Mills 1991; Pratt 1992; Blunt 1994). Travellers' informal diaries and reports such as those written by Gertrude Bell 'provided the British Empire with the raw material that enabled it to govern large parts of the globe' (O'Brien 2000: xi). Women with detailed regional knowledge such as Bell and Stark were employed by the British state during wartime and evidence suggests that others such as Bird and Cressy-Marcks may have worked covertly for the state at some point during their travels. More all-pervasive than the role of individuals was the colonial discourse which, combined with the trend to systematise knowledge within European natural sciences, resulted in a Eurocentric framework of analysis which reflected dominant social and cultural values. The imperative for travellers to describe potential raw materials and markets 'supported the colonialist venture' (Mills 1991: 159); Pratt (1992: 38) argues furthermore that discursively 'natural history asserted an urban, lettered, male authority over the whole of the planet; it elaborated a rationalizing, extractive dissociative understanding which overlaid functional, experiential relationships among people, plants and animals', and was often based on and reinforcing of a capitalist conquest narrative. As indicated in Chapter One, there has been much debate concerning a 'woman's way of knowing' as expressed in women's travel writing. Pratt suggests that women travellers were often more self-reflexive, prepared to subordinate themselves to local knowledge and tended to more ethnographic, urban, centripetal interpretive forms of 'social exploration' compared to their male counterparts, but they could also contribute to north European expansionism through expressions of 'female imperial authority in the contact zone' (Pratt 1992: 170). However, with a few notable exceptions women paid relatively little attention to the women who lived where they travelled (Birkett 1989). Even where gender-based insights of women travellers could disrupt elements of racialised analysis such as Orientalism, accounts could also remain mediated by and reinforcing of class-based prejudice (see Secor (1999) on Lady Mary Wortley Montagu). Women's travel writing was frequently both socially transgressive and conforming, a personal negotiation of permissible and credible experiences and writing styles (Mills 1991). The tropes of the imperative for scientific ordering, race, class and colonial rule as well as gender bring nuance to our understanding of the nature and workings of geographical knowledge as an emerging discipline, as well as the relationship between that geography and the changing forms and expressions of state and Empire.

As recent anthologies of women travellers testify, there are many women travellers who could lay claim to inclusion in this chapter, but space only allows the consideration of a handful of that number. The eight women selected are: Mary Kingsley, Kate Marsden, Gertrude Bell, Freya Stark, Isobel Wylie Hutchison, Violet Cressy-Marcks, Grace Dibble and Thora Halsem (Isabella Bird's work is discussed in Chapter Two). They have been selected to represent a spatial and temporal spread, a variety of writing styles and varying degrees of incorporation within geographical societies, as well as different perspectives on women's role in British society and Empire.

Mary Kingsley

Born on 13 October 1862 in Islington, London, daughter of Mary (née Bailey) and George Henry Kingsley, brother to authors Henry and Charles Kingsley. Her mother had been her father's housekeeper and they had married only four days before Mary's birth, something which only became public knowledge long after her death (Birkett 2004b). During his travels as a private physician, George Kingsley began a comparative study of sacrificial sites around the world and Mary grew up sharing his interests in anthropology and travel. She spent most of her childhood in the family home in Highgate, London, where her younger brother Charles was born in 1866. Mary never went to school but read widely from her father's library and was taught German to assist him in compiling his amateur anthropological findings when he returned home. Most of Kingsley's young life was centred round her invalid mother and her usually absent father. She clearly suffered from a high degree of social isolation during her childhood, but also proved resourceful, learning basic plumbing and sailing (Simpson 1950). The family moved to Bexleyheath in Kent in 1879 and then to Cambridge in 1886 when her brother Charles became an undergraduate at the university. This move was a great opportunity for Mary as she could mix in academic circles and she became friends with Francis Burkitt and Agnes Smith Lewis (Birkett 2004b).

From 1888 to 1892 Kingsley nursed first her mother, who eventually became paralysed, then both parents when her father returned home ill after a bout of rheumatic fever. When her parents died within six weeks of each other in early 1892 and her brother departed to the East, Kingsley found herself with a £4000 share of the family inheritance and without any domestic obligations (Blunt 1994). She took a ship to the Canary Islands in August 1892 and took stock of her position. It was in the ports of the Canaries that she first came into contact with the people and goods from Africa, which were to engage her for the rest of her life. While the debate concerning women's right to take the title of 'geographer' raged at the RGS in 1893, Kingsley made her will, sent letters of introduction ahead and departed for West Africa. Her correspondence suggests she was motivated by continuing her father's work, that she was suffering from depression and maybe had a death wish (Claridge 1982) – perhaps all three were true at some point. Her first journey was something of a reconnaissance and 'Few women of any period have stepped so abruptly from conventional circumstances into the unknown' (Claridge 1982: ix). She landed at Freetown in Sierra Leone on 17 August and travelled south to Luanda, visiting missionaries, officials and trading stations. She also collected or acquired scientific specimens, which were of interest to British scientists on her return. Arriving in the UK in January 1894 she immediately began plans for a longer trip. Determined that it should be more rigorous in purpose and outcomes, she contacted Dr Albert Günther, keeper of zoology at the British Museum, who, impressed with the specimens found on her first journey, equipped her with collection and storage materials. She was also commissioned by George Macmillan to write a book on her return (Birkett 2004b).

Having travelled from Liverpool with Lady Ethel MacDonald, wife of the Commissioner General of the Oil Rivers Protectorate, whom she had met on her first visit, Kingsley then stayed four months with the MacDonalds at the Residency. The stay was far from uneventful as Kingsley helped nurse Europeans suffering from smallpox, accompanied the MacDonalds on an official visit to Ferdinand Po and made short journeys inland and upriver before embarking on her longer independent travels (Birkett 2004b).

The diaries from her two trips were used as the basis for her book *Travels in West Africa* (1897). The account of her first journey included a description of the bights along the West African coast, trade, transport and navigable rivers, the colonisation of mangrove swamps and the work of earlier explorers such as Mungo Park. The account of the second journey included her cross-country trek across uncharted ground from the Ogooué River to the Ramboë River, her ascent of Mount Cameroon by a new route for Europeans, details of inland trade, including the local trader factory system, indigenous social and religious customs; and the landscape, flora and fauna. She traded as she went, partly in exchange for goods she wanted en route, but partly to fund her journey. Of the 65 specimens of fish Kingsley brought back to Britain, many were rare and three were named for her. She is also credited with having made 'the most extensive ethnological field studies there to date' (Robinson 1990: 138).

Ambiguity best describes Kingsley's own attitude to her work and status, and her writing is riddled throughout with her contradictory relationships to gender and colonialism (Mills 1991, Blunt 1994). Without formal training in ethnography beyond assisting her father, she felt that she was perfectly capable of undertaking the collection and tabulation of field data, but that the more important work of analysis and interpretation should be left to someone who was better qualified. She wrote didactically, that after collecting facts together '... then I advise you to lay the results of your collection before some great thinker and he will write upon it the opinion that his greater and clearer vision makes him more fit to form' (Kingsley 1897/1982: 165). The gender attached to this expert is not accidental. However, she also satirised the adult male as the scientific yardstick (Mills 1991) and challenged masculinist representation of length of journey as a measure of success and worth: 'The "arm-chair explorer" may be impressed by the greatness of length of the red line route of an explorer, but the person locally acquainted with the region may know that some of those long red lines are very easily made ... In other regions a small red line means 400 times the work and danger, and requires 4,000 times the pluck, perseverance and tact' (from *Travels in West Africa* cited by Birkett 1989: 131).

As an independent woman traveller going to West Africa there was a limited range of categories into which fellow Europeans placed her. When on ship it was assumed she was a missionary or a botanist, but 'Naturalist' was the label Kingsley chose for herself when completing travel documents (Birkett 1989). In a letter to Ethel MacDonald Kingsley identified herself as separate from the typical RGS traveller. 'I am beginning to think that the traveller – properly so called – the person who writes a book and gets his FRGS etc. is a peculiar sort of animal only capable of seeing a certain set of things and always seeing them in the same way, and you and me are not of this species somehow. What shall we call ourselves?'[1] In *Travels* Kingsley positively repudiated being a geographer and therefore the necessity of following the sequential discursive structure of 'geographical' writing: 'I am not a geographer. I have to learn the geography of a region I go into in great detail, so as to get about; but my means of learning it are not the scientific ones – Taking observations, Surveying, Fixing points, &c., &c.,. these things I know not how to do' (1897/1982: 100–1). Her comments were partly in response to 'being informed on excellent authority that publishing a diary is a form of literary crime' and insufficient to the task. Diaries were a 'permitted' literary form for women as a personal and provisional voice, but women's diary-based writing was also criticised for the same characteristics and not considered rigorous in literary or scientific circles (Mills 1991). Kingsley disregarded these warnings, defending the diary form which allowed every-day things to be recorded that might go unnoticed in a more formal text: 'for the reader gets therein notice of things that, although unimportant of themselves, yet go to make up the conditions under which men and things exist' (1897/1982: 100). These extracts illustrate both her self-deprecating humour, and her sensitivity to ethnography and what we would now call 'everyday geographies'.

Kingsley's return to Britain was greeted by fanfares in the *Telegraph*, where her travels were compared to those of Speke, Stanley and other respected male explorers. She was also described as a 'New Woman' and promptly denied this title and its feminist associations, stating that all her accomplishments had been achieved with the assistance of the 'superior sex' (Birkett 1989). It is not clear whether she is referring to white colonials or the African porters who guided, carried her luggage and cooked for her. If it was the latter then Kingsley is articulating a hierarchy in which gender subservience takes precedence over perceived racial superiority. Despite the social freedoms she had achieved through her independent finances and her power as an indirect colonial 'ruler', at home she was a social conservative in many respects. She did campaign from public platforms for the Colonial Nursing Association and the professionalisation of nursing, but was an anti-suffragist. Birkett (2004b) suggests Kingsley's 'relationships with other professional women were often competitive and confrontational'; there was certainly friction between Kingsley and the colonial editor of The Times, Flora Shaw (ibid.).

In some ways Kingsley's attitudes to her own achievements were remark-ably progressive. When she travelled across territory unknown and uncharted by Europeans (such as the Lake Ncovi) 'She never accepted that she had "discovered" them. It was her contention that those who lived there knew about them all the time' (Wilcox 1975: 111) – but this did not stop her from stressing these achievements in her writing. She was also prepared to revise her views whether it was about European traders or black Africans. Despite an early encounter with black Africans on her ship in 1893 which set her cleaning her cabin with carbolic acid,[2] and although far from uncritical of the Africans she came into contact with, the ways in which Kingsley wrote and spoke on her return to Britain challenged the British domestic impression of the 'child-like savage'. 'You cannot associate with them long before you must recognise that these Africans have often a remarkable mental acuteness and a large share of common sense; that there is nothing really "child-like" in their form of mind at all. Observe them further and you will find they are not a flighty-minded mystical set of people in the least' (Kingsley 1897/1982). In defence of the cannibalistic tribe who guided her on her exploratory journey to the Ramboë, Kingsley deliberately misspelt their name, referring to them as the 'Fan' rather than the 'Fang', to avoid cheap jibes in the British press at the expense of a people she generally respected. Kingsley used intelligence, commonsense and gifts rather than coercion in her travels (Blunt 1994). Despite not seeing Africans as 'child-like' *per se*, she took on a parental role in relation to her porters and cook. Kingsley wrote 'I was a Father and a Mother to them'(1897/1982: 605), but 'Treating Africans as children, and, in her case, taking on both paternal as well as maternal roles, was still an imperial gesture' Kearns (1996: 468). Kingsley's style was also to downplay dramatic encounters. The occasion when she fell into a camouflaged animal trap is typical, where in her tongue-in-cheek style she stresses 'the blessings of a good thick skirt' rather than the likelihood of impalement. It is not surprising that Mary Russell (1986) adopted this phrase as the title for her anthology of women travellers. Kingsley initially omitted some of her more extreme experiences, such as fighting off a hippopotamus with her umbrella, for fear that she wouldn't be believed. Such a position was hardly surprising in the light of her publisher's disbelief of her piloting a boat and the incredulity expressed in regard of some other women's contemporary travel accounts (e.g. Alexandra David-Neel in Tibet (Mills 1991) and Kate Marsden in Siberia, see below).

Kingsley's *Travels in West Africa* was widely read and she was in demand as a speaker on the overseas travellers' circuit (Frank 1986; Blunt 1994). There appears to have been a geography of Kingsley's preferred engage-ments, influenced by her relative social ease in different places. She pre-ferred speaking in northern manufacturing towns, but was pragmatic about other speaking engagements as a means to an end: 'for the rest every pound means five miles in West Africa next time out'.[3] She was also sensitive to criticism of her accent and tendency to drop her 'aitches' and 'gees' when she spoke, criticism most likely to occur in the home counties (see Blunt 1994: 130). The book also placed her at the forefront of the current political battle regarding the extent of colonial powers in West Africa and she made her support for the traders and trade expansion perfectly clear. She may have been against female suffrage, but she felt she was particularly well informed to influence this particular political debate and did so through her papers to the Liverpool Geographical Society with its strong trade member-ship, through stimulating debate in the press, encouraging her trader friends to speak out and using social networks open to her. Birkett (1989) goes so far as to credit Kingsley with provoking parliamentary debate, government enquiries and orchestrating a press campaign. Her article in *The Spectator* (19 March 1898) was another foray into colonial politics, this time on the subject of hut tax in Sierra Leone, which caused Joseph Chamberlain, colonial secretary to secretly seek her advice (Birkett 2004b). All of this was quite an achievement for a woman who felt that politics was the business of men and exemplifies women's indirect participation in the politics of Empire. In ongoing lecture tours she continued to speak out against settler colony policy and in support of laissez-faire trade in West Africa, including trade in alcohol, which had been criticised by missionaries (ibid.). For Kingsley her own evaluation of what she had observed for herself *in situ* outweighed any dominant wisdom.

Her second book *West African Studies* appeared in 1899 and was more serious in content than the first. Here she outlined her 'alternative plan' for the area, based on British trading interests and incorporating African views (Birkett 2004b). Kingsley felt that as a woman she could relate to African culture and experience and thought that both government and missionary attempts to westernise local practices were harmful, whereas existing customs of polygamy and the slave trade could be justified (ibid.). 'Her appeal for an understanding of African social and legal systems and the importance of commercial interests forged a new pressure group in colonial politics called the Fair Commerce Party, Third Party or simply "Kingsleyis"' (Birkett 2004b, also see the *Glasgow Herald* 27 December 1902). This lobby group was supported by a variety of leaders in British civil society but had insufficient coherence of public appeal to have long-term or significant political impact (Blunt 1994).

Despite her successes, Kingsley's last years were not happy ones. The strains of repeated public appearances and her desire to be back in Africa are thought to be contributory factors in Kingsley's breakdown in 1898. In 1899 she became besotted with the acting governor of Sierra Leone, Matthew Nathan, but her feelings were not reciprocated. She sailed for South Africa in March 1900 and offered her services as a nurse on arrival. She was posted to Simon's Town Palace Hospital to nurse Boer prisoners of war where typhoid was raging. She died of typhoid on 3 June 1900, after which she was buried at sea, at her request, with full military honours. Sympathetic to pantheism, she wanted her spirit to drift north to the West Africa she loved (Robinson 1990). The Liverpool School of Tropical Medicine established a medal in her honour and the Congo Reform Association and African Society were founded in her memory and became vehicles through which her politics could continue (Birkett 2004b). Kingsley's obituaries appeared in many journals and as a female subject she was unusually honoured with an obituary in *The Geographical Journal*. The latter celebrated her achievements while distancing the RGS from her and her views by stressing her status as an 'incidental' contributor to geographical knowledge and a holder of unusual and unlikely to be agreed with opinions on the government of West African colonial lands. 'Devoting herself to scientific studies, she chose the West African coast lands as a field for original research in the subjects which exercised a special attraction for her, zoology and anthropology. Her work did not, therefore, lead her into any entirely unknown countries, and her contributions to geographical science were but incidental to her more definite labours in other fields. Still, she did much to bring before the public a clear understanding of the nature of the West African countries, for which, in spite of their many drawbacks, she continued to feel a surprising enthusiasm, her powers of observation and description rendering all she wrote unusually valuable and suggestive, although many might dissent from the particular views which she held' (*GJ* 1900: 115).

Kingsley's books were widely read in her own lifetime and generated reviews in a correspondingly wide range of publications, especially when she had become a household name. The second book was reviewed in 25 different publications ranging from the *Daily Telegraph* to the *West African News* and *African Mining Review* (see Blunt 1994: 166–7) – although Flora Shaw refused to review her books in *The Times* (Birkett 2004b). The two books, different in tone, received different responses: 'Reviews of *Travels in West Africa* focused on the novelty of a woman traveller, while reviews of *West African Studies* focused on the book's style and content' (Blunt 1994: 117). Reviewers stressing her femininity highlighted her courage and stamina and her refusal to compromise on feminine dress; reviewers of the second book were split between the many criticising and the few praising her feminine style of writing (ibid.). Writing in the *SGM* one reviewer described *Travels* as 'quite a new departure in African literature, and after reading it, one is not surprised at its popularity and extensive sale, for such a sprightly, interesting, vivid, and in some respects audacious, account of travels in Africa, it has never been our lot to read – and the author a lady!' Other newspapers, which emphasised her love of the landscape and described her journey as a 'walk' and the Ogowé as a 'stream', feminised and 'minimised her achievements compared with a masculine tradition of exploration' (Blunt 1994: 118).

Kingsley continued to be read in the twentieth century and her books were abridged for children. She almost certainly would not have approved of the title of Helen Simpson's (1938, 1950) retelling of her story under the title *A Woman Among Savages*, a perfect example of a racialised heroic myth superseding authorial intention 40 years after Kingsley's death. Nor would Kingsley have liked being reclaimed as a proto-feminist in the 1980s and 1990s (Birkett 2004b) or been likely to have chosen Virago as her publisher.

Kingsley saw most of her work as anthropological and she did become a member of the Royal Anthropological Society. She resisted involvement with the women's fellowship debate at the RGS, seen in her 1899 letter to RGS secretary J.S. Keltie disclaiming any association with Marian Farquharson's 1899 campaign to have women admitted to the Linnaean and Royal Geographical Societies: 'I do not wish to alarm you but I feel it is my duty as a friend to warn you that there is a dangerous female upon you! ... I herewith do send you a copy of this which I sent to her in case she goes and says I want to be a FRGS'.[4] In her letter to Farquharson she stresses her own lack of scientific education and experience of *surveying*: 'I have never found a point or taken an observation or in fact done any surveying work that entitles me to be called a geographer'. She went on to articulate a position which did not deny women's abilities *per se*, but argued that it was better for women not to force themselves on learned societies and that 'if we women distinguish ourselves in science in sufficiently large numbers at a sufficiently high level the great scientific societies like the Royal and the Linnaean will admit women on their own initiative or we shall form similar societies of our own of equal eminence'.[5] Despite these protestations, Kingsley's correspondence with Keltie on other occasions emphasises her speaking engagements and invitations at other geographical societies, which might suggest she would have been open to an invitation from the RGS. Keltie encouraged her to submit a paper for the Council's consideration for an evening meeting (as was the normal practice) and requested a short factual account of her journeys for the *GJ*, but she never addressed the Society.[6] She did have papers read at and published by the Scottish and Liverpool geographical societies in 1896, and at the Tyneside society in 1897,[7] when her papers were read for her by men; it is difficult to know whether this was because of her gendered deference or her insecurities about her accent.

Kate Marsden

Kate Marsden was one of the cohort of women elected as fellows of the RGS in 1892–3. She was born on 13 May 1859 in Edmonton, Middlesex, daughter of Sophia Matilda (née Willsted) and Joseph Daniel Marsden, a solicitor. She was the youngest of eight children and when her father died in August 1873 the family was left financially vulnerable (Middleton 2004c). She trained as a nurse during 1876–7 at Tottenham Hospital, Edmonton, where nurses were also schooled as an evangelical 'deaconesses' for the Christian faith, something which suited Marsden, who had once considered becoming a nun (Keay 1989; Baigent 2008). After a short training Marsden was one of a group of volunteer medical staff who went to nurse Russian war-wounded soldiers in Bulgaria in 1877. She was awarded a Red Cross medal and had her first contact with leprosy patients, which influenced her

Figure 4.1   Kate Marsden.
Source: courtesy of RGS-IBG picture library

sense of personal 'mission' and was to shape much of her later work (Robinson 1990).

On return to Britain, Marsden nursed in London and Liverpool for four and a half years, until she retired on the grounds of ill health in 1884 (Middleton 2004c), the first of numerous breakdowns (Baigent 2008). Subsequently she and her mother sailed for New Zealand to nurse her sister who was dying of tuberculosis. During her stay Marsden took a post as Lady Superintendent of Wellington Hospital and made significant improvements there, but her employment was foreshortened by a 'serious accident' (Middleton 2004c). She appears to have had another nervous breakdown, a 'black night of the soul' which challenged her faith as well as her mental well-being (Johnson 1895; Robinson 1990). On her return home she decided to spend her life in caring for those with leprosy. Having gained the support of Florence Nightingale and Louis Pasteur and an introduction to the Empress of Russia through the Princess of Wales (solicited when she was presented at court in 1889), Marsden set off to investigate the extent of the leprous population in Russia, the Near and Middle East, calling at Alexandria, Jerusalem and Scutari before travelling overland to Moscow (Robinson 1990; Middleton 2004c). In February 1891 she set off for Siberia where leprosy patients were in particularly acute need of care; she was accompanied by a Russian-speaking friend Ada Field, but Field was forced to give up at Omsk after travelling by sledge in the severe winter weather. Marsden continued to Irkutsk, travelling by barge to Yakutsk and then 2000 miles by horseback to Vilyuysk. Her long journey began in snow storms, ended in summer storms and went through forests, swamps and burning peat grounds where there was a danger of her horse sinking through the thin crust covering the smouldering earth. During her journey she gave what immediate nursing she could in prisons and villages, and to the lepers she found banished to destitution in the deep forests. She lobbied local officials and organisations to take action and lend support for a larger project, using her letters of introduction and/or her Bible as authority as suited the situation (Keay 1989). On her return to Moscow, she was met by her friend Ada who supported her on the last leg of her journey. Marsden was physically the worse for wear, with bones protruding and skin chafed raw, and she described herself as resembling one of the lepers she had seen. She was never well enough after this to undertake her planned journey to Kamchatka.

Marsden wrote an account of her travels in *On Sledge and Horseback to Outcast Siberian Lepers* (1893). The book included copies of letters from Queen Victoria and Countess Tolstoy among 30 pages of testimonials; it was published in London and New York and was in its fifth edition by 1895. In Middleton's (2004c) view, Marsden 'who charmed the tsarina and her ladies-in-waiting and blasted her way through the embattled bureaucracy of imperial Russia' must have been 'impressive'.[8] From Moscow she campaigned for funding and raised some £2400 through a London committee to build a hospital in Vilyuysk, which was opened in 1897 (Middleton 2004c). Her reputation at home, however, was fatally damaged by an attack on her in *The Times* by Alexander Francis, pastor of the St Petersburg British-American church, who headed a committee investigating her work in Russia. The British Charity Organization Society also wrote a critical report on her work, and she was, at best, guilty of poor accounting (Baigent 2008). However, the reasons for these criticisms have never been clear: 'Hints of financial impropriety and lesbianism cannot explain why Isabel Hapgood, an American translator, worked so remorselessly for her destruction' (Middleton 2004c). Marsden began libel action against Francis and *The Times* in 1895 but abandoned it, possibly in the face of investigations into her sexual orientation (Baigent 2008). A sympathetic but rather stilted biography by Johnson appeared in 1895 and Marsden wrote her own account in 1921 – *My Mission in Siberia: a Vindication*. Marsden's reputation remained largely intact in Siberia and at home she was presented at court a second time in 1906, which suggests she was far from universally discredited. She converted to Catholicism and used this avenue to promote work with lepers, spending some time in the USA to this end (see Baigent 2008). Blighted with illness, she had ongoing financial problems and appears to have relied upon the Norris sisters for hospitality for many years (ibid.). She was never fully vindicated and died in poverty on 26 March 1931 (Middleton 2004c).

Robinson (1990: 142) suggests that 'It was Kate Marsden's misfortune that people always found it hard to take her seriously', suggesting 'even the title of her book sounds like some elaborate and gothic spoof'. For many of the British public her sensational journey with its Dante-like landscapes and extremes of climate and hardship simply seemed incredible for a woman to have undertaken; some disbelieved that she had travelled in Siberia at all, others that she could not have ridden the distance, or if she did, there must have been sexual impropriety en route. Interestingly, the committee investigating her work included the veracity of her travel account as one of its four key concerns, as doubts about the authenticity of her geographical account were coupled with her financial probity and reliability as a charity fundraiser and worker, despite the establishment of the hospital in Siberia. In the event, the committee considered the word of a single man who had travelled part of the way with her as sufficient verification.[9] In the short notice of her book in the *GJ*, the reviewer, Peter Kropotkin,[10] described her book as 'A lively account of the hardships' and went to lengths to stress the accuracy of Marsden's travelogue. 'The account is quite correct, without any tendency to sensationalism or exaggeration. The description of a forest land on fire, not when the trees but the whole peat soil is burning, is very good and makes a valuable addition to the description of the scenery of Siberia' (*GJ* 1893: 376). Whether representing the solidarity between a royal society and Marsden, who might be considered a royal 'favourite charity worker', or the RGS defending 'one of their own' against outside accusers, or even support for a pilloried woman, through the vehicle of this review the RGS vindicated the veracity of Marsden's geographical account. Marsden was given the unusual honour of an honorary life fellowship of the RGS in 1916, but this may have been a self-interested act on the part of the officers of the Society who regularly received letters pleading time to pay her annual subscription (Baigent 2008). Marsden's fellowship of the Society gave her status in a male-dominated public world (ibid.); her account was recuperated by Dorothy Middleton's (1965) book on Victorian women travellers and subsequent writers, but Baigent's archive work (2008) shows a complex and unresolved picture, including the RGS distancing itself from Marsden when she sought endorsement for her later book (Hinks wrote to the president that it would be 'awkward for the society to pledge itself') (ibid.).

Gertrude Bell

Gertrude Margaret Lowthian Bell was born at Washington Hall, County Durham on 14 July 1868, daughter of Mary (née Shield) and Hugh Bell, iron foundry magnate (baronet from 1904). Her industrialist grandfather Isaac Bell had been a Liberal member of parliament and associate of Darwin and Huxley (O'Brien 2000). She grew up at the family home of Red Barns near Redcar where she had a privileged and enlightened upbringing, but her childhood was inevitably influenced by her mother's death after childbirth when Gertrude was only three years old. She had a close relationship

Figure 4.2   Gertrude Bell.
Source: courtesy of RGS-IBG Picture Library

with her father and a reasonably good relationship with her stepmother Florence, but her formative experiences produced a mixture of resilience and vulnerability in Bell's character, as well as leadership as the oldest child of five siblings (Lukitz 2004).

Bell attended Queen's College in London and not yet 18 years of age went on to Lady Margaret Hall, Oxford in 1886, where her physical prowess, broad reading and self-confidence impressed fellow women students. The status of women students at the University of Oxford was ambiguous at this time and Bell recorded one lecturer putting the small group of women attending his class in seats with their backs turned to him (Goodman 1985; O'Brien 2000); women were able to study at Oxford, but not able to use all the university facilities and libraries; they could take examinations but not receive degrees. She made friends and worked solidly and gained first-class marks in her finals examinations in modern history after only two years (Lukitz 2004). David Hogarth commented in her obituary that 'Her examiners never forgot either the keen features or the studiously smart attire of the nineteen-year-old candidate who asked them across the viva voce table more searching questions than they asked her' (D.G.H. 1926). Bell's Whiggish view of progressive political development, found in her later writings, can be traced to the type of history taught at Oxford when she was a student (O'Brien 2000).

The years immediately following Oxford were spent in a social whirl, but one which in many ways was circumscribed by gendered social conventions. The years 1890–2 were spent in a conventional way for her gender and class, teaching her siblings, supporting her stepmother's charity work and spending the social season in London. Her family sent Bell on a European tour 1888–9 and in 1892 she stayed with Florence's sister in Tehran. Bell went on two round-the-world trips with her brother Maurice (1897–8) and with her half-brother Hugo (1902–3), when they attended the Great Durbar in Delhi as guests of Viceroy Curzon (Keay 1989). She also took up mountaineering and climbed regularly in the Alps (1899–1904) with the Fuhrer brothers as guides. Her risk-taking on and off the mountains has been linked to the possibility of a death wish and/or sublimating sexual tension (O'Brien 2000), but mountain climbing was a common activity for the less conventional and more active subset of wealthy women in the late nineteenth and early twentieth century (e.g. Fanny Bullock Workman, Freya Stark (see below)). Bell summited the Meije, Mont Blanc and the Matterhorn, as well as surviving a three-day ordeal on Finsteraarhorn in 1902, but she declined an invitation to join a pre-1914 expedition to the Himalayas (D.G.H. 1926). During her European travels Bell made many friends among literary and diplomatic circles, largely through her stepmother's brother-in-law, Sir Frank Lascelles, British minister to Romania. This network would serve her well in future years, particularly her acquaintance with Charles Hardinge, later Viceroy of India.

Although romance had failed to blossom at home, the bright and striking Bell found love when visiting the Lascelles in Persia in 1892, but her parents refused to give permission for her engagement to Henry Cadogan, who died of a fever not long after she left Tehran. After her return to Britain she was persuaded to publish her travel sketches, which she did anonymously under the title *Safar Nameh: Persian Pictures* (1894). The book contained many interpretations of Persian life consistent with the dominant British orientalist discursive construction of the 'East' as mysterious and fatalist, but it also demonstrated her ability to write well and she eased into her career as an author with the benefit of anonymity. She developed her linguistic skills by translating Hafiz' poems from Persian to English. Nearing 40, Bell met the married soldier-diplomat Dick Doughty-Wylie in 1906; they began a passionate attachment in 1913, conducted mostly by letter and apparently never consummated. Bell, who had finally determined to overcome her fear of physical intimacy and public knowledge of their relationship, was devastated when Doughty-Wylie was killed at Gallipoli in 1915 (Wallach Scott 1996). Subsequently she was to admire one or two of her colleagues and there were rumours about a relationship with diplomat Kinahan Cornwallis and Haji Naji, an attractive merchant in Baghdad, but she never had a long-term partner (ibid.).

Bell's first visit to Jerusalem in 1900 afforded the opportunity to travel in the desert to historic sites such as Petra and was followed by a return visit in 1905 when she travelled from Syria to Asia Minor, including the forbidden Jebel Druze region. The area included lots of archaeological ruins and Bell collected evidence, writing 'I began the day by copying inscriptions – I found several Greek, one Cufic and one Nabathean – Lord knows what it means but I put it down faithfully and the learned shall read it'.[11] This may echo Mary Kingsley's demurring to experts, but reflects more an awareness of her limitations (as a trained historian rather than archaeologist) and that she was an empiricist at heart, a gatherer of data; but her confidence in interpreting data did grow over time. She published the first part of her journey under the title *The Desert and the Sown* (1907), which was well received and described as 'brilliant' by *The Times*. The *Times Literary Supplement* commented that women were perhaps better travellers than men, 'for when they have the true wanderer's spirit they are more enduring and, strange to say, more indifferent to hardship and discomfort than men. They are unquestionably more observant of details and quicker to receive impressions. Their sympathies are more alert, and they get into touch with strangers more readily'.[12] Bell undoubtedly underwent rigours in her desert travels, but she also took a retinue of servants, table silver and a canvas bath, in much the same way that men of her class going on an armed forces campaign or safari travelled complete with specialist furniture. David Hogarth described Bell's account as one with 'with rare sympathy and literary power ... whether for its matter or its veracity or for its style, it ranks among the dozen best books on Eastern travel' (D.G.H. 1926: 365) and compared it with Charles Doughty's[13] classic *Deserta Arabia*. It became 'a classic of pre-First World War travel literature' (Lukitz 2004). In late 1906 and 1907 Bell worked with archaeologist Sir William Ramsey in Anatolia, when she dated the Byzantine churches in the region, producing a chronology of their spread, and they co-authored *The Thousand and One Churches* (1909) (Lukitz 2004). Her expedition to Ukhaidir and return via Baghdad and Mosul is considered her most important exploratory journey (ibid.) and her account of the day-to-day lives of local people and the impact of changes brought about in Ottoman rule by the Young Turks since 1908 in *Amurath to Amurath* (1911) was widely read. Although not as poetic as her first book, which was a disappointment to some readers, the book was judged by *The Times* 'a serious contribution to Mesopotamian exploration'.[14] This was some consolation for the fact that the French archaeologist Louis Massignon had just published on Ukhaidir. While she lost the heroic moment of revelation to Massignon, she returned to the site to extend her researches and the resulting monograph *The Palace and the Mosque of Ukhaidir* (1914) was her most important scholarly contribution to archaeological literature (Wallach Scott 1996). While in southern Turkey in 1911 she met T.E. Lawrence and they became friends with many shared interests and opinions. At home she was a leading figure in the Women's National Anti-Suffrage League and was president of the northern branch. Her interests were very much with her friends and family who represented the ruling classes (ibid.) and she only supported the idea of women's public service in *local* government as promoted by Violet Markham;[15] her father and stepmother were both anti-suffragists, as was her friend Lord Cromer, but her younger half-sister Molly supported the constitutional suffrage campaign (Lukitz 2004).

Frustrated by her unfulfilled relationship with Doughty-Wylie, Bell embarked on a circular journey around the Syrian Desert, leaving Damascus in December 1913 and travelling to Hayyil (Ha'il) and the Nefud Desert where the Turkish-allied Rashid and British-allied Saud tribes were fighting. She set off with 17 camels, supplies, personal effects, gifts, staff and surveying instruments supplied by the RGS (O'Brien 2000). She was the first westerner to cross the Nefud and charted the rare water sources as she went (ibid.). When she reached Hayyil she was kept captive by the Rashids in the harem for over a week, which made her engage with harem life in a much deeper way than ever before, reflecting on the women's insecurities and frequent status as war booty. After leaving Hayyil, she went on to the Euphrates, Najaf and Baghdad, before returning to Damascus via Ramadi. En route home she saw the British ambassador in Constantinople to report her observations (ibid.). Bell's best account of these experiences were recorded in the diary she wrote for Doughty-Wylie, based on her daily notes; these have now been edited and combined with their correspondence by Rosemary O'Brien (2000).

Bell was elected to the fellowship of the RGS in 1913, and received the RGS Gill Memorial prize the same year 'for her many years' work in exploring the geography and archaeology of Syria, Mesopotamia, and other parts of Asiatic Turkey, and for the excellent works in which they are recorded' (*GJ* 1913: 359). Her award was presented by RGS president Nathaniel Curzon, Earl of Keddleston. Curzon, who had originally opposed but then overseen women's admission to the RGS, commented on Bell's award: 'This is the first occasion on which we have made an award to a lady fellow … for many years we have been familiar with her travels and most admirable writings, and I feel we are honouring ourselves rather than her in asking her to accept this award'.[16] Bell asked for a tiny three-inch theodolite for her prize, which she took on subsequent journeys and is now found in the RGS-IBG archives.[17] In 1918 Bell was awarded the prestigious Founder's medal. Her citation notes her 'valuable' books, her archaeological work and her government office (thought to be unique for a woman) and her various journeys dating from 1905, with special reference to her 1913–14 journey to the Nefud Desert: 'It was a truly astonishing journey for a lady to make, and as Miss Bell takes her travels most seriously, speaks Arabic with fluency, has the great gift of making friends with the people of the country, and is an accomplished antiquarian scholar, it is small wonder that our Government have found her services so useful in connection with the Mesopotamian campaign that she cannot be spared for to-day's ceremony' (*GJ* 1918: 66).

At the outset of the First World War Bell worked for the Red Cross in Boulogne and London. She had already been informally supplying data gathered on her travels to the intelligence arm of the Admiralty[18] and O'Brien (2000) goes so far as to suggest that Bell's growing interest in archaeology was developed in part as a cover for her observation of Arab, Ottoman and German influence in the region. In 1915 she was asked to give her unpublished maps of Syria to the War Office and then called upon to give a briefing on the 'East' to government officials (Wallach Scott 1996). After Doughty-Wylie's death Bell was invited by archaeologist David Hogarth (brother to her university friend Janet) to join the military intelligence department in Cairo. Bell worked as a staff officer for the Arab Bureau (which was under the direction of the India Office) with Kinahan Cornwallis, T.E. Lawrence and others, developing and mapping networks of Arab support for the British. She was then sent to Baghdad to evaluate the efficacy of the Arab Bureau's policy on the ground for the India Office, and was attached to the Mesopotamia Expeditionary Force (MEF) under the leadership of Sir Percy Cox, whom she came to admire greatly. Bell had to cope with sexist marginalisation by the political officers in Basra and suspicion of her social network, which included the Viceroy of India, and therefore conflicting loyalties between the MEF and the India Office, especially when David Hogarth started citing her personal letters in bulletins (ibid.). Unpaid and without official title she spent much of her time mapping the tribal groups between the Tigris and the Euphrates and monitoring their affiliations, but had to beg work space and secretarial support. Eventually, after writing a cross letter to Hogarth, she was rewarded with an official post and pay. 'Major Miss Bell was now the only female Political Officer in the British Forces' (Wallach Scott 1996: 180). In 1916 her official reports received high praise in Britain and her work in trying conditions was acknowledged (ibid.; Maddrell 2008). Despite her honorary status as a 'man' in many Arab circles and her official position, Bell still felt limited by her gender, especially in the face of other political officers' field action: 'One can't do much more than sit and record if one is of my sex, devil take it; one can get the things recorded in the right way and that means, I hope, that unconsciously people will judge events as you think they ought to be judged. But is small change for doing things, very small change I feel at times'.[19] Bell's assiduous information gathering and mapping culminated in *The Arabs of Mesopotamia* (1918), a vital source for strategists in the area and she was soon promoted to Oriental secretary to the British High Commission in Baghdad. Once there she faced sexist opposition again from General Maude, the new military commander of Mesopotamia, but was defended by Percy Cox (Wallach Scott 1996). Bell also became curator of antiquities and established the Baghdad Museum. Using her Arab contacts from earlier visits she was soon meeting and greeting and mapping the political loyalties of local groups. She was known as El Kathun, 'the Lady' or 'the lady who served the state' and succeeded by her social and linguistic skills, intelligence and common sense to persuade Arab leaders to side with the British (Wallach 1996). Fahad Bey claimed he was persuaded by her arguments but told his men 'She is only a woman but she is a mighty and valiant one. Now we all know that Allah made women inferior to men. But if the women of the Anglez are like her, the men must be like lions in strength and valour. We had better make peace with them'.[20]

Bell was appointed CBE[21] for her work in 1917, and after the war was sent by Arnold Wilson (Cox's successor) to the Versailles Conference to represent the interests of Mesopotamia; but working from an Arab perspective, she was increasingly frustrated by Anglo–American policy to support the foundation of a Jewish state in Palestine and came to the view that Arab states should largely be self-ruled. The importance of Bell's contribution to strategic maps was recognised by Cubbitt in his post-war evaluation of the RGS's contribution to the campaign of the 1914–18 war, principally through the preparation and indexing of maps of the theatres of war and strategic areas (see Heffernan 1996): 'The original work, that is to say, the reduction to map-state of what had existed only in surveys of travellers, many of them unpublished, was confined principally to Arabia; and the Arabian sheets prepared by Mr. Douglas Carruthers and his assistants stand in a class apart as original contributions to cartography. They owe much to the unpublished observations of the late Captain Shakespear and of Miss Gertrude Bell, as well as to the good offices of Commander Hogarth' (Anon. 1919: 336–7).

In post-war years Bell and Cox persuaded Churchill as Secretary of State for the Colonies to maintain British rule in Iraq with a view to supporting self-government. Bell was instrumental in Feisel ibn Hussein's ascent to the throne in 1921; they had met in Paris and became close confidantes, resulting in her being known as the 'Uncrowned Queen of Iraq'. Bell did much to formulate what would constitute Iraq's borders and coordinated a united Iraqi front to the border commission when their boundaries were regulated with Jordan, Saudi Arabia and Turkey (Lukitz 2004). She consolidated a core collection for the National Museum in Baghdad established in 1923 and contributed to policy on education. Although never much of a participant in female Arab culture, Bell did organise social events and lectures by a woman doctor for the women of Baghdad (ibid.), which echoed the lectures her stepmother had organised for ironworkers' wives in the northeast of England.

When Iraq's new constitution was established in 1924, Bell's position was redundant, she was no longer 'Queen' and began to fade. She spent the summer of 1925 on sick leave in England, but when author Vita Sackville West visited in early 1926 she commented on Bell's vitality (Keay 1989). However, appearances can be deceptive and four months later Bell took an overdose of sleeping pills in her home in Baghdad and was found dead the following morning on 12 July 1926. She was buried the same night in the British military cemetery and a memorial service followed at St Margaret's in London (Lukitz 2004). Whether she took the overdose deliberately is unclear, it was certainly not made explicit in obituaries which described her 'dying in her sleep'. In the preface to the 1947 third edition of *Persian Pictures* Arberry described her as 'the maker of Iraq', the country being a 'monument to her genius'. Always aware of the unresolved border issue with the Kurds in the north Bell would not have been surprised by ongoing problems there. We can only speculate what she would have made of the current political situation in Iraq, but she would almost certainly have sought to blend Iraqi and British strategic interests. Lukitz (2004) concludes that Bell is best remembered as 'one of the greatest chroniclers of Britain's imperial moment in the Middle East'. A heavily edited selection of her letters was published by Florence Bell in 1927 and her papers were ultimately deposited at the University of Newcastle. Bell never published in a geographical journal, but was given the highest medal by the RGS, where a bronze bust of her resides. In 1927 Hogarth, then president of the RGS, used Bell's diaries to present an account of her 1913–14 journey. He praised her mapping of wells, previously unplaced or unknown, her insight to desert frontiers of Syria and tribal relations in the area, all of which became more significant during the Second World War. The significance of her work, both as a traveller and as a political officer, were attested to by Sir Percy Cox, who described the honour of working with her for 10 years. Brigadier General Sir Gilbert Clayton spoke of her as a great traveller and explorer; about the importance of her ground work to the success of T.E. Lawrence's wartime exploits; and the opening up of major transport links on two routes she travelled. He concluded: 'I have had the privilege, and have the privilege, of the acquaintance of most of the leading Arab rulers and of the chiefs in Arabia. I have never met one who has not either known Miss Bell personally or by repute, and I have never met one by whom she is not held in affection, in honour, and in reverence' (*GJ* 1927: 19). They referred to her 'untimely death', obfuscating her suicide, just as her obituaries referred obliquely to Doughty-Wylie and her broken heart. Her representation of Arabs was criticised as an exemplar of collective formulaic Orientalism by Said (1978), but recent biographies demonstrate continued widespread interest in her work.

Freya Stark

Freya Stark was born in Paris on 31 January 1893, daughter of Flora and Robert Stark. She had a younger sister, Vera, and an older sibling had died

Figure 4.3   Freya Stark and companions in Jebel Druze.
Source: courtesy of RGS-IBG Picture Library

soon after birth in 1891. Her parents were first cousins, had a fraught relationship and moved frequently around Devon for her father's work renovat-
ing houses, spending summers in a rented house in Asolo, Italy. In 1903 the
couple separated with Flora and the girls staying in Italy and Robert returning to Devon, later emigrating to Canada. Flora became an investor in an
artisan silk factory owned by Mario di Roascio (they had some sort of
attachment but commentators are unclear about its nature and duration).
These relationships became more complicated when Freya suffered a hor-
rific accident when her hair was caught in a loom at the factory pulling off
part of her scalp and eyelid. She subsequently sought to sue di Roascio who
later married her sister Vera (Geniesse 1999; Hansen 2004). Her period of
convalescence after the accident represented one of the few times in her life
that Stark felt she had her dominant mother's full attention and this has
been linked to her tendency to take to her bed when she was unhappy in
later life (Geniesse 1999; Hansen 2004). She was left emotionally sensitive
about her injured face and wore her hair over one side of her brow; the
effects were reduced by cosmetic surgery in middle age, but her looks were
always an issue for her and she overcompensated with elaborate hats and
clothes.

With such an itinerant childhood Freya was taught at home, completed a
correspondence course and then went to Bedford College at London
University to study history 1911–14. During this time she met Professor

William Paton Ker, who adopted her as one of his many 'honorary god-children', took her mountaineering and generally took a supportive interest in her life. During the war years Flora trained as a nurse and became engaged to a doctor called Guido Ruata. Stark was mortified when their relationship foundered, and at Ker's suggestion she joined a wartime ambulance unit in Italy in 1917. Eager for emotional reassurance all her life she was to have several episodes of unrequited love and made a disastrous late marriage to wartime colleague Stewart Perowne in 1947.

After the 1914–18 war she continued her climbing and travelling when her health allowed, and she was particularly proud to be only the second woman to climb Mount Rosa by its east face (Hansen 2004). Ker died on a climb in 1923 when Stark was in the party and she greatly missed his fatherly presence in her life (Geniesse 1999). Around this time in the mid-1920s Stark began to study Arabic with a view to travel and possible employment as a governess in the Middle East (Hansen 2004). When a family friend gave her a house in Asolo in 1926 it seemed to be a turning point in her life; she and her mother moved there and Freya dropped her lawsuit against di Roascio. This was soon followed by her first visit to Lebanon in 1927, which included journeys in Syria with her friend, railway heiress Violet Buddicom. She visited her father in Canada and went to live in Baghdad in 1929. On a tight budget, Stark lived in lodgings in a poor district and there was mutual shunning between herself and the British community. She got work as a journalist and using Baghdad as base, made adventurous expeditions into Lurestan and Mazandaran, where she often played cat and mouse with both British and Ottoman authorities in order to avoid travel restrictions. In the event, the British government was appreciative of all the information she provided to update maps and intelligence on the area. She found that when she gained some fame for her journeys the British expatriates were more solicitous of her company and she had many entrées to social and intellectual circles when she returned to London, including Sydney Cockerell of the Fitzwilliam Museum in Cambridge (Geniesse 1999). She published her travel account *Valley of the Assassins* in 1934.

Stark's next journey (1937–8) was an expedition to the Yemen. She gained sponsorship from the RGS and Lord Wakefield, a benefactor who had previously funded adventurous projects undertaken by women. The all-women team was made up of Stark, archaeologist Gertrude Caton-Thompson and her Bedford colleague the geologist Elinore Gardner,[22] with whom she had previously done fieldwork in Egypt and the Saharan oases. Stark and Caton-Thompson were both graduates of London University, but there the similarities ended. Stark was short of stature, of ambiguous social background, flushed with success from her first publications and the receipt of an RGS award; Caton-Thompson was statuesque, patrician, with a proved academic record, and at this time sat on the governing bodies of Newnham and Bedford Colleges, the RGS, BAAS, Royal Anthropological Institute and was president of the Prehistoric Society; Gardner, a scholarly Oxford graduate, had less of a dominant personality (Geniesse 1999). Stark's serendipitous approach to travel and work did not sit well with the academics' ordered scientific way of doing things, not least as Stark was commissioned to undertake people-centred ethnographic study. Stark and Caton-Thompson were soon at loggerheads, exacerbated by illnesses, with Gardner attempting mediation. Stark wrote to Jock Murray that she did not like travelling with women and was an 'Anti-Feminist';[23] in turn she tried the patience of her team members and British authorities twice asking to be evacuated on the basis of medical emergency, only to refuse to go when Harold Ingrams despatched a plane for the second time! Geniesse (1999) argues that it was the status Stark gained from being on an RGS expedition that saved her from being recalled by the Aden government after this incident. The archaeologists had made an important discovery of a pre-Islamic moon temple and an ancient irriga-tion system. On their return Caton-Thompson and Gardner wrote their own account of their findings and Stark a separate one in which she recorded her discovery of ancient Cana on her solo journey after the others had left. They also lectured to the RGS separately. Stark received lavish praise for her lecture from the RGS chair Admiral Sir William Goodenough and from Lord Wakefield. When travelling Stark used her social skills to cultivate friendships and assistance, sometimes as a means to an end (such as hospitality), but also because she enjoyed meeting people. 'She made the most too, of a woman's privilege of being allowed behind the harem cur-tains. She helps adorn the bride, she shares the meals, she even collapses with measles in the distant Hadhramaut, and has to ward off ancient but interesting remedies' (Middleton 1993: 368)

During the 1939–45 war years Stark was posted to Aden as assistant information officer, exploiting her knowledge of south Arabia. Then she organised propaganda in Cairo, Yemen and Baghdad. She founded and fostered the Brotherhood of Freedom, which worked as an informal net-work of those open to supporting Britain in the war and democracy in the post-war period (least successful in Baghdad) (Hansen 2004). Ever the maverick, she got into trouble for selling a government car and giving the proceeds away – she was not disciplined, but this may have cost her a diplomatic career in the longer term (Geniesse 1999). As a gifted public speaker she was asked by the Ministry of Information to speak in the USA counselling caution regarding plans for a Zionist state in Palestine. She did not meet with success and there was a small diplomatic incident around the status of her message in relation to British policy when she felt abandoned by the government who had sent her. Speaking to women's groups in India in 1945 was a less fraught endeavour (ibid.).

In later years she travelled to Greece, Turkey, Afghanistan, India, Pakistan and China. In 1979 she was commissioned by the BBC to do a programme on the Euphrates, and the following year she trekked through the Himalayas by pony (Middleton 1993).

Stark had much in common with Gertrude Bell: their love of Arabia, their contribution to war work in the Middle East, their inability to find a fulfilling marriage and their passion for clothes. Yet Stark was deeply resentful of these inevitable comparisons and rather ungenerous with regard to acknowledging Bell's achievements (Geniesse 1999); she sneered at the comforts of Bell's caravans and thought her aloof from the people she travelled among (Middleton 1993). Stark, like Bell before her, was one of a group of women who were very much adopted by the RGS as one of their own. Interestingly, she spoke at the RGS and published six articles in its journal before she was elected a fellow in June 1936. Throughout her career, Stark received several honours. The RGS awarded her two prizes in recognition of her travels: the Back grant in 1935 for her journey in Lurestan in northwest Iran and the Founder's medal in 1942 for her travels in southern Arabia. Stark was only the fifth woman to receive an RGS gold medal and wouldn't be followed by another woman until 1989 (Middleton 1993). Doreen Ingrams had shared the Founder's medal with her husband Harold in 1940, in part as a result of a piece written by Stark in *The Times* praising their achievements (Geniesse 1999). The RSGS honoured Stark with the Mungo Park award in 1936 for her work on Arabia; she was the first woman to be given the honour of the Burton medal by the Royal Asiatic Society, which they followed with the Percy Sykes Memorial medal; the universities of Glasgow and Durham awarded her honorary degrees; and the Order of St John of Jerusalem made her first a Sister in 1949 and then a Sister Commander in 1981. In 1958 Stark received state recognition with a CBE in 1958 and was made Dame of the British Empire in 1972. She died in Asolo, her Italian home town, on 9 May 1993 (Middleton 1993; Geniesse 1999; Hansen 2004).

Publications

Middleton described Stark as an 'outstanding traveller' as a result of her engagement with people and considered her travel writing as 'incomparable'. 'She was observant, sympathetic, witty and articulate, recording what she saw in page after page of lively and literate prose ... Her ear for a phrase, her eye for a garment, her speedy summing up of a character are what makes her books so enjoyable' (Middleton 1993: 368). By the end of her life Stark had published eight papers and three shorter pieces in *The Geographical Journal* (1931–58), but the bulk of her output was personal, namely four volumes of autobiography and six volumes of letters; this might be dismissed as autobiographical, but as noted in Chapter One, different textual forms are revealing not only of varied expressions of self-representation, but also of everyday cultural geographies, social and political networks.

Valleys of the Assassins (1934) was an 'immediate success and known for its elegant prose, lively wit, and observations of people' and *The Southern Gates of Arabia* (1936) 'is often considered a classic of travel writing' (Hansen 2004). Gertrude Caton-Thompson was offended by Stark's book *A Winter in Arabia* (1940), even though most of its prickly content had been forcibly excised by Jock Murray (Geniesse 1999). Still making amends to the Ingrams, she dedicated the book to them. Middle East diplomat Sir Kinahan Cornwallis wrote in the book's preface 'the average Englishman is not blessed with an exaggerated sense of imagination in his dealings with other races, but it is to be hoped that all who read Miss Stark's pages will learn the difference between the right way and the wrong, and profit thereby'. Most of her post-war publications were autobiographical or collections of correspondence, but she continued to write on her European travels and historical ponderings out of financial necessity. Stark was celebrated within the RGS from the 1930s to the 1950s at a time when a growing number of academics looked to the newly formed IBG as a home and source of scholarly geographical work.

Isobel Wylie Hutchison

Isobel Wylie Hutchison played a significant role in the RSGS, serving on its Council from 1936 to 1940 and between 1944 and 1953 she was the only woman ever to be honorary editor of the *Scottish Geographical Magazine*; she went on to become vice president of the Society from 1958 to 1970.[24] She was born on 30 May 1889 at the family home, Carlowrie in West Lothian, Scotland. She was the third of five children of Jennie (née Wylie) and Thomas Hutchison, a wholesale wine merchant, who had worked in India. Thomas was a keen gardener and this passion was passed on to his children who were educated at home. Carlowrie was a self-sufficient world for the children where, in addition to her lessons, Isobel had her own garden plot from 10 years of age and learned how to develop and print film from her brother Walter; she also relished in sports and physical activities (Hoyle 2001). From 1904 when Isobel was about 15 years old, she and her sister Hilda began attending Miss Gamgee's private school in Edinburgh. However, the idyll of Isobel's adolescence and young adulthood were punctuated by the deaths of male members of her family: her father died of a sudden illness in 1900; her brother Frank was killed in a climbing accident in the Cairngorms in 1912; and her brother Walter reprieved from the battlefield by a broken ankle, then died in an accident while on duty in North

Figure 4.4   Isobel Wylie Hutchison.
Source: Courtesy of the RSGS

Berwick during the First World War. An introspective character even before
her father's death, Isobel's young life was characterised by an enjoyment of
long walks, which she was to develop into a serious occupation. Her mother
looked to marriage rather than higher education for her daughters, but
Isobel preferred to explore poetry. During the 1914–18 war she and her
sister signed up for Red Cross duty, which exposed her to the living conditions of the urban poor in Edinburgh (ibid.).

Building on her love of gardening, Isobel began studying at Studley
Horticultural College for Women in Warwickshire from 1917. It was at
Studley that Hutchison met her life-long friend Medina Lewis, daughter of
Welsh landowners, and Hutchison seems to have become conscious of her
lack of sexual interest in men (Hoyle 2001). She successfully completed her
certificate in horticulture before ploughing into a variety of part-time theol-
ogy courses at King's College, London where she had a period of 'fevered
activity and emotional instability' which led to a breakdown (ibid. 28–30).
One of the things Hutchison had noted in her cathartic and partially autobiographical novel *Original Companions* (1923) was a desire to travel. In the

year the novel was published she studied Arabic and set off for Palestine and Egypt, travelling independently but on well-worn tourist routes. The following year she spent six weeks on an organised tour of Spain and Morocco with a wealthy friend, but they argued and she swore never to travel under someone else's direction again. By 1924 she had also undertaken several walking tours in the Highlands, including the length of the Outer Hebrides. The last journey provided the copy for her first article submitted to *National Geographic* for which she received $250 (far in excess of the ten guineas Kew was to pay her later for 308 plant species!). Ironically the account would not be published for 30 years when a photographer accompanied her on a repeat visit, but the acceptance allowed her to think of herself as a serious writer (Hoyle 2001). Hutchison used the money to visit Iceland, when she walked 260 miles from Reykjavik to Akureyri, which she wrote about in *The Field* magazine and the *National Geographic* (1928). She returned in 1933 for a walk across north-eastern Iceland. At home, her membership of the Tory Party and the Scottish Vernacular Society is indicative of her political and linguistic conservatism.

Described by the RSGS as an 'explorer-botanist' (*SGM* 1983: 54), Hutchison made four major expeditions to Greenland, North Alaska and Arctic Canada in the late 1920s and early 1930s. The North Pole had been reached in 1909 and the South Pole in 1911 and there had only been a few academic post-war expeditions into the Arctic before Hutchison undertook her travels (Hoyle 2001). She travelled to eastern Greenland in 1927 and returned to spend a year in Umanak in north-west Greenland in 1928, an area closed to visitors without permits. Here she learned Greenlandic and accompanied the local pastor, doctor or anyone else making local journeys near where she thought she would find the flora specimens she was seeking. In 1931 the Antarctic explorer and academic Frank Debenham invited her to speak to a geography class at Cambridge University; she was initially petrified but he wrote to her afterwards that she 'held the class enthralled' by her vivid style and passion for Greenland and its people.[25] In 1934 Hutchison made her most spectacular journey (Robinson 1990) when she travelled from Vancouver to Alaska, travelling by steamer to Skagway and overland to Nome, a small ship took her along Alaska's north coast and she completed her journey with a 120-mile dog sleigh ride to Herschel Island accompanied by Gus Masik, because the ice had closed in; she returned by a mere three-and-a-half-day ride in a mail plane along the course of the Mackenzie River back to Alberta. She made a return trip to Greenland in 1935, plant collecting and maintaining friendships, funding a bell for the new church at Umanak, paid for from her fees for lectures given on Greenland (Hoyle 2001). In 1936 she gained some funding from the British Museum and permission to visit the Aleutian Islands; and being based in Unalaska, also home to a US coastguard team, she managed to hitch a ride on their ship the *Chelan* while on a local tour. Her presence as a woman was quite irregular, but the crew were duty bound to accommodate representatives of government or scientists, so she was permitted on board and the crew christened her 'Admiral Hutchison'! She took every opportunity to land and search for specimens and found interesting Asiatic types of plants on Attu (Wylie Hutchison 1937b; Robinson 1990; Hoyle 2001; see Figure 4.5). She concluded her expedition in Japan and travelled back to Britain overland with her sister. Whilst on these journeys Hutchison's principal purpose was to collect botanical specimens and she carried a letter from Kew authorising her as a plant collector. She also gathered local folklore and tales, cultural artefacts and information on the cultural, political and economic relationships of the indigenous peoples (Inuit, Inuvialuit and Aleutians), settlers and traders she met. Her cine films recorded subjects as diverse as seal hunting from a kayak to the governor's coffee party and Greenlanders dancing Scottish reels introduced by whalers a century earlier.[26] Her botanical specimens were commissioned by the Botanical Gardens at Kew and the British Museum's department of natural history. The Royal Scottish Museum in Edinburgh and Cambridge University's Museum of Anthropology and Ethnology commissioned the collection of cultural artefacts, which were acquired at a vital time before the tourist trade affected the production of such items (Hoyle 2001).

During the 1939–45 war Hutchison worked for two years as a censor of correspondence in Danish and Norwegian; she and sister Hilda were initially posted to Liverpool, then Glasgow, and their home became an RAF billet. After the censorship work finished she gave lectures for the Ministry of Information and delivered a talk on BBC radio when Japan invaded Attu, which marked the beginning of her phase as a radio broadcaster. A timely paper on Attu was accepted by the *National Geographic* (1942) and she was commissioned to write articles on wartime in Scotland and Wales, again initiating a series of publications in that journal, including accounts of her long-distance walks through the UK and Europe undertaken in her late fifties and early sixties.

Even from the perspective of the twenty-first century, Hutchison's 'achievements are astonishing' (Kaye 2002). An article in *The Scotsman* described her as 'an enigmatic combination of the wilfully determined and the near mystical, seeing the hand of God in the sublime and awesome Arctic environment, and who did much of her travelling alone, or in the rough and ready, all-male company of sailors, trappers and hunters'.[27] It also cast her in the frame of the hardy independent Scottish woman explorer (see Munro 2002).

With the 'eye of an artist and the soul of a poet' (Hoyle 2001: 113) Hutchison made for an exceptional traveller and observer; this was helped by the fact that she was well organised but also flexible and opportunistic.

Bog cotton on Kodiak Island   *Elymus arenarius used for basket weaving,*   *Epilobium angustifolium, Unalaska*
Dutch Harbour

Figure 4.5   Photographs of plants observed *in situ*.
Source: Wylie Hutchison 1937b; GJ

She was always a collector rather than a 'discoverer' of plants, ever ready to acknowledge the frequent local assistance she had to find plants; but that did not make her collections any less valuable to the scientific institutions she supplied. Although a trained horticulturalist she considered herself an amateur botanist. The originality of her authorial voice came in her recording of sense of place in all its physical, social and cultural richness, in prose, poetry and art. Hutchison was in great demand for public-speaking engagements after the publication of each book and she gave presentations in village halls, on the BBC radio and at the RSGS, RGS and the Scott Polar Institute in Cambridge, amounting in total to over 500 speaking engagements (paid and unpaid) (Hoyle 2001).

In 1934 Hutchison was awarded the RSGS Mungo Park medal in tribute to her explorations; she wrote expressing surprise to a friend: 'I certainly never expected that my feminine researches would gain a medal'.[28] As Hoyle (2001) points out, Hutchison was relatively isolated in her work; the Arctic Club, founded in 1932, which might have provided a natural research network for her, was only open to men who had been on Oxbridge expeditions. She was also honoured with the Freedom Medal of Denmark for her work in Greenland and in 1949 she received an Honorary LLD from St Andrews University for her work as botanist, explorer, writer and artist of the north. She was the only woman to receive an LLD from St Andrews that year and considered it to be her highest accolade (Hoyle 2001). She bought a new hat for the occasion, but had recently had to sell jewellery and cameras to make ends meet and had converted Carlowrie's enclosed lawn into a market garden in order to finance her large home.

Hutchison spent her later years at Carlowrie with sisters Hilda and Nita, increasingly crippled by arthritis. She died in 1982, aged 92, in the home where she had been born. Hutchison was a strong believer in reciprocal obligations, sent money in times of need to the communities where she had travelled and was remembered fondly by her tenants. Her papers were preserved thanks to her friend Medina Lewis and resulted in a memorial exhibition which was held at the National Library of Scotland in 1987.

Her writing career began with poetry and fiction, but as Hoyle's (2002) bibliography testifies, her writing was prolific and varied. It included articles in the Kew Bulletin, the *Scottish Geographical Magazine* and 12 articles in the *National Geographic*. She also became a self-taught but nonetheless accomplished cine film maker and exhibited a number of her paintings. Her collected specimens and artefacts, correspondence, films and writings are held in 17 different archives in the UK and North America. Hutchison's writing has two characteristics: first that she didn't sensationalise the difficulties she experienced on her journeys; and secondly, that she was not judgemental of others (Hoyle 2001). Her observations included the domestic practices of the different communities she visited, somewhat in contrast to the heroic narratives of male Arctic explorers, but echoing earlier women explorers (ibid.).

For the most part Hutchison was not excessively constrained by her gender, but gendered social mores surfaced on occasion. In contrast to Greenland, Hutchison experienced some resistance to her presence as a single woman travelling without 'proper reason' in northern Canada – female suffrage was a volatile issue at the time which may have contributed to her reception. She also had to be careful in reporting her journey to the press and public when it came to giving account of her months alone with Arctic trader Gus Masik (Hoyle 2001). David Munro noted of her writing style: 'With Hutchison you get to meet the people and you encounter the landscape, she notices everything, whereas with many another early travel writer it's all about them …'[29] In contrast Hutchison made few claims for herself and made light of the rigours she underwent, but the result is relative obscurity: 'Her success as a traveller, combined with her essential modesty, has caused Isobel Hutchison's name to be almost forgotten. The public imagination is captured by tragedy and failure' (Hoyle 2001: 6). In contrast, Marian Harvey, a woman who had befriended her at Shingle Point mission, wrote to Hutchison: 'You little know the amount of silent tribute you received from people here who knowing the country could more readily understand a few of the discomforts, and the true courage that many of your lonely experiences called for'.[30]

Her paper 'Flowers and farming in Greenland' was first published in the *SGM* (1930) and reprinted by Edinburgh publisher T.A. Constable in the same year. Her travels, including the many people she met and who assisted her, are recorded in three major books. A manuscript based on her first visit to Greenland was not considered sufficiently novel to be financially viable by publishers, but additional material from her year's sojourn saw the quick publication of *On Greenland's Closed Shore: the Fairyland of the Arctic* (1930 (Hoyle 2001)). *North to the Rime-Ringed Sun*: being the record of an Alaska–Canada journey made in 1933–4 (1934) was in sufficient demand to be reprinted in 1935 and published in New York in 1937; it received almost uniformly good reviews except in the *Saturday Review of Literature*, whose description of the book as 'consistently drab' suggests the reviewer did not get past the initial leg of the journey! (Hoyle 2001). In contrast F.S.C. (1935), the *GJ* reviewer of *North to the Rime-Ringed Sun*, commented after reading the book that it was no surprise that the RSGS had awarded its Mungo Park exploration medal to Hutchison after this journey; and the *SGM* (1935) maintained her journey probably had no comparison with any made by another woman in modern times. The book was chosen as the Scientific Book Club of New York's 'book of the month' (Hoyle 2001) and Hutchison was the subject of a chapter in Marjorie Tiltman's (1935) *Women in Modern Adventure*, which placed her travels in the context of other women's remarkable achievements. She later used Tiltman's book herself as the source of information when giving a talk to the Lyceum Club in Edinburgh in 1955, with the telling title 'Some women did not stay at home – women's place in exploration' (ibid.). In *Stepping Stones from Alaska to Asia* (1937a), her account of the people was appreciated as much as the information about the flora and landscape (A.J.W. 1938). The book was reissued by Blackie in 1942 under the more politically topical title *The Aleutian Islands: America's Back Door*. Her account was criticised as dotted with 'whimsy' by the *Times Literary Supplement*, but here lay the rub: in order to generate now-needed income she had to write a *popular* book that would be readable and generate sales (Hoyle 2001). Hutchison also transcribed an account of Gus Masik's years of trading in the Arctic, based on their many shared evenings of story telling, and translated Knud Ramussen's book of Alaskan fairy tales from Danish to English (Robinson 1990; Hoyle 2001). Hutchison's article in *Nature* on plant collecting in the Pribilof Islands (1937) is taken by Hoyle as an indication that she was 'allowing herself' to take her botanical work seriously. Being taken seriously was not simply a matter of negotiation with herself – Hutchison joined the RGS in January 1936 and offered to give a paper on Greenland on her return from Alaska; she returned to find no paper had been scheduled and had to enter into correspondence with the RGS, who wanted an original paper which could be published in the Society's journal. This caused complicated negotiations between the RGS, Hutchison and her publisher, who didn't want her new book to be undermined by significant material appearing elsewhere. The result was that Hutchison showed some of her cine film at an RGS film night in April 1937 and gave a paper in May, but her published article (1937b) was a short extract from this paper on the new reef discovered off Attu, the most westerly of the Aleutian islands, i.e. largely reporting the findings of the US coastguard who provided photographs and a map for her. When she addressed the RGS, Professor Henry Balfour introduced her as 'a botanist who, for the purpose of collecting plants for the British Museum, made a somewhat adventurous journey into a part of the world little visited by British travellers' (*GJ* 1937: 544).

In later years Hutchison still travelled in Scotland, including leading a cruise to the Shetlands for the Scottish National Trust; she also holidayed in Europe in the 1950s, often with Hilda, and continued to write travel and nature pieces as a source of income. Famous in the 1930s and 1950s, Isobel Wylie Hutchison's name and achievements have largely been forgotten in the interim (Kaye 2002) beyond the Edinburgh societies and specialist Arctic botanical circles to which she belonged. Her work is recorded in Robinson's (1990) *Wayward Women*, but Hoyle's (2001) detailed biography is the first to give a full account of her life and work.

Figure 4.6   Violet Cressy-Marcks.
Source: Cressy-Marcks 1930

Violet Cressy-Marcks

Violet Olivia Cressy-Marcks (née Rutley) was born on 10 June 1895 at the High Street, West Wickham, only daughter of Olivia Ada (née Leake) and W. Ernest Rutley. She had two brothers Reginald and Allan. On her birth certificate her father is recorded as a butcher, but by the time of her first marriage he is listed as a 'merchant' and living in Grant Road, Addiscombe, Croydon.[31] Best known by the family name of her first husband, Captain Cressy-Marcks, Violet had one son before they divorced, going on to marry Francis Edwin Fisher in 1932, with whom she had two further sons, Ocean and Forest.

Credited with travelling in every country of the world, Cressy-Marcks was keen to have a 'scientific' grounding to her travels and was a fellow of the Royal Geographical, Royal Zoological and Royal Archaeological societies. Her membership of the first two is inscribed on the title page of her first major work, *Up the Amazon and Over the Andes*. Having been trained in surveying at the RGS and surveyed part of the north-west Amazon basin (1929–30), Cressy-Marcks also acknowledges the tutelage of E.A. Reeves of the RGS (Maddrell 2004h). She was a capable cinematographer and photographer, bringing films and photographs from many of her travels, including politically sensitive areas. She studied in Arabia and undertook widespread archaeological studies, including Egypt, Syria, Palestine, Persia, Java, China, Ethiopia, Afghanistan, and the Khmer, Inca, Aztec and Mesopotamian peoples. Considered primarily as an archaeologist (Robinson 1990), collecting contemporary ethnological data on ethnic groups little known in the west was also part of her remit.

Elected to the RGS in 1922, Cressy-Marcks, then living in Buxton, Derbyshire, was described by her nominator as of independent means, already having 'travelled extensively' from Alaska to Java and having made private 'explorations in Tibet, Kashmiri etc.'[32] She went on to travel overland from Cairo to the Cape in 1925; to Albania and the Balkans in 1927–8; and spent the winter north of the Arctic circle travelling by sledge from Lapland to Baluchistan in 1928–9. She travelled through the Amazon and Andes to Peru by canoe and foot (1929–30) and gave an account of her travels in *Up the Amazon and Over the Andes* (1932). Her fourth journey around the world took place during 1931–2; she revisited Spain in 1933 and travelled through India, Kabul, Tashkent and Moscow in 1934. (*Who's Who* (1947) describes her as flying from England to India, over the Hindu Kush and from Moscow to England and her photograph in the National Portrait Gallery shows her in a leather flying suit (see Birkett 2004a: 35).) In 1935 she took the first motor transport from Addis Ababa to Nairobi during the Italian invasion of Ethiopia when she visited the warfronts taking cine film; in 1937 she travelled from Mandela to Peking over land and in 1938 she travelled from Turkey to Tibet by motor, yak and mule.[33] The latter included a study of the war conditions in China and an interview with Mao at Communist headquarters in the cliff town of Yenan, as well as ethnographic studies and reaching Lake Koko Nor as her end point. At home she led a society life (giving an 'at home' at the Lyceum Club before travelling to Lapland for example), was well known at The Ritz and Claridges, and had a taste for hunting and game shooting as well as collecting East European icons and rugs (Maddrell 2004h). Her obituary in *The Times* stated: 'There is good reason to believe that no contemporary British woman equalled or indeed approached her record of adventurous journeys in unfamiliar lands with modes of transport so varied and calling for determined and fearless achievement of her objectives'.[34] Although taking her husband with her as a 'passenger' on her journeys to Kabul in 1934 and China in 1938 she relished travelling without a European companion, echoing Isabella Bird's dictum that adventurous travel was 'no place for a man'.[35] However, she would still create 'space' for herself by walking ahead of the party: 'I liked the solitude which is to me the kindest of all travelling companions' (1940: 29).

Up the Amazon and Over the Andes is dedicated to Violet's mother 'the bravest and most noble human I know', who is also credited with both encouraging and financing her journeys. However, her will requesting that a copy of her biography be shown to the chief of MI5 'for his appreciation', as well as entrusting that biography to (the then deceased) Bernard Rickatson-Hatt, who had spent three years in intelligence in Istanbul (as well as being editor-in-chief for Reuters), suggests she may have had sponsorship from the secret service for some of her many travels (Maddrell 2004h). The co-executor of her will, William Cuningham, reported that he 'hadn't the faintest notion that she had any connexion with espionage',[36] as yet this is unsubstantiated, but Cressy-Marcks certainly managed to coincide her travels with periods of international political sensitivity on a number of occasions. In Russia she achieved largely unfettered travel and visited most of the Foreign Office officials (as she reported to Mao). During the Second World War she was an ambulance driver for the British Red Cross abroad, as well as special war correspondent, based in Chungking, for the *Daily Express* 1943–5; she was also accredited to the War Office as war correspondent at the Nuremburg trials, Paris peace conference and the Greek elections (*Who's Who* 1947). If Cressy-Marcks was engaged in intelligence gathering during some of her travels then she would have been in the company of Gertrude Bell and Freya Stark among others, adding weight to Middleton's (1965) view that several women travellers were on the fringes of the intelligence 'game'.

In post-war years her travels were still frequent, travelling in Indo-China, Katmandu and Japan 1953–4 and completing her seventh and eighth journeys around the world in 1955 and 1956. Her husband Francis died at Nassau in 1956 on this last trip. When in the UK, Cressy-Marcks divided her time between homes at Princes Gate, Kensington, London and Hazelwood, Kings Langley, Hertfordshire. She died on 10 September 1970, at the age of 75, and her request was to be buried with her husband at Langleybury church. Her net estate was £288,799 and her will, as is the nature of such things, says much about her character and the things she sought to perpetuate after her death. It included a bequest in her name providing a RGS travelling scholarship for women's geographical research in the field, as well as bequests to the Royal Central Asian Society, Chatham House (International Relations), and the China Society. The latter was an endowment to fund the annual party which she usually hosted. There were also individual bequests to waiters at the Ritz and Claridges and her gardeners, housemaids and secretaries, and to Nora Norcliffe, her companion of several years (Maddrell 2004h).[37]

Publications and reception of her work

Cressy-Marcks only published accounts of two of her many journeys: *Up the Amazon and Over the Andes* (1932) and *Journey into China* (1940). Both were well received by critics, who realised her serious and determined journeying under often difficult and even life-threatening conditions. *Up the Amazon and Over the Andes* topped the list of recommended reading in the

British Journal of Nursing (December 1932) and was quickly abridged to appear in *Women in Modern Adventure* (Tiltman 1935). Although writing the *Amazon* book primarily as a travel account, she included supplementary appendices for those seeking practical and reference information. This included advice on equipment to fellow travellers, a map of her journeys, lists of survey instruments, anthropological instruments, 'natural history requisites' and photographic equipment as well as camping gear, navigation tables and suggested trading goods. A list of Latin and common names for a selection of South American flora and fauna and statistics on the Amazon, including current flows and volume of water were also provided. With a dismissal of economy travellers as 'not conducive in furthering the prestige of our country' (Cressy-Marcks 1932: 331), she argued that if a journey is to be of any scientific use then money must be spent on equipment, but was equally happy to suggest that samples of interesting rocks can be taken merely by use of a geological hammer and analysed at home. Despite all the equipment she took with her, her book was criticised by a reviewer in *GJ* for its useless map, but the appendices were considered useful and interesting. The same reviewer thought her prose style inadequate to the task of describing the wonders of the journey and landscape, but that the 'vivid touches fully compensate for some common-place observations on the civilized parts of the country' (D.C.S-T. 1933: 555). In contrast, the reviewer in the *Times Literary Supplement* complimented her spare style and emotional control: 'In the Andes part of the book there are many such unelaborated touches which illuminate the character of the people and also her own. In these comments there is individuality without eccentricity and feeling without illusion. They abundantly justify all the F's after her name on the title-page'.[38] Needless to say Cressy-Marcks slashing her own leg to spill the venom after a snake bite and fellow traveller Antonio Vlhesek's fever-induced suicide in her canoe were the most dramatic moments in her 'strange tale ... of endless obstacles and hardships surmounted by iron determination' (ibid.), but the sheer slog of parts of the journey with reluctant mule drivers were the aspects of the journey which called for the most resolve.

Perhaps in response to comments on her first book, *Journey into China* is a more structured volume, and to forestall criticism she explains that although her account could be delivered in a more exciting manner she chooses to keep the 'frills and effects' out of her book (1940: 2). She sets out her own criteria for success when she states that she wishes to convey the happiness she derived from her journey; if this was achieved then she would 'feel I have done my job, and that writing about it was worthwhile', (ibid.). Once more travelling in a war zone, Cressy-Marcks' personal target was to reach Lake Koko Nor; the explicitly political aims of her sixth journey to China included studying Chinese communism, Japan's chances of crushing China and Russia's influence on China. It is at Koko Nor in Tibet that her descriptive writing style is at its best and most reflective, not least when she noted how reaching her much aspired to destination mapped onto her own biography as 'one of Asia's ancient landmarks, but a milestone in my life' (1940). She writes in an understated style on culture, war and politics. Her texts also include some humour at her own expense (her embarrassing unwillingness to sing for locals when requested during a bus ride) and her husband's (his staring out a leopard in the consulate garden for half an hour before he realised it was chained). She also stressed with humorous irony the gender inequalities of social conventions which inverted British practice: when they galloped through dangerous areas their escort surrounded Frank, 'I might go on ahead or behind, taking photos as I pleased, but never had one of the guards' (1940: 32). Ultimately her work on China was valued principally for its contribution to understanding China in general and particularly the importance of China's 'back doors' to Russia and Burma, and Chinese means of resistance to Japanese invasion, all of which was stressed by soldier-diplomat Percy Sykes in his introduction. Cressy-Marcks interviewed Mao and reported on the status of the Red Army, including their ammunition and warm clothing shortages – they resorted to targeted guerrilla tactics and sentries handed over their coats to the relief guard at the end of duty (1940: 168). She notes the importance of unity in the new China, but also the labour camps. The troop train she travelled on out of Sian was fired upon and she articulates the impact on civilians of modern warfare. She helped a doctor treat the injured in a field station for a while, but recognised that she was not brave enough to go to the front. After visiting a military hospital she wrote 'I was dazed and sickened by the suffering. My stomach is not weak but I stumbled out for air …[an air raid sounded] but I felt it mattered little what happened. A bird sang sweetly and the notes hurt, and the sun mocked' (1940: 249).

The *GJ* reviewer thought the book needed more time taken over editing, but appreciated the urgency of publication in terms of world affairs. The insight and detail she brought to British understanding of China's geography and political relations, as well as her own courage and perseverance, were praised. 'With a woman's acute observation of detail, she gives a rich tapestry of sane and humble lives under the widely varying geographical and climatic conditions on her route' (A.M. 1940: 214). The *Times Literary Supplement* reviewer considered her visit to Command headquarters of the Red Army to be one of the most remarkable sections of her book (a view shared by Carter 1943). Although her actual interview with Mao recorded in the book revealed little, her description of the living conditions at Yenan, with its offices in caves and improvised sinks from petrol cans, was considered fascinating. Perhaps the highest compliment for her account, which was published when Britain was at war, is that which notes her lack of affect in retelling the experience of being under fire or dodging bombs. 'These and other adventures are all related with the implication that they are no more dreadful, or remarkable, because they are happening to her, a European, than to the philosophical Chinese with whom she is sharing them. Indeed one of the most welcome deficiencies in this long book is the entire absence of patronizing facetiousness'.[39] It is in the everyday life and developments in politics and the war that Cressy-Marcks excels. Allowing for its weaknesses, Dagny Carter begins his review 'This extraordinary book should have had a less ordinary title' and concludes '*Journey into China* is an exciting firsthand account of the new and strange world of motor roads, mechanic shops, airdromes, and fast-growing cities which is now emerging in the heart of ancient Asia' (1943: 71–2).

Despite being a fellow of the RGS, and the validation given to her books by leading RGS figures such as Sir William Goodenough and Brigadier General Sir Percy Sykes, Cressy-Marcks was not accepted into its inner circle and her travels gained limited recognition. A letter from Captain E.W. Fletcher, R.A., from the British Legation at Kabul to RGS secretary Arthur Hinks set the tone for her reception in the Society, which seems in no small part to be based on a combination of social snobbery and a related desire to protect the Society's status. 'We had Mrs. Fisher, who you probably know as Mrs. Violet Cressy Marks (sic) here. She was armed with letters of introduction from the Afghan Minister … describing her as belonging to "High London Society" which in addition to making me smile rather complicated things, for her behaviour was hardly a good advertisement for "high society" which she would probably pronounce " 'igh society". She was an unpleasant person and gave everyone a lot of bother and I pitied her husband, whom she derided in front of strangers, and who is a real nice old British farmer. I am telling you this as I fancy she means to write a book and possibly to lecture and I think you should be warned, especially as Sir W. Goodenough apparently wrote a preface to her book on South America. I am not being jealous or obstructive and in many ways I admire her toughness of spirit but there it ends. It is a pity such people are allowed to come to these parts. We don't feel proud of them'.[40] Hinks replied, without naming Cressy-Marcks, that he appreciated the advice but 'I have never had any fear that she would offer us a lecture, because she never seems to have anything to say after her travels, which have been very extensive and should have been very interesting. Her *performance* in getting to the head of the Amazon and up over the escarpment of the Andes was quite extraordinary' (my emphasis).[41] Ironically, Fletcher's comments about social status echo Cressy-Marck's own remarks about another unnamed woman traveller 'who has travelled a great way and is well known for being extremely resourceful, that she travelled around the world for one year with one suitcase and for £100, though this form of travelling should not, I feel, be encouraged, as it is not conducive in furthering the prestige of our country' (1932: 331).

In the event there *was* a protracted correspondence in 1937 between Cressy-Marcks and Hinks on the subject of a paper. Cressy-Marcks had been invited to show the RGS president Sir Percy Cox her cine film on Abyssinia for consideration for an afternoon meeting. Cressy-Marcks subsequently maintained that Cox had also invited her to give an evening paper on Ethiopia the night before she left for China. Cox had died while she was away and without a copy of Cox's letter to her, Hinks stood by what he had on file, which was that Cox thought her draft notes were only suitable to accompany an afternoon film.[42] This leaves the question of whether Cressy-Marcks would have opportunistically lied on finding out that Cox had died, in order to secure herself a lecture spot at the RGS. In the event she appears not to have given either paper or film and no papers by her were published in the *GJ*. This did not stop Cressy-Marcks from leaving a bequest to the RGS, perhaps pointedly stipulating it should support *women's* travel. The *GJ* reciprocated with an obituary, but one in which she was discursively constructed as not quite in their own mould: it stressed her impetuosity in addition to her determination and described her as 'very much in the tradition of the intrepid lady travellers of an earlier, and less scientifically-minded generation' (*GJ* 1970: 670). In contrast to this obituary Robinson argues: 'she strenuously collected as much scientific data as she could, and confronted any little difficulty ... with stern courage. She was one of the most committed and thorough of women travellers, constantly trying to prove and improve herself, and, it seems quite fearless' (1990: 43). Sir William Goodenough commented in his introduction to her 1932 book on the Amazon that Cressy-Marcks was 'more mindful of scientific work than may appear in her pages' and respected diplomat explorer Sir Percy Sykes noted the fact that she had studied surveying at the RGS and that she 'ranks high among English explorers' (1940: 11). These widely differing views of Cressy-Marcks' character and work underscores ways in which the extent of geographical work may not always be fully represented in published accounts, and that obituaries should be seen less as a panopticon and more as a viewing point onto a biography.

Other Twentieth-Century Women Travellers

The post-war years brought growing opportunities for independent travel, generated by increased affluence and improved communications, especially air transport. This created a new genre of travel writing, seen for example in the writings of Dervla Murphy. Such travel accounts can be found as traces in GA branch meeting programmes such as that of the Central London Branch, which included talks in 1953 by Miss Bennetton 'Slides from East Africa', and in 1955 Miss Tyrrell's 'Informal talk on New Zealand'. Personal papers or the minor publications of school teachers also reveal different sorts of 'professional development' travel. **Grace Dibble** was educated at Maidstone Girls' Grammar School, trained as a teacher at Homerton College and gained a University of Cambridge Diploma in Geography, followed by a honours degree in geography from London University. She had her first experience of European travel through the Cooperative Holiday Association and joined trips organised by the Thanet Geographical Association (with Alice Coleman) and vacation courses offered at universities such as Uppsala in Sweden (Dibble 1993). Dibble began teaching in Kent in 1931, and then worked for a year in Canada and eight years in India. She spent a further eight years teaching in West Africa before being appointed as principal of Ilesha Training College in Nigeria in the 1950s (Robinson 1990; Dibble 1993). Throughout her working life and after her retirement in 1965 Grace Dibble travelled in over 100 countries, mostly by everyday means, preferring buses, carts and other forms of public transport (Dibble 1993). She published a series of travel books under the title 'Return Tickets for Armchair Globetrotters' with Stockwell in Ilfracombe. The publishers were unable to provide information on the volume of publications but recalled that Miss Dibble had gained 'great pleasure' from publishing her volumes.[43] She was clearly tickled (see 1993: 5) by her inclusion in Robinson's (1990) *Wayward Women. A Guide to Women Travellers*, where Robinson described her as a 'traveller of the old traditional school: the cheerful maiden whose inveterate world-wending habits are harnessed to a clear sense of Christian duty, and whose prodigious books edify and entertain' (1990: 88). For Dibble, her training as a geographer was central to her understanding of and engagement with place: 'Travelling was made so much more interesting for me, because I had read Geography at Cambridge, and for the London Honours Degree. Thus, I could better understand the topography, because of a little knowledge of geology. One knew what climate and flora and fauna to expect. In our anthropological classes we learned about the various indigenous peoples'. Reflecting strong themes in the London degree, she continued, 'I found Economics especially interesting and so visited factories where possible – pineapple in Honolulu; scent in Grasse; silk in Japan and glass in Sweden. I was always thankful for the good training we received in map making and reading' (Dibble 1993: 5–6). In contrast, **Thora Halsem** travelled less widely and did not publish her experiences, but created her own illustrated diary accounts. Halsem graduated with an honours degree in geography from the University of Liverpool in 1939,[44] became a geography teacher at Clitheroe Royal Grammar School and spent her summer holidays travelling by bicycle (and later by car) in the 1940s to 1970s. Her personal travel accounts cannot quite be considered as vernacular – she is a graduate of geography – but neither are they 'professional' in terms of their mode of production or intended audience; they perhaps more accurately

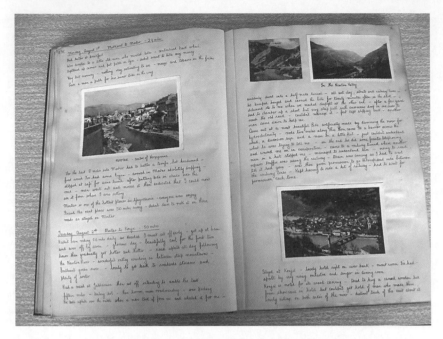

Figure 4.7   One of Thora Haslem's travel accounts held by the GA (author's photograph).

fall into the category of Rojek's (2000) notion of 'serious leisure': personal, pleasurable, experience-based, identity-shaping, but also worthy, enhancing her professional role as a teacher. Mrs Gulliver, a former pupil, who serendipitously recognised and rescued Haslem's books at a local auction sale, wrote: 'I remember the talks given on the return from her tours. They were lively, interesting and humorous – real highlights in an otherwise dull curriculum' (*GA News* 1992: 2). Halsem's collection of European and Scandinavian travel accounts (see Figure 4.7), now lodged at the GA, represents a fascinating repository of personal encounters of the way of life of the people she met and stayed with. This is particularly interesting in the case of those areas that were revisited after a significant hiatus, which allowed Halsem to analyse social, economic and political changes.[45]

Conclusion

As can be seen from debates such as that between Domosh (1991a, 1991b) and Stoddart (1991), as well as within the private critiques and public reviews of travel literature discussed above, the epistemological relationship between 'geography', 'travel', 'fieldwork' and 'data collection' is a contested one. As the opportunities for 'pioneering' exploration decreased in the early twentieth century, so the demands for detailed empirical study of place increased (Younghusband 1917) and geography in its professionalised academic form shifted its discursive emphasis from exploration to fieldwork in the early twentieth century. During the inter- and post-war years the Le Play Society and later the Geographical Field Group and GA played significant roles in offering interested parties opportunities to undertake field study in Britain and abroad. These field trips, which varied in scope and demands, ranged from excursions to expeditions, but mostly fell in the middle ground of serious fieldwork. Single people and married couples made up these parties and it was not unusual for women to be the majority of group members. The influence of these avenues for fieldwork will be highlighted in successive chapters, notably in the biographies of Joan Fuller and Gladys Hickman.

Victorian women's travel writing underwent something of a renaissance within both literary and geographical work in the 1980s and 1990s. This has served to bring their contributions to geographical knowledge to the attention of a wider audience within the history of geography and in doing so linked them to theoretical debates about empire and gender, including demonstrating that the female gender and service of Empire were not necessarily mutually exclusive. Archive material also shows how individual women could be simultaneously published contributors to the geographical canon, while being excluded from membership of the RGS; and later women fellows had their work excluded from meetings and journals, while being endorsed by its president or other geographical societies. The geographical work of post-Victorian women travellers was acknowledged by publishers (notably John Murray) and national geographical societies such as the RGS and RSGS, as well as smaller groups such as local branches of the GA, but is largely neglected in the historical canon of the subject. The combination of the creation of the IBG as a professional body for academic scholars, coupled with the early-to-mid twentieth-century shift from exploration to fieldwork within the geographical discourse left generalist travellers marginalised from the academic disciplinary framework. This was exacerbated by the decline of post-war regional courses taught in universities as that paradigm was eclipsed by systematic approaches to geographical enquiry. Yet, as these brief examples illustrate, women travellers in the Edwardian and interwar years were nonetheless engaged with charting the uncharted, in part the physical landscape and to a greater degree the social, cultural and political landscape of the regions where they travelled, both of which proved of significant value to the British state, particularly during times of war. Women travellers do not conform to a particular stereotype (Birkett 1989, 2004a), but this makes them and their work all the more interesting. Each travel account shows different and often evolving strategies to negotiate gendered literary structures and social expectations. Members of geographical societies often incorporated increasing amounts of geomorphology, geology, population, climate and trade statistics, and they evidence women's ability to serve the country's strategic and economic interests through their travel writing. However, there was sometimes tension between addressing the perceived requirements of geographical societies and their canon, and the interests of their wider income-generating readership. Although issues of credibility persisted in women's travel writing (possibly through internalisation (Mills 1991)), women's authorial voices did become more authoritative, especially those like Bell and Stark, who, with the benefit of university educations, had a more reciprocal relationship with geographical societies. Analysis of twentieth-century women travellers' geographical work (published and unpublished) merits a more detailed engagement within the history of geography, especially in the light of the 'cultural turn' in geography and biographical approaches to historical subjects.

Chapter Five

Women in Geographical Education: Demand for Geography Teachers and Teaching by Example

Introduction

There was enormous change in access to education from 1850 to 1970 for girls of all classes and for the working classes in general. Although there was a tradition of formal schooling in Scotland, in England, education for middle- and upper-class girls in the mid-nineteenth century was limited to the 'feminine' skills of drawing-room music, water colour painting, fine embroidery and domestic management, which were taught at home by governesses or tutors. There were 12 endowed schools for girls in England and Wales in 1846, and Queen's College and Bedford College were founded for women students in 1848 and 1849, respectively, principally for the training of governesses. This was followed by the foundation of pioneering schools for middle-class girls, notably the North London Collegiate School (1850) and Cheltenham Ladies' College (1853). In 1866 girls were able to sit Cambridge local (schools) examinations and in 1868 non-collegiate students were admitted to the University of Oxford. By 1891 there were 36 public day schools for girls across England (Barnard 1961). As the century progressed opportunities increased for girls across the social spectrum, especially with the Elementary Education Act in 1870, which provided free primary education for all (although this was not necessarily taken up until it was made compulsory 10 years later), provided partly through funding the existing network of church schools and partly through increasing local government Board of Education schools. The 1870s also saw the foundation of women's colleges in Oxford and Cambridge and the admission of women to take degrees at the University of London in 1878. At the same time the university sector was expanding with university colleges gaining full university status (e.g. the Federal University of Wales 1893, and the universities of Sheffield 1905, Bristol 1909, Nottingham 1948, Southampton 1952 and Keele 1962). Gradually women gained access to equal educational opportunities but pockets of inequality persisted according to class and gender. Social differentiation was a vital part of the increasing demand for formal education of girls in the late nineteenth century, with schools being chosen as much on the basis of social segregation as by curriculum (Sutherland 1990). However, in a patriarchal socioeconomic system, many middle- and upper-class girls continued to be 'educated for dependence', with employment being considered a 'last resort' if marriage did not materialise (David 1987). Women were only able to take their degrees (as opposed to sitting the examinations) at Oxford and Cambridge universities in 1920 and 1947, respectively, reflecting particular localised institutional barriers, and widely circulated views that higher education was too demanding for women and might affect their reproductive capacity. As late as 1912, respected academics J. Arthur Thomson and Patrick Geddes (who worked with many geographers) argued in *Problems of Sex* that intellectual work could damage a woman's health and cause premature death (David 1987; Maddrell 1997). Christina Bremner's (1897) survey *The Education of Girls and Women in Britain* showed girls' experience of elementary education to be different from boys, and these differences were further concretised by early twentieth-century legislation intended to assist the post Boer War drive for national efficiency (Thane 1990). The 1902 Education Act initiated state-provided secondary education and co-education was favoured where resources were limited, especially in rural areas; co-education subsequently became the norm for state secondary schools during the post-war years. The government's 1904 Education Code stipulated that the purpose of elementary education was concerned with 'assisting both girls and boys, according to their different needs, to fit themselves, practically as well as intellectually, for the work of life' (i.e. productive labour or reproductive labour) and this statement shaped the education of 80 percent of British children for the following 40 years (Eaglesham 1967: 53).

When the government charged local education authorities with the responsibility for secondary schools in 1914, post-elementary education became more accessible for the poor, but the pupil–teacher system was abolished which had been a route to continued education for bright children from poor backgrounds. In the 1918 Education Act half-time exemption for children over 10 years of age was abolished and the school-leaving age was raised to 14. After 1918 children in Scotland took the Qualifying Examination at 12 years and the results determined whether they would proceed to full secondary education or short technical courses. The Board of Education of England and Wales moved to a similar model and working-class children from elementary schools could compete for free places at secondary schools (Sutherland 1990). State provision for secondary education was codified and strengthened in the 1944 Education ('Butler') Act, but the quality of provision was far from even and scholarships to selective grammar schools played an important part in providing access to academic secondary education and therefore a chance to apply for university entry. However, university places remained very limited until the introduction of comprehensive secondary education and the expansion of the university sector in the 1960s.

Although often neglected in the historiography of geography, there was an important link between the establishment of geography as a university discipline and teacher education. The place of school geography within the history of the discipline has begun to be explored in relation to school texts (e.g. Maddrell 1998; Ploszajska 1999) and in more general terms (e.g. Walford 2001), but much of the specific detail of this relationship remains unexplored empirically and to a degree conceptually. An important issue is the extent to which gender was significant in the production of geographical knowledge for consumption within schools: both in terms of the gender of the subjects producing that knowledge and whether gender influenced or figured in that work. Many of those availing themselves of early teacher training courses in geography were women, a number of whom went on to make significant contributions to the subject through their texts and pedagogic writing. The Geographical Association (GA; founded in 1893) and its journal *The Geographical Teacher* (*GT*)[1] were significant avenues for women to participate in and contribute to geographical debates. Although initial demand for the GA to be established was based on the premise that 'the present state of knowledge of the subject throughout the schools was unsatisfactory and far inferior to that possessed by *boys* in foreign schools' (my emphasis)[2] the remit was soon expanded to include teachers in girls' schools and elementary schools. Women were admitted to full membership from the foundation of the GA in 1893, with the Association defining its membership in terms of all who taught or were interested in geography teaching. Several women, including leading educationalist Miss H. Busk, were noted as present at the GA's first annual general meeting in 1894[3] and by 1905 just under one-third of the 486 members were women (Balchin 1993: 11). Throughout the twentieth century women played important roles in both local branches of the GA and its national Executive Committee and Council. Teaching was an important means of access to post-school training and education in the early twentieth century, especially for bright working-class women (and men); the same route has been shown to be an important avenue for women from a variety of social and ethnic groups in the USA (Monk 2004, 2007). As teacher training became institutionalised with a growing number of residential colleges, these were organised on gender lines and represented opportunities for women to find employment in higher education, other than the university sector. As the careers of women discussed here and many of those discussed in subsequent chapters illustrate, a number of women geographers moved between university and teaching or teacher training posts during their careers. Archives relating to school geography, including those of the GA, are very patchy and I cannot do justice to the history of the large number of women working in geographical education in this chapter, however, I feel that it is important to provide examples of geographical work as expressed in both teacher training and classroom teaching. I am sure that other studies will follow which will address this important subject in the detail it merits. Many of the women university teachers of geography discussed in the following chapters also contributed directly or indirectly to school teaching and/or teacher education through their publications, contributions to summer schools for teachers and their roles within the national or local branch organisations of the GA. Alice Garnett who served as honorary secretary of the GA for over 20 years is the most obvious example, but Eva Taylor, Dorothy Sylvester and Alice Coleman also made significant if differing contributions to supporting school geography, as did numerous others. This chapter provides evidence of the influence of women educationalists' geographical work, through training geography teachers and writing school texts, which were then used in teaching large cohorts of children across the country (and abroad in some cases); the importance of the GA in shaping and updating the geography curriculum in schools, and women's roles within the Association; the centrality of fieldwork in school geography; and occasions of political engagement through school geography.

Ellen Rickard

Ellen Jennifer Rickard, born c. 1873, had worked as a school teacher after attending the Cambridge Training College; she taught at St George's School in Edinburgh (1895–1903) and St Saviour's and St Olaf's School in Southwark, London (1904–7). She then returned to higher education spending one year at Somerville College, University of Oxford (1907–8), where she completed the Diploma in Geography. She moved into teacher education, becoming a lecturer in geography at the Maria Grey Training College, London (1910–16) and it was during this period that Rickard became a demonstrator on the nationally influential summer vacation courses in geography at Oxford, organised by the Herbertsons (Oughton 1963). Rickard then moved to a post at Badminton School c. 1918 until her retirement in 1934.

Rickard joined the GA in 1908 and was soon elected to its Council in 1912 – 'and was therefore among the first women members to join in the administration of the Association's affairs' (Oughton 1963: 418). During the difficult years of the 1914–18 war Rickard stood in firstly as honorary correspondence secretary (1914–19) after the resignation of J.F. Unstead, then as head of the organisation as honorary secretary between Andrew Herbertson's death in 1915 and the election of H.J. Fleure to the post in 1917. During this time she was invited to speak at GA branch meetings, including the North London group in 1915, where she spoke on teaching geography to junior classes (GT 1916: 220). Although teaching was a feminised profession, women took few leadership roles outside girls' schools and colleges and Rickard's obituarist Marguerita Oughton notes that she was not only a figurehead but also represented the GA at meetings with other national associations, as well as in discussions on educational reform 'at a time when few women undertook or were elected to such office' (1963: 418). Whilst 'warm tribute' was paid to her 'eminent contribution' during the wartime difficulties (Fleure 1953), and Balchin (1993) notes her role as correspondence secretary, her work has been obscured by the years and the general sweep of history. Whether by choice or not, Rickard was not to play a leading role in the GA after Fleure was elected honorary secretary and a new correspondence secretary was in place. It is quite possible that Rickard, who was described as 'devoted to' Herbertson, may well have considered this was appropriate, that she had helped to hold the GA together and now the role was handed on to a fitting successor, while she took on her new post at Badminton. She appears to have published only one co-written piece in the GT on the use of statistics in geography teaching (1912).

F.D. Herbertson

Frances (Fanny) Dorothea Herbertson (née Richardson) was a goddaughter of Dorothea Beale, pioneer in women's education and head of Cheltenham Ladies' College, so it is not surprising that after attending the King's High School in Warwick, she went on to study at the University of London and was remembered as 'an exceptionally cultured woman' of numerous gifts (GT 1916; Walford 2001). She became a teacher of classics at Cheltenham Ladies' College and met A.J. Herbertson at one of Geddes' summer schools in Edinburgh. They married in 1893 and she went with him to Montpelier that winter as he pursued his meteorological work (Jay 1979). F.D. Herbertson is described as 'intimately associated with her husband's work' (GT 1916: 211) particularly in Oxford and during his 15 years as honorary secretary of the GA (1900–15). During his tenure at the School of Geography at the University of Oxford (1899–1915) she is characterised as 'assisting' her husband in his varied and great workload, particularly working with him on publications and summer schools for teachers (Jay 1979; Scargill 1999).[4] Rudmose Brown (who was head of the geography department at

Sheffield) described her as 'busy, energetic … a driving force in the summer schools, always everywhere, keeping everyone up to scratch (Rudmose Brown 1948, cited by Scargill 1999: 12). She was both an educationalist and a gifted organiser, and 'those who had the good fortune to meet her, especially at summer vacation courses at Oxford, are aware how keen was her interest in everything that was done and how difficulties melted away when submitted to her sympathetic yet searching judgement' (*GT* 1916: 211). Described as a 'brilliant and sympathetic co-worker' to A.J. Herbertson, she died on 15 July 1915, only 15 days after her husband, succumbing to the long-standing heart problems that had limited her in her last years. The couple had two children, Margaret born c. 1899 and Hunter five or six years her senior; Hunter served in the 1914–18 war and was reported as 'missing in action' at the battlefront.[5]

As geography emerged as a university discipline, school teachers were the main constituency for geographical training and most of the first generation of academics wrote school textbooks to meet demand for new accurate geographies. Scargill describes A.J Herbertson as having written his school texts 'with the assistance of his wife' (1999: 12). Writing to Walter Freeman in 1948, Joan Reynolds recounted how Andrew Herbertson had stressed his wife's contribution to his writing of the best-selling *Man and His Work*: 'even pointing out whole passages she had written'.[6] In addition to co-authored books F.D. Herbertson wrote several independently for the Oxford Geographies series (edited by her husband). These books were written for elementary schools, with illustrations and exercises at the end of each chapter intended to engage pupils with the material; a series of *Descriptive Geographies* followed, which used a compilation of carefully selected travel extracts and other sources to create Geographical Readers. According to Reynolds, these books reflected Frances Herbertson's belief in the need to engage younger pupils especially with a greater sense of individual place: 'Mrs Herbertson told me how, while entirely agreeing with his Regional System, she felt there were too many names in his [Andrew's] books without any space left for adequate description of the places, or geographical features mentioned' (ibid.). In Walford's (2001: 91) opinion her books were well pitched, written 'with admirable clarity, and a feeling for appropriate shortness of sentence and vocabulary. Though they appear disarmingly simple, the books probe fundamental geographical issues'. The Herbertsons' school books sold 1.4 million copies (Gilbert 1965; Scargill 1999) and were an important means of spreading and popularising the new regional approach to geography. In essence they constituted the core of geographical knowledge imbibed by millions of British schoolchildren in the first half of the twentieth century. Both Herbertsons subscribed to the comparative method in their texts as can be seen in their joint work *Man and His Work. An Introduction to Human Geography* (1899); this book, one of the first to use 'human geography' in its title, moves from simple to complex society with all the underlying assumptions about hierarchy of lifestyles and its relation to environment. While the comparative method was pedagogically sound in principle, taking pupils from known to unfamiliar phenomena, it was prone to Eurocentrism and hierarchical classifications which mimic those of the natural sciences (Maddrell 1998). In her 1914 text on the Southern continents, F.D. Herbertson was quite unusual in allowing that Bushmen [San] cave paintings might be historical in nature at a time when African societies were commonly held to be ahistorical or without history, but her text did perpetuate the representation of racial hierarchy in Africa, with a continuum which moved from forest pygmies to more civilised coastal dwellers. These representations were concretised in pupils' minds by exercise questions such as 'Can you suggest reasons why the natives of northern Nigeria are more civilised than the natives of southern Nigeria?' (1914: 76; Maddrell 1998). Frances Herbertson also wrote a biography of Frederic Le Play, republished by the Le Play Society in 1950, which remains a key reference on his life and work.

J.B. Reynolds

Joan Reynolds was one of the first cohorts of students to complete the formal qualification in geography offered at a British university, when she achieved the University of Oxford Diploma in Geography in 1901. A photograph was taken to mark the occasion and Reynolds can be seen sitting next to Halford Mackinder (see Figure 5.1).[7] Reynolds had attended the King's High School at Warwick (with Frances Herbertson), completed a BA and then presumably started teaching before going to Oxford. At Oxford she experienced warm hospitality from her former co-pupil and her husband and enjoyed long discussions about geography: 'in their home, we spent hours discussing the regional method of teaching geography and the humanistic aspect of the subject, so well explained in *Man and His Work*'.[8] In addition to the kindness of the Herbertsons, Reynolds also benefited from individual coaching from Halford Mackinder on the correlation between history and geography.[9] Reynolds was particularly appreciative to Andrew Herbertson for his introduction (and presumably recommendation) to the Civil Service authorities, resulting in her setting and marking the geographical papers (principally for the Post Office candidates) for many years. Joan Reynolds spent several years in London after her time in Oxford and taught at the University College School (see her 1904 paper). By 1918 she was assistant examiner for the London University Matriculation Examinations, acting in this capacity as an unseen but important gatekeeper to geography at university level as well as a provider of geographical

Figure 5.1   University of Oxford Diploma in Geography students and staff 1901 (Joan Reynolds seated next to Halford Mackinder, A.J. Herbertson on the right).
Source: Courtesy of the School of Geography and Environment, University of Oxford

knowledge through her widely used texts. She also marked papers for G.G. Chisholm's London pupil–teacher classes when she moved to Bristol,[10] which implies a portfolio of income streams including teaching, publications and mass examining similar to that of Marion Newbigin. In 1948 Reynolds sent a copy of an A.J. Herbertson letter to Freeman from an address in France, suggesting that she may have been domiciled there. In her correspondence with Freeman, Reynolds drew attention to the fact that she was the only woman in her 1901 Diploma cohort and when he requested information on A.J. Herbertson's work as an early pioneer in geography, she responded but drew attention to and provided more detail on Mrs Herbertson's geographical work, clearly indicating that she thought women played a significant role in the early and formative years of the Oxford department.

Prior to her Oxford Diploma Reynolds had published a report on teaching the regional method in Switzerland and northern Italy in 1899 (based on her tour of Swiss schools in 1898) and was already working on a book entitled *World Pictures* for A&C Black Publishers, which Herbertson helped her to edit. After completing the Diploma Reynolds published several papers in the *GT*. Her first paper on class excursions (1901) was read to the 1900 Cambridge University Extension Summer Meeting on Teaching Geography, organised with the GA (i.e. before Reynolds started her Diploma). It stressed engaging of younger pupils with their *heimatkunde* (local area) and giving older pupils specific data collection tasks to contribute to a larger set of combined findings, i.e. active and analyical fieldwork. In 1904 Reynolds published a paper on 'The regional method of teaching geography' (1904a) and a two-part summary of official geographical literature on the colonies (1904b): in both she reflected the geographical ethos of Andrew Herbertson, who sought to promote the interests of Empire through geography every bit as much as Mackinder who is more commonly labelled the imperialist. Map work was at the core of her work, as she stressed the lessons to be learned from maps in the field and in understanding regions. This theme was developed in her (1910) paper on 'Map reading and the imagination' where she related to her own childhood experience of drawing imaginary lands with her brother and sister to the innovations some teachers were using with sand models and maps of imaginary places such as Robinson Crusoe's island, which in turn were linked to the imaginative leap of reading topography and place from printed maps. She also wrote two series of textbooks on Regional and Junior Regional Geography for her publishers A&C Black, and many of these books demonstrate her commitment to teaching through visual media, as she noted in her 1927 preface: 'One of the chief ways ... by which children easily learn many facts is by means of pictures'. True to her principles her books are rich in illustration, with those for younger children having whole-page plates on every double-page spread and those for older children having a least a sketch map, photograph or table. Her *Regional Geography: The World* (1915b) is of necessity a denser text, but still contains about 150 maps and diagrams in its 335 pages. These books sold very well, for example her 1905 book on *Europe and the Mediterranean Region* was in its eighth edition by 1918. The frontispiece for this volume listed her other books and her qualifications as validation for her work as well as her post as 'assistant examiner for the Matriculation Examination of the London University'. The implication of the latter being that the book might be a route to success in those examinations and this was articulated more explicitly in the preface: 'The book will be found to contain sufficient information, with regard to the area described, for candidates for geographical examinations of the standard of London Matriculation' (1927: ii). The preface also acknowledges George Chisholm's help in refining the text – later, at his request, she revised a late edition of his *Commercial Geography*. Reynolds also produced a set of themed books called the Quotation and Picture Series, which were a precursor of the coffee table picture book. In her letter to Freeman Reynolds describes something of the process in the production of these books: 'They [the publishers] had a lot of coloured pictures in books on certain places and I could choose some pictures and get poems and brief descriptions to describe the scenes e.g. "London", "the

Lake District", "Scotland" and "Sussex".' She notes that she 'rather borrowed this idea' from F.D. Herbertson's *Descriptive Geographies*, but this obviously coincided with the availability of the pictures in the publishers' possession, including some she had used previously in earlier texts. Reynolds also noted that their production ended with the war. The volume on Scotland (1915a) includes tasteful watercolours of key Scottish features as might appear in a well-illustrated travel guide, such as Edinburgh and Holyrood castles and Staffa rocks with accompanying poems and prose from authors such as Burns and Wordsworth. The ultimate meshing of her work with that of the Herbertsons occurred when she revised their *Man and His Work*, the book she admired so much, for its sixth edition in 1931, which was reprinted in 1934.[11]

Rachel Fleming

Rachel Mary Fleming is one of the category of people who are often invisible even in institutional histories, one of those who do their work with quiet devotion, playing a pivotal structural role in the 'back room' rather than the 'front room' of the discipline (in Goffman's (1959) terms). Rachel Fleming was born in 1882, daughter of a Congregational minister and his wife (Fleure 1968). Her mother had grown up in a wealthy family and found it difficult to cope with the daily demands of her impoverished married family life, leaving her daughter Rachel to manage the household while she used the skills she had been trained in to make beautiful lace and fine needlework, but which sold for very little money.[12] Fleming became a pupil-teacher and attended a summer school in geography at Aberystwyth in 1917, when she was recuperating from a serious illness. In his obituary Fleure simply states 'At that moment, the office of the Geographical Association was being established at Aberystwyth, and Miss Fleming became its assistant secretary' (1968: 327). A contemporary describes it in more dramatic terms: '[She] had a hard life, and much illness until she was rescued by Fleure – that would have been just after the First World War, when a lot of women, who had been doing war work etc. came to Aberystwyth as Adult Students. With the cruelty and callousness of inexperienced youth the students who were just up from school called them "The Monkey Tribe".'[13] Fleure describes the GA as reaching a low ebb after Herbertson's death but that 'largely because of her devoted work and enterprising efforts, the membership of the Association grew a little and it was possible to give her a salary, though it was always quite inadequate'. This is typically self-effacing of Fleure's own efforts, who is described by Walford (2001: 115) as 'the ceaselessly working guiding hand' at the GA, but this should not detract from the credit he gives to Fleming for her role in this growth. In his words, 'Under her care' the membership grew to 3000 and she gave talks to teachers which were 'a source of much enthusiasm' (ibid.). Fleming was involved with converting *The Geographical Teacher* to *Geography*, helped to found the GA Library and responded to the many queries which came in from teachers at a time when they had access to few specialised publications in geography. In 1930 when Fleure moved to Manchester University, the GA office went with him; Fleming then moved to London and became Librarian of the Royal Anthropological Institute. A linguist by training, she specialised in Russian, and was a member of Sir John Russell's group visits to Russia. She drew on her visits, linguistic and geographical work when she wrote the entry on the geographical features of Russia for the *Encyclopaedia Britannica* (14th edition). In addition to journal articles Fleming wrote *Ancient Tales from Many Lands and Round the World in Folk Tales* (Fleure 1968). When she left Aberystwyth she was awarded an honorary MSc by the University of Wales in recognition of her work for geography; later, when her health failed, friends were successful in petitioning for a civil list pension for her (which Fleure attributes to the support of General Smuts) (ibid.). The pension allowed her to retire to the Isle of Wight where she died on 23 March 1968. In Fleure's words: 'Our association owes to her much more than can be conveyed to readers in a short notice. She gave us inspiration, wide knowledge and unceasing attention to detail and to issues of policy' (1968: 328). Rachel Fleming's geographical work was largely invisible, but made a difference to the day-to-day teaching of geography as well as strengthening the institutional framework and resources of the GA at a time when it was quite vulnerable.

Charlotte A. Simpson

Charlotte Simpson was born in Rugby in 1879, daughter of a doctor and his wife. It is known that she went to Oxford in 1908 to study with Herbertson and gained a distinction in the Diploma in Geography in 1910 (Beaver 1962). The quality of her work is indicated by the fact that her paper on the upper reaches of the Warwick Avon, based on the regional study component of the Diploma, was presented to Section E at the 1913 meeting of the BAAS at Birmingham.[14] The paper was subsequently published in *GT* (1913–14) and followed by another on processes of valley formation (*Geography* 1923–4). She worked at the Hydrographic office of the Admiralty in the 1914–18 war, which was followed by two years' teaching (ibid.) It was at Oxford that Simpson became involved with the Le Play Society, taking part in numerous Le Play House Educational Tours including a surveying course with Geddes at Montpelier in 1928, and field study became the central tenet of her work. She completed a BSc Research degree at Oxford 1924–7 and was appointed Lecturer in Geography at Warrington Training

College, Liverpool in 1927, a post she held for five years. At 53 years of age she went into semi-retirement, organising residential courses in field survey at her home at Cranham in the Cotswolds; and numerous courses were run for Le Play, teacher training colleges, the GA (which formed a Regional Survey Committee in 1926 (Evans 1986)) and other groups (1932–40). This work had a tremendous influence on a host of geographers, as well as producing rich local data (Beaver 1962). Simpson published several books and papers centred on field study: *Rediscovering England* (1930) (based on a series of adult education classes); *The Study of Local Geography – A Handbook for Teachers* (1934); 'A venture in field geography' (*Geography* 1945) (based on data collected at Cranham) and *Making Local Surveys – an Eye for Country* (1951) – described by Beaver (1962: 315) as the 'consummation of her life's work'. The interwar years were characterised by increased public access and participation in the countryside, which was fostered through civil groups such as the Rambler's Association (formed from constituent groups in 1935) and state initiatives to promote a healthy and informed citizenry (Matless 1998). Geography, fieldwork and leisure overlapped at this time (Livingstone 1992; Matless 1992) and Simpson's work contributed to the culture of observant citizen (Matless 1998). Simpson wrote a pamphlet on 'Village survey making' for the Ministry of Education (see 1951 title page) and in her other writing encouraged engagement with the landscape on family holidays as much as through formal education, in ordinary as well as spectacular landscapes.[15] For Simpson local survey channelled and developed children's natural curiosities, as well as giving them worthwhile occupation and stimulating new hobbies, but this 'eye for country' was closely linked to informed and active citizenry, a means of encouraging young people to contribute to debates about future policy and decision making in post-war years. While acknowledging the necessity of secondary data, she argued the benefits of first-hand observation were overwhelming.

> It will be found that personal observation, if accurate, is more valuable as a training and in the making of records than blind reliance on the opinion of others, however eloquently expressed, and such a training should enable young people to make careful study of a setting before they decide what new developments they would like to see. One of the objects of ... local study ... is to help those young enough to take a hand in the shaping of a new world to use their powers of observation and discrimination (1951: 2).

D.M. Preece

Dorothy Mary Preece taught at Crewe County Grammar School from 1916 and was promoted to 'Senior Geography Mistress' in 1938. In a personal account of her early years, university lecturer Dorothy Sylvester represents Preece as a model teacher: 'No one could have been more fortunate in their Geography teacher. Her lessons were always so clear and impressive that I never forgot one … It would be impossible to evaluate adequately the work she did and the influence of her teaching and I know that I am one of a great body of old students of the school who owe her a great debt, though none greater than I. She was tireless in her efforts, endlessly resourceful in explanation and illustration'.[16] Preece tutored Sylvester for university examinations and they became friends as Sylvester moved into adulthood and a career of her own in geography. Sadly Preece died suddenly 'at the height of her powers' just as her publications were becoming well known (ibid.). Preece co-wrote the influential text *The Foundations of Geography* (1938) for the Modern Geography Series; the text was still in print in revised form in 1966, but the other planned volumes in the series had to be left for others to write. Sylvester (1939: 136), who wrote her obituary, described Preece as 'one of the finest teachers of Geography in this country' but also as a woman of great loyalty, humanity, kindness, charm and personality. Preece was involved with the GA and helped to reactivate the dormant Crewe Branch, subsequently becoming its Secretary and no doubt played a significant role in it quickly gaining a large membership and a lively programme of events (ibid.).

Gladys Marten

The history of Manchester High School for Girls illustrates how geography became significant within the secondary school curriculum in this and other schools as a result of the Education Act (1902). The 1902 legislation made geography a compulsory subject to matriculation standard for all those intending to become teachers, the result being 'of late years it has become a much more important subject at the top of the school. The new methods have been introduced by Miss Beatrice Leach, a mistress since 1890, who is responsible for geography throughout the school' (Burstall 1911: 184). No doubt these changes in the curriculum caused an increased demand for teachers specialising in geography, resulting in a knock-on effect for summer schools and supplementary qualifications in geography. Gladys Marten was appointed as geography teacher only three years after the new legislation.

Gladys Maud Marten (see Figure 5.2) was born on 9 August 1878. She was educated at Berkhamsted Girls' Grammar School 1890–7, when she took her Cambridge Higher Local Certificate. Between 1897 and 1905 she completed an honours degree in history, mathematics and geography but the name of the university is not recorded on her employment file. She went

Figure 5.2   Staff of the Manchester High School for Girls, c. 1910; Gladys Marten far left bottom row.
Source: Manchester High School for Girls Archive

on to attend the Cheltenham Ladies' College Teacher Training Department (1897–8) and was awarded the Cambridge University Teachers' Certificate in 1898. This was followed by her first teaching appointment at the Quadrant School, Coventry (1898–1904) and her subsequent post at the Manchester High School for Girls (MHSG) (1905–38), which was to be the lynchpin of her career.[17] However, Marten's time at MHSG was not unbroken, as at the age of 33 she took leave of absence to complete a Diploma of Geography at Oxford (1911–12), in which she gained a distinction. She took a further break from school teaching when she took a temporary post as assistant lecturer in geography at the University of Manchester from 1914 until 1922. She was appointed as a short-term measure in 1914 to assist the Reader in Geography and was originally paid a very low salary of £85 per annum. She was given specific responsibility for teaching geography to Teaching Diploma students, although in all likelihood she would have taught more than this.[18] In 1921 Marten gained an MA from the University of Manchester and she returned to MHSG in 1922; she became *de facto* head of the school's Geography Society (the headmistress being the president of all school societies) and organised many Geographical Club expeditions, including to the Lake District and Derbyshire. These expeditions were an important part of her teaching and the girls' experience of geographical education, and they figured prominently in *MHSG School Magazine* accounts of her work on her retirement and death. In retirement Gladys Marten lived in Devon with another ex-MHSG teacher, Grace Bradshaw. She died in Dorset in 1955, leaving some of her geography books to the school library. Her obituary in the magazine (1955) recorded:

> Old girls will remember her as a form mistress who not only taught them but also entered into their interests and had their confidence and whose vivid and interesting lessons and geographical excursions roused enthusiasm for what became a living rather than a text book subject. Those who were with her on the staff remember her as a talented colleague with broad intellectual interests, a good friend whose wise advice was often sought and one whose repartee and quick wit delighted all who were privileged to know her'.

She wrote two articles for the *GT* 1911–12 (one on geography in the middle school, the other on the influence of the environment on Arabian and Syrian peoples), probably generated during her Diploma year; these publications combined with her Diploma were presumably significant factors in her appointment to the post at Manchester University. However, her biography is interesting less as a geographical author than as an example of a mature teacher of geography gaining subject accreditation and moving between school, university and school again. Initially switching into the university sector during the first year of the First World War, she continued there for four years after its cessation. It is not clear whether she returned to school teaching in 1922 from choice, but her school record certainly shows she was an engaging and committed teacher throughout the 16 years of service before retirement.

Later MHSG school magazines detail numerous geographical excursions and expeditions in the 1940s to 1960s, principally organised by geography teacher **Miss Leigh**. These trips to the 'field' included school visits to both rural and industrial areas of northwest England, including day excursions to Edale to study limestone features, Merseyside to study industry at Port Sunlight (1928), the Formby oilfield (1947, in conjunction with the Manchester Geographical Society) and Grassington (1955) to study historical geography. There were also field weeks around the UK (1927 study of glacial features in the Lake District and 1969 study of glaciation, soils and land use in North Wales), as well as a school visit to Switzerland (1953), which incorporated geomorphology for the geographers (see Figure 5.3). Evidence from these accounts in school magazines and reports demonstrate the consistent practice of field visits as part of the geography curriculum in this private school, and although little detail is given of work undertaken, even the look-and-see approach gave an opportunity to compare text-based

Figure 5.3   Manchester High School for Girls field trip, 1953.
Source: Manchester High School for Girls Archive

learning with the real landscape, and later reports suggest active study of themes by students. Children from state schools may not have had regular opportunities for excursions in Switzerland but articles in the *GT* and *Geography* suggest that local fieldwork was a normal part of the early to mid-twentieth-century geography curriculum in state secondary schools and week-long field courses within the UK were common practice in the 1960s and 1970s.

Jessie Watson

Jessie Black was educated at the University of Edinburgh and whilst there she and fellow student Wreford Watson formed the Geographical Society – he became president and she secretary. The couple were married in 1939 and went to Canada, in part a response to Wreford's frustration at not being fit for active service in the war. He studied for his PhD under Griffith Taylor, worked as an instructor and then became first professor in geography at MacMaster University. While employed as chief geographer in the Department of Mines and Technical Survey, he also worked part time to establish a geography department at Carleton; at this time Jessie had a full-time lectureship (Collins 1991) and they worked together to found what became a thriving department (Crosbie 1991). When they left Canada to return to Edinburgh a Jessie and Wreford Watson Award was established in their honour in the geography department at Carleton. When Wreford took up the chair in geography at the University of Edinburgh, Jessie was appointed temporary lecturer in geography (1963–7) at Moray House College of Education (later Edinburgh University's Department of Education), where she continued teaching until 1978.[19]

The couple, who had two children (Margaret and Jamie), 'shared everything' (Crosbie 1991). They were constant companions in their continued trips to Canada and had a common interest in their geographical work. They were warm hosts to local students, colleagues and visitors, especially those in the Centre of Canadian Studies established at Edinburgh and acted as a social focus for the department. '...together they became an institution in both Canada and Edinburgh. They worked very closely together and their academic collaboration is most clearly illustrated in their co-authored book, *The Canadians: How they Live and Work*' [1978] (Collins 1991: 230). It was this book that Wreford always maintained gave him the greatest satisfaction (ibid; Robinson 1991), reflecting the couple's view of themselves as very much as a team, which 'was a beautiful blend of Wreford's quiet diplomacy and Jessie's open frankness' (Collins 1991: 230).

In 1987 Jessie and Wreford Watson were honoured by Edinburgh University's Geography Society on its 50th anniversary and before Jessie's death in 1989 the Edinburgh department established a prize in her honour. The prize was awarded to a student who not only achieved high academic performance but also gave most to the 'life' of the department: 'this perhaps reflects most accurately the true nature of Jessie Watson' (Collins 1991: 230). In addition to Jessie's own work in teacher training Crosbie (1991: 72) credits her with making a major contribution to Wreford's work on social geography 'with perceptive insights into the character and lives of people in the many places covered in their travels'.

Other women who played important roles in teacher education and local geographical societies include Edith Coulthard and Beatrice Swainson. **Edith Coulthard** was a graduate of the Manchester department and was very involved with the Manchester branch of the GA; she also assisted Alice Garnett as honorary secretary of the GA in the early 1950s. Coulthard published three short papers in *Geography* (1942, 1943, 1946) on what she described as 'experiments in teaching'. The 1942 paper describes a particularly interesting account of how the girls in the current affairs group at the Bishop Auckland School gleaned, synthesised and presented geographical, historical and political information on the major theatres of the war, including projects on Russia and Asia, as well as discussion of emigration and the White Australia Policy on immigration. The 1943 'experiment' was based on an exchange with a school in Quebec and was concerned with local games relating to place identity; and the 1946 piece outlined a local study intended to provide geographical understanding of the local area 'fit for real life' for those likely to pursue non-academic careers within Bishop Auckland. These examples demonstrate a willingness on the part of a classroom teacher to experiment with teaching methods and to engage with the actual interests and needs of pupils, creatively applying geographical knowledge to a range of issues; and a 1938 GA report cited her work as a good example of 'education in citizenship' (Marsden 1999). The editor's note under her 1943 title to the paper illustrates the editorial role in encouraging Coulthard to move from private correspondence on the matter to bringing this material to publication for wider consumption; she, of course, may have used correspondence with the editor as a deliberate strategy of testing response to her write up of a topic that might have appeared as less mainstream to geography.[20]

Patricia Pemberton

Pat Pemberton had studied geography at Queen Mary's College before completing her teaching qualification at the London Institute of Education. Her first post was at Bromley High School, and this was followed by a bursary to teach English in Sweden at the English Centre in Sundsvaal. When she returned to the UK Pemberton moved into teacher training and held posts at Wall Hall, King Alfred and Hereford Colleges. Her career culminated in her appointment as vice principal of Sunderland College of Education in 1970, but her life was tragically cut short by a road accident in May 1973 (GA 1973). Whilst remembered for her teaching and administration, her legacy lay more in her combination of travel, research, publications, teaching and community service. She had a lifelong interest in historical and cultural geography and applied these interests to less economically developed countries, producing textbooks and running vacation courses to the Cameroons, as well as fund raising for a school in Lesotho (GA 1973). She also served on the GA Executive Committee and running Council and was an important member of the GA Colleges and Departments Section Committee, as well as active in her local branch. In the years prior to her death Pemberton had been appointed to the Geography Committee of the Schools Council. Pemberton researched on spatial learning and edited *Geography in Primary Schools* (1970), a leading guide for teaching geography in the 5–11 age group.

Olive Garnett

Olive Garnett, born c. 1899, sister to Alice, the better known of the two in geographical circles (see Chapter Seven), also devoted her life to geographical education. She was deeply involved with the GA, representing and developing the Training College and Primary School sections. She was a lecturer and then deputy principal of the Froebel Educational Institute (where Eva Taylor also taught in the 1920s). Her publications in *Geography* are limited to book reviews and an obituary for J.F. Unstead (1966), but her text *Fundamentals in School Geography* was in print for over 30 years (1934, 1949, 1965) and has been described as 'the most significant and widely read [school geography] book of the 1930s, 1940s and 1950s' (Walford 1991: 139). However, Olive Garnett's impact on the discipline was as much through influencing and supporting generations of geography teachers as via the printed word, not least her advocacy of active learning through models (Ploszajska 1996) and field education, for example, visits to observe road construction and fishing boat arrivals to understand the reality of industry and urban change (Marsden 1999). When she died in 1982 she was remembered as someone who 'pioneered teaching methods in geography in junior and primary schools' (*Geography* 1982: 349).

Gladys Hickman

Born 17 April 1912, Gladys Hickman was the daughter of Rose Minnie (née Green) and George Hickman. Her father had grown up on Romney Marsh near Rye in East Sussex, where his father kept a sheep farm.[21] She was educated at a state school in north Kent, where she remembered the staff as committed and gifted subject teachers who were mostly war-widows or -fiancées. Although the school was newly started in 1922 and Hickman was the first of her generation to attend university, several other girls from her school were at University College London (UCL) at the same time, indicative of the academic drive of the school. Hickman won a scholarship to university from her school and studied geography at UCL (1930–3). She was taught by the youthful but dynamic staff of Fawcett, Ogilvie, Buchanan and Dickinson. Although enjoying her classes the university was a social world away from what she had previously known:

> It took me about two months to go into the women's common room. I saw people going in and out, but I didn't know that I could go in and out. Nobody told me [laughter] ... I thought it was a fantastic place. Coming out of an ordinary girls' school, that had really only been started just after the war ... Well, it was out of this world. I mean I'm not joking, it was a world I just didn't recognise, I had no academic contacts, no academic relations really, though some very clever relations, they hadn't had anything but what was then called an elementary education .... the lacrosse team at UC had 5 or 6 members from my school, so it was all very amusing. But we also had playing lacrosse with us Flinders Peach's daughter, you know coming from the sort of background I'd had it was all rather unbelievable. I don't mean that I was impressed by it necessarily, but it was all new, new, new.[22]

Figure 5.4   Gladys Hickman (centre) with Bertie and Molly Roberson, on a pre-war skiing trip.
Source: Courtesy of Richard Hickman

Hickman enjoyed the regional courses, Asia in particular caught her imagination, and she relished being set a local study which she opted to do on Rye, her home area. She particularly enjoyed lectures in physical geography by Owen Garwood: 'a real character who was a geologist-cum-physical geographer who was always talking about the north face of the Eiger and things like that [laughter]'.[23] Studying in London also allowed access to the annual GA conference where she heard Christie and Edna Willats and Dudley Stamp talk on the Land Utilisation Survey and other 'names' from textbooks. Most fieldwork was undertaken at weekends on a voluntary basis, but a week's field course to the College Des Ecosses at Montpellier with Fawcett was particularly memorable for Hickman as the first time she had been abroad, the occasion of her 21st birthday and an opportunity to put text-based learning alongside field observation. By the end of her third year she had saved up sufficient money to go with the Student Group of the Le Play Society on a field study week to Poland and revisited shortly before the war. It was on one of these visits that she met Lieutenant Ryszard Rusiecki, of the Polish army, the man she was to marry in 1941. They met again when he escaped to Britain in 1939 and was stationed at Blairgowrie in Scotland.

After graduating with First Class Honours in 1933, Hickman got a job teaching geography at the Barnsley High School for Girls in Yorkshire, where many of the pupils were on free school meals because of the high rate of unemployment in the area at the tail end of the nationwide economic depression. Hickman initially thought she was 'at the end of the world' but the job taught her a great deal and she stayed four years, being promoted to Head of Geography on the retirement of the senior geography teacher. She found the girls wanted information that was socially and politically relevant: 'I had a group of girls who said they weren't interested in learning about the three main rivers of China [chuckles] and they wanted to know why the Japanese were invading China. Most of all they wanted to know why the Chinese had broken the walls – the dykes – on the Pang Ho and flooded, killing 300,000 Chinese – their own people. They wanted to know that. That was the geography they wanted to know. Fifteen year old girls – I ask you'.[24]

Hickman was appointed as lecturer at Goldsmith's College, University of London in 1938. Goldsmith's was evacuated to Nottingham University (Maddrell 2008; see Chapter Seven) in 1941 and she married there that year; her son Richard was born in 1942. After the war Gladys separated from her husband and later divorced, returning to her maiden name – a rare thing at this time – and she brought up Richard as a single parent. Encouraged by Molly Bex-Humphries to apply for a post at Bristol University in 1948, Gladys, who had now acquired an MA, spent 25 years at the university. She is recorded as having a PhD after 1951,[25] which was no mean achievement for a single parent holding a full-time academic post and it is not surprising that she was known for her determination. Richard was to accompany her on some Bristol field trips and Le Play Society field weeks abroad (including those she led) and went on to study geography himself.

During this post-war period British colonies were rapidly gaining independence and Bristol University fostered connections with the indigenous educated classes in East and West Africa. These connections led to Hickman, always known to friends as Hickie, being seconded to Makerere College in Uganda in 1955, where external London University degrees were taken (only a few years before it became part of the University of East Africa in 1963), and the University College of Ghana in Accra (1958–61). She was a socially adventurous traveller to other people's everyday worlds. 'During this time she travelled to over 30 different countries, often to places and by means that would be considered very hazardous today, including from Ghana to Timbuktu by VW Beetle. Gladys travelled by public transport as much as possible during her trips in order to meet the local people and much of her research was based on interaction with local residents' (GA 2006). This was an approach she started when first exploring the social life of the Alps with friends on skis in the 1930s, simply going off-piste and meeting local farmers etc.[26]

On her return to Bristol Gladys became director of the Schools Council Geography 14–18 Curriculum Development Project (1970–3). Hickman used both her former students and GA network to recruit schools to pilot the scheme, which was to become influential on both 'O' and 'A' level geography courses in the 1970s and 1980s.[27] Through her work at Bristol she made 'notable contributions to curriculum innovation and development as well as to teaching and learning' (GA 2006). Gladys was promoted to senior lecturer in education in 1971 and retired a few years later in 1974.[28] She was made an honorary member of the GA in 1990, only the 20th person and third woman to receive this honour, given in her case in recognition for her work in geographical education and the Bristol branch of the GA.[29] Gladys endowed GA initiatives to study China and Africa and was one of its longest standing members at the time of her death (GA 2006). In retirement she moved to Edinburgh where she became an active member of the China Group, visiting China in the 1970s and 1985. In the 1980s and 1990s she wrote one book on China and co-wrote another. When I interviewed Gladys Hickman in 2005 she was 92 years of age and briefing RSGS officials to petition for the needs of refugees in Darfur and elsewhere at the Edinburgh G8 Summit. Gladys died on 20 September 2006.

At Makerere Hickman co-wrote *The Lands and Peoples of East Africa* (1960) with William Dickins (who sadly died before its publication). The book was written principally for African pupils sitting the Cambridge Overseas School Certificate and 'young teachers in Africa and elsewhere' (Hickman & Dickins 1960: v). After Dickins' untimely death, the production of the final text was greatly helped by the advice of S.J.K. Baker and R.W. Steel (ibid. vi). Apparently there had been a need for such a book for 12 years,[30] so it is not surprising that by 1970 the book was in its seventh impression (furthermore it was reprinted in 1986). Hickman went on to write *The New Africa* (1973), principally for a British market. This text, written near her retirement, was based on her experience of living and working in both West and East Africa and travelling widely in the rest of the continent. She stressed the collaborative input of all those who assisted her on those travels and studies, arguing 'there is no doubt that if this book carries something of the real Africa, it is because of the unstinting and friendly contributions of numerous Africans' (1978: 3). In the text she combined a regional and thematic framework and challenged the view held in Britain that Africa was an 'easy' continent to study. She believed in engaging pupils through case studies and other sources, writing that 'the intention is to take students inside situations so that they begin to understand how cultural as well as physical factors affect the development of a country and a continent … the emphasis is on the processes of change and on understanding based on sound critical study' (ibid.). In her author's note Hickman logged that Professor Michael Wise had contributed much to the shape of the book, not only through his editing but 'also for his encouragement and support for aspects that may be regarded by some as unconventional for a book at this level' (ibid.). This 'unconventional' content may have included Hickman's frank discussion of apartheid under a heading 'Separation and discrimination', where she discusses what apartheid means to ordinary non-white people. She articulated the political and daily reality, ranging from limited voting rights and Pass Laws to examples that pupils would readily understand such as segregated bus seats, noting that:

> In Cape Town bus seats are reserved separately for whites and non-whites and the latter stand if all their seats are occupied even though white seats are empty. In Cape Point Nature Reserve non-whites are directed to a 'non-white picnicking area'. Even a few metres from the Transkei Legislative Assembly building in Umtata the Chief Minister (perhaps akin to a Prime Minister in a newly independent country) may not sit down in a nearby café, though he may buy coffee at the counter and drink it standing up (1978: 146).

Hickman noted the relative economic well-being of South Africans compared to other African states but stressed that 'Africans in other countries have said over and over again that they would rather manage their own affairs in discomfort or poverty, than live more comfortably under dictation and "tyranny"' (1978: 147). It is quite clear where Hickman stood on the issue of apartheid and why. As colleagues testified, she had a remarkable ability to engage young people (GA 2006).

> All through her life, she had an insatiable curiosity about places, why they were how they were, and what it was like for the people who lived there. She loved to travel. And she loved to stop and talk to local people about what went on and why it was organised how it was. She was a pioneer of this technique of making detailed local studies of a particular village, farm, or factory, and using a carefully chosen selection of such studies to show what went on within a much larger area, set against a background of more general information.[31]

At Birmingham University **J. Helen Murray** was Assistant Lecturer (1925–8) and **Ethel David** (née Fisk) was Assistant Lecturer (1929–31): both held *joint* posts in the departments of geography and education. Fisk/David published in *Geography* on the regional geography of the world and map work (1931, 1940, 1941). **Phyllis A. Nicklin**, a graduate and postgraduate of Birmingham University, was a temporary lecturer at Nottingham University during the 1939–45 war (see Chapter Seven), then pursued a career in teaching and teacher training before being appointed staff tutor in geography at the extra-mural department at Birmingham University. She was described as 'the indefatigable Honorary Secretary of the Birmingham [GA] branch', which had a strong membership, allowing them to organise a GA Spring Conference in 1965 and publish excursion guides to the West Midlands. She was also a popular speaker at other GA branches, for example, speaking at the Crewe and Nantwich branch in 1959 and again in 1963.[32] Within the national organisation of the GA, Nicklin served on the Council and the Executive Committee. On a personal level she was remembered for her kindness, efficiency and her stimulating extra-mural course and field excursions (GA 1970). 'Miss Nicklin was considered a great rock in the world of geographical education and spent over twenty years as the driving force behind what is possibly the most successful liaison between an academic department, an extra-mural department, and a local branch of the Association that could be found in any University city' (GA 1970: 232–3). **Beatrice Saward**, a Reading University graduate who had been a WREN meteorological officer during the war was appointed assistant lecturer in geography (1946–8) and lecturer (1948–54) at Birmingham, where she primarily taught climate-related courses. She left in 1953 or 1954 to take up a post at Westhill Teacher Training College (Giles 1987). Her move is thought to coincide with her marriage when she changed her name to Eckstein. She left Westhill on the birth of her first child but returned to become head of department there.[33] Given the number of women starting as school teachers it is not surprising to see the combined roles of subject and teacher training posts, but Saward's move on marriage and her ability to return after a career break to head the department she left suggests much greater flexibility for women in teacher education rather than subject-based departments.

I.M. Long

Molly Long (née Jackson) was born in London, educated at Bedford College and later trained as a teacher at the University of London Institute of Education (ULIE). She joined the ULIE staff in 1946 as a temporary assistant tutor, and was upgraded to permanent lecturer of education, with special reference to geography from January 1948. She was promoted to Senior Lecturer in 1972 and retired in 1976.[34] She worked at ULIE with Neville Scarfe and Reginald Honeybone, with whom she co-authored the *Geography for Schools Series*. She undertook a wide range of duties in the department including lectures, fieldwork activities and supervising school practice; she was also an active researcher studying the relationship between children's interests and hobbies and their geographical knowledge, as well as collaborative research with GA colleagues on perceptions of photographs (Graves 1997). She published widely in *Geography* and edited the long-standing and much-used publication *Handbook for Geography Teachers*[35] where she argued for a greater emphasis on the study of people in geography and for a place for geography within the humanities (1964: 4). Long also co-wrote *Teaching Geography* (1966) with ULIE colleague B.S. Roberson, in which they continued to support the regional approach to geography, arguing 'those who would not place the main emphasis on regional geography in school must justify themselves with some other philosophy'. This attachment to the regional method and a rather aloof attitude to junior members caused the younger generation of teachers such as Sheila Jones to view Long as socially and pedagogically conservative.[36] Her papers included two addressing the use of visual materials in teaching geography (1953, 1961) and she sat for many years on the GA Executive Committee as Representative for Teaching Aids. Long and Garnett were the only long-serving women on the GA Executive, which acted on behalf of the GA Council between meetings of that body. Long also represented an important link between university and school geography, reviewing recent developments in research for the readers of *Geography* in 1964. Ultimately Molly Long became president of the GA in 1970, two years after Alice Garnett and only the second woman to hold this post. She died on 12 January 1997.

Sheila Jones

Sheila Jones was the third woman to hold the presidency of the GA, which followed in 1974, quite soon after Molly Long's tenure. Jones was born in 1929, an only child and daughter of Jessie (née Dawkins) and Leslie Jones, a technical representative. Her parents sacrificed financially to send her to Redland High School in Bristol from four and a half years of age, where Jones was always taught in the year group above her own age cohort. After three years in the sixth form, completing the Higher School Certificate in geography, physics and pure mathematics, she went on to Bristol University to read geography: 'it was less easy to get away from home towns in 1947'.[37] Jones' school had assumed that she would continue to higher education, but this was a significant moment for her parents as she was the first member of her family to go to university. Although recognising her own naivety as a fresher, in retrospect Jones and her peers were not impressed with the quality of the teaching they received in the immediate post-war years at Bristol. 'The Department at that time was very poor and keeping in touch … with my contemporaries we agree that [our] successes have been in spite of our Uni. Background. Our only field week was in the Third Year which was a tour of SW Wales. My subsid. Subject was Geology and although most of the group were ex Service it was much more enjoyable (although more difficult) course, with two hard working field weeks'.[38] After university Jones considered becoming a meteorological officer in the Wrens or a map maker in the Colonial Survey, but to her 'amazement' found her earlier resistance to a career in teaching completely undermined by a three-week trial teaching practice after graduation. She graduated with a Lower Second ('all I deserved at the time') and began a year's postgraduate teacher training course at Bristol, what Jones describes as 'the most enjoyable of my Uni. Years'. The lynchpin of that formative year was having Gladys Hickman as her tutor: 'Gladys Hickman was a revelation to students then. Full of new ideas, field work to the fore, etc.'[39] Jones continued to be influenced by Hickman's student-centred approach and ongoing work, and worked on the pilot of the 14–18 Geography Project (see below).[40]

Jones went on to take up teaching posts at the Girls' Grammar School in Falmouth for three years, where she found the work straightforward and 'the textbooks were God' – she did a great deal of local field work with her pupils. After a year in Dorchester she decided to return to Bristol, applied for a post in the new comprehensive school opening in 1956, but in the absence of a response applied for a post as head of geography at Colston's Girls' School. Just after being appointed at Colston's she was called for interview for the comprehensive, as Jones noted: 'In this way a path is defined!'. She was to stay at Colston's for the rest of her career. With a supportive head teacher (Sarah Dunn), Jones was able to make the most of opportunities arising for professional development and became one of the first teacher tutors in the country under a scheme devised by Gladys Hickman and two other colleagues at the Bristol University School of Education. Being in Bristol also provided the first opportunity for real involvement with a local GA branch and she joined its Committee in 1959. Bristol hosted the GA Easter conference in 1961, which meant much preparation and work for local members. Jones was invited to join the national GA Field Work Committee and then the Models and Quantitative Techniques Committee (at Hickman's behest) and she became the Secretary of the latter group. As a result of this involvement, Jones attended the famous Madingley lectures on quantitative techniques given by Chorley and Haggett. Although increasingly integrated into the local and national GA network, Jones records her astonishment at being invited to become president of the GA: '… no one was more surprised than I was when Robert Steel, a member of the inner circle who then ran the GA, invited me to be the President 1974/5'.[41] After her year as president Jones spent a period as branch officer and continued to be involved with her local GA branch.

Conclusion

The work of the GA cannot be underestimated in shaping geographical knowledge in British schools in the twentieth century, arguably second only to the influence of state legislation and associated examinations, inspections etc. As can be seen from repeated references to the Oxford Diploma in Geography in the biographies of women geographers and geographical educationalists in the early twentieth century, the Oxford School played a formative role in shaping geographical knowledge and how it was taught, not least with Andrew Herbertson heading the GA, but following his death and with the availability of full degrees in geography elsewhere after 1917, Oxford's influence on geographical education declined. Like Herbertson, Fleure and Garnett combined academic and GA leadership, but gradually as the twentieth century progressed the GA became increasingly independent of university geography, if not university geographers. GA sub-committees, such as those on visual aids and fieldwork, were important in producing resources (and thereby a publication outlet through in-house publications), staff development in new techniques, and opportunities for field experience. The GA annual conference and its journals were key elements in the success of the GA network, not least in keeping school teachers up to date with new developments in pedagogy as well as within the discipline. Local branches, although sometimes headed by well-known academics, were for the most part run for and by local teachers, as can be seen in the obituary of **Evelyn Brown**, teacher at St Swithun's School, Winchester. Brown served on the GA Secondary Schools Section Committee, was a founder member of the Winchester GA branch set up in 1955 and represented the branch at the GA Council. Hence she is described as 'a true representative of the geography-teacher member of the Association' (*Geography* 1962: 423). Whilst recent commentaries have bemoaned the withdrawal of academic geographers from the GA in the Research Assessment Exercise era (Castree, Fuller & Lambert 2007) and the issue of the estrangement between school and university geography deserves addressing, it should not be forgotten that the great majority of the work of the GA has been done by 'ordinary' members, a great many of whom were women like Evelyn Brown.

However, that is not to say that the hierarchies of the GA represented the gender balance of the geographical teaching profession. Balchin's (1993) institutional account of the GA, with its emphasis on national GA office holders and the 'bigger picture', is a largely male story prior to 1970, with Alice Garnett as the notable exception and other women such as Ellen Rickard, Edith Coulthard, Marguerita Oughton, Olive Garnett and D.M. Forsaith mentioned in passing. The archives of the GA are patchy, with no extant record of past members or even the full membership of the GA Executive prior to 1962. However, analysis of the Jubilee Appeals Record Book 1946–9 shows that approximately 46 percent of donations were made by women and a 1953 analysis of membership showed 11 of 44 branches to have female honorary secretaries.[42] The Minute Book of the Central London Branch of the GA 1936–58 shows a male figurehead as President (Rodwell Llewellyn Jones from LSE) and men acting as chairs of meetings and providing the majority of speakers; but women were nonetheless regular speakers and played full and active roles in the running of the branch, including organising field excursions. The minutes of the annual general meeting of 1936 record: 'The annual report for 1935 was presented.

The adoption of the honorary secretary's report was moved by Miss Burton and seconded by Mr. Scarfe and carried unanimously. The adoption of the balance sheet was proposed by Miss Smith, seconded by Miss Harnes and carried unanimously'. By 1957 the roles of treasurer, secretary and assistant secretary were held by women, but men continued to hold the chair.[43] Records for the Crewe and Nantwich branch of the GA, which had Dorothy Sylvester as president, show that during 1958–70 it was typical for only one of the invited speakers to be female in the course of the eight to nine meetings each year.[44] Whilst as has been noted above, numerous women represented local GA branches and/or interest groups such as primary or secondary schools on the Council, the Executive Committee, which acted on behalf of the Council on a day-to-day basis, was overwhelmingly dominated by men in numerical terms, although Garnett was to have supreme influence as honorary secretary from 1947 to 1967; the only other long-term woman member was Molly Long. Other women were included as representatives from the GA Council and briefly represented training colleges in the early 1960s and secondary schools in the late 1960s; one woman represented the recently introduced Models and Quantitative Geography in 1970 and Sheila Jones became secretary then chair of this sub-group. The trustees of the GA were all male from 1951-to 1970, as were the presidents (until 1968), honorary treasurers, honorary editors, honorary librarians and those representing primary schools, public schools, further education, field studies and sixth form/university (1962–70). Of the 72 people co-opted or invited to join the Executive Committee 1952–70 only three were women,[45] demonstrating how informal routes, reflecting circulating views as to who is 'useful' or 'influential', can reinforce masculine dominance. Even in the 1970s the power structure within the GA operated within an inner circle or cabal, who in effect nominated presidents of the Society, as was experienced by Sheila Jones when put forward for the 1973 presidency.[46] In this case, the informal power block favoured a woman, but such quangos, by their nature, are neither representative nor accountable, and as shown elsewhere informal mechanisms tend to favour the dominant group. This brief analysis of GA sources cannot purport to be anything other than a selective commentary, but it does show that while the officers of the Executive Committee of the GA were overwhelmingly male, women were nonetheless active on the GA Council and within certain roles on the Executive, as well as being central to the organisation of many local branches prior to 1970. The proportion of individual GA branch officers by gender varied by branch and over time.

There is evidence of fieldwork being central to the practice of school geography, principally in the local region, but also nationally and abroad where finances allowed. Women such as Charlotte Simpson, who took fieldwork to their hearts, played important roles in facilitating fieldwork, through fieldwork guides or centres. Women seem to have been able to move in and out of teacher training posts during years of childcare responsibilities with more ease than in university departments. This facilitated their ongoing long-term careers. Several women working in teacher training departments wrote influential school texts or guides to teaching geography in schools. The impact of these works should not be underestimated given the large number of school teachers and pupils potentially influenced by such work.

Chapter Six

Diplomas, Degrees and Appointments: The First Generation of Women Geographers in Academia

Introduction

The 'first generation' broadly encompasses women taking geography courses and beginning careers between 1900–20 and the work of five women will be discussed in detail: Nora MacMunn, Catherine Raisin, Blanche Hosgood, Hilda Ormsby and Eva Taylor. The intersection of their lives with the beginnings of geography as a university discipline will be examined, with reference to the women's individual institutional status and career development, their research, publications and teaching. This reciprocal relationship between the emerging new university discipline of geography and women's access to university courses and careers as academics is a recurring theme in this chapter; as is their application of the regional method as the prevailing discourse in this period, with its associated grounding in field and map work.

The creation of Queen's College and Bedford College (1848–9) in London, largely through the gifts of F.D. Maurice and Mrs. E. Reid, respectively, were important milestones in women's higher education in England. Both of these colleges were open to middle-class fee-paying women students and geography was taught at both institutions from the outset, as was also the case in a number of other smaller women's colleges, for example, Crosby Hall, Finsbury and Hyde Park (where German exile Kinkel taught geography) (Ashton 1986). Queen's College, initially founded to educate governesses, included in its alumni such pioneers of women's education as Frances Buss and Dorothea Beale (Watts 1981). Frances Buss founded the North London Collegiate School for Girls in 1850, which was innovative in only using trained teachers and having an academic curriculum for girls. Buss also used her connections with Queen's College to invite subject specialists as occasional lecturers at the school. Henrietta Busk, who went on to become a leading figure in the campaign for women's access to university degrees and in the Teachers' Guild (through which she worked to promote the teaching of geography among other subjects), gained her geographical training at Bedford College, 1861–4. In the early years of Bedford College, the Reverend Dr James Booth taught physical geography and later mathematics and astronomy. Booth does not seem to have been successful and the nineteenth century saw a succession of nonspecialist teachers of geography, with the subject seen merely as part of the general education of women students (Hilling 1994). From 1869 to 1874 Professor William Hughes of King's College taught geography at Bedford. He started from the premise that physical geography was *foundational* to all other aspects of geographical study and understanding, and he argued for the nomenclature of 'social' rather than 'political' geography (Freeman 1961; Hilling 1994). E.G. Ravenstein, cartographer and historian of exploration, as well as RGS Victoria Medal holder (for studies of population distribution based on the 1871 and 1881 census) taught geography at Bedford from 1882 until 1885, after early retirement from the Intelligence Department at the War Office (Hilling 1994). John Scott Keltie, in his report to the RGS on geographical education, described a lecture he heard on political geography at Bedford as 'one of the most instructive and most interesting lectures I have ever listened to' (Keltie 1885: 500). Despite this accolade, geography seems to have been marginalised at Bedford College after this, appearing on the prospectus, but not actually taught for a number of years (Hilling 1994) and even when taught, courses did not lead to recognised qualifications, apart from general certificates awarded by Queen's College. Both Bedford and Queen's were incorporated into the University of London in 1901. In this and following chapters it becomes clear that Bedford College and the North London Collegiate School played an important part in the lives and careers of a number of the women geographers discussed in this book.

The University of London played a crucial role in giving women access to university qualifications from 1878. This influence extended over a wide area as the university also awarded degrees to a number of validated courses around the country (Marion Newbigin, for example, gained her degrees at Aberystwyth as an external London candidate). However, after an early success with the appointment of Captain Alexander Machonochie to the chair in geography at University College London (UCL), 1833–6, geography was not to be established as a permanent department within the university until the turn of the century. Keltie (1885: 31), in his study for the RGS on the nature and scope of geographical education, reported that 'It has been objected to the recognition of geography by the universities that it is not a "manly" subject, but one fit only for the elementary classes', representing a feminised view of both geography and the working classes.

Geographical societies and academics lobbied for geography to be established in the nation's universities, and gradually lectureships and then departments were founded, often with subsidies from one of the geographical societies. The tenuous status of new geography courses and departments in British universities in the late nineteenth and early twentieth centuries resulted in geography lecturers welcoming women students to their courses, if initially only to supplement otherwise small class numbers. Women were one of the groups that took advantage of the opportunities afforded by these geography courses, including those in new universities (Stoddart 1986). School teachers were a natural constituency for the emerging courses in geography, demonstrated by the large numbers attending the vacation courses initiated by Arthur Geddes in Edinburgh and Mackinder and the Herbertsons in Oxford. A significant proportion of these teachers were women, teaching being considered an appropriate profession for middle-class women. Teaching also represented an opportunity for bright working-class young women and men to work as pupil-teachers in elementary schools. In this reciprocal relationship, university geography departments were able to demonstrate demand for their courses and women were able to gain professional accreditation as geographers (Maddrell 2004a). The Oxford geography department was the first to offer formal university qualifications in academic geography and Joan Reynolds was one of the first cohort of three students to take the University of Oxford Diploma in Geography in 1902 (see Figure 5.1). The first degrees in geography in the UK were offered at Aberystwyth University College and Liverpool University in 1917, soon followed by London in 1919 (Stoddart 1986).

Nora MacMunn

Nora Eileen MacMunn was born on 25 June 1875 in Chelsea, third daughter of John Alexander MacMunn, an army doctor. She was educated privately and then studied at Oxford as a member of the Society for Home Students (non-residential); women were not allowed to be members of the university or to be awarded degrees prior to 1920 but she passed her examinations in modern history (with marks equivalent to a third-class degree) in 1903.[1] She gained her University of Oxford Diploma in Geography in 1904 and was appointed by A.J. Herbertson as demonstrator in geography at the Oxford department in 1906. Her appointment was generated by the increased demand for certificate and diploma courses, despite Halford Mackinder's move to the London School of Economics (LSE). MacMunn's was a junior post in which she principally assisted the Reader with practical classes, reading seminars and geographical excursions. She gave her first full course of lectures in 1909 on the subject of 'Northern and Central

Europe'.[2] Her career is briefly recorded within Scargill's (1999) study of the department: having been appointed as demonstrator, 'she continued to serve the School [of Geography] until her retirement in 1935'. This hints at, but obscures her relatively junior standing throughout her near 30 years of service. After 16 years as demonstrator, MacMunn was promoted to assistant lecturer in 1922 (she matriculated in October 1920 and had been awarded an MA by decree)[3] and then lecturer in regional geography in 1924.[4] Interestingly these promotions follow Hilda Ormsby and Eva Taylor's appointment to lectureships in geography at the LSE (1920) and Birkbeck College (1921), respectively (see below). The news of these women's posts in London University must have circulated within the small British geographical community and perhaps had an influence on MacMunn's promotion. MacMunn held the lectureship in regional geography to which Herbertson had first been appointed, nonetheless, the hierarchical lists of lectures appearing termly in the university *Gazette* show how her male juniors such as H.O. Beckit and J. Cossar overtook her to play leading roles in the department (both MacMunn's promotions came when Beckit was Reader and director of the school). Lecture lists show the range of courses taught by MacMunn, which covered regular practical classes in topographical and historical maps; lectures on the history of geography, the historical geography of India, Canada and Central and Northern Europe; the history of exploration and regional geographies. Sometimes courses were shared with colleagues, for example, with Beckit on the British Isles, but for the most part she was the sole lecturer for a wide range of regional courses: France, Italy, the Rhine Basin, Inter-Tropical Africa and Extra-Tropical Africa, The Guineas, Canada, China and Indo-China, and the West Indies. The only other woman lecturer mentioned in relation to the geography department was Miss A.M. Ramsay whose informal lectures on the Historical Geography of Greece and Asia Minor were open to geography students in the summer of 1921, however, Eva Taylor worked as Herbertson's private research assistant 1908–10 (see below) and would have been part of the geographical community, as were Mrs. F.D. Herbertson and Mrs B. Beckit.

Photographs and attendance lists from excursions during these early decades of the department indicate the high number of women students. A manuscript for a 1913 student magazine describing staff includes a verse about a woman, presumably MacMunn:

> A maiden there is strong of limb,
> Who's surely done [?] and gym –
> With projections she copes
> And she goes up the slopes
> With a wonderful vigour and vim (*Geophil*)[5]

In the same magazine, stickperson caricatures of staff show a tall woman wearing a hat brandishing a copy of Vidal, suggesting MacMunn frequently referred to and was influenced by his work. Another sketch shows a geography excursion group, mostly made up of women, all with their bicycles (often used on local field trips), indicating a cohort of active female students.

MacMunn's affectionate obituary of A.J. Herbertson, found in *The Geographical Teacher* (1915–16: 144–5), is indicative of the influence he had on her life and work. For MacMunn and others he had been an 'eminent geographical authority' who had done much to transform dry geography, but was also a 'helpful adviser and friend'. Writing of his relation with students she seems simultaneously to be speaking for herself and others when she writes: '… his help was not confined to geography as a school and examination subject only, but extended to the problems of life generally and to our personal difficulties, and his far-reaching philosophy was open to all who enquired of him … Another thing that acted as a powerful stimulus to his pupils was that he always thought they could accomplish anything if they put their mind to it – he expected good work, and so he usually got it'.

Towards the end of MacMunn's career the Honours School in Geography was established at Oxford in 1932, with one woman and two men making up the first student cohort in 1933 (see photograph in Scargill 1999: 19). At this time, Head of School Kenneth Mason gained sponsorship from the Drapers' Company (to which he belonged) to fund three travel studentships, with the intention of attracting strong graduate students to geography. The first prize (1934) was awarded to Mary Doveton to undertake research on the political and economic geography of Swaziland (Scargill 1999).

Publications

MacMunn's first publication appeared in *The Geographical Journal* (1906) and was a map and table summary of the orographical regions of England and Wales, based on painstaking calculations. After her appointment as demonstrator of geography at Oxford, MacMunn went on to publish a detailed regional study entitled 'The economic geography of a county. Illustrated from Essex and Cumberland' in *The Geographical Teacher* (1907), edited by Herbertson. Her study reflects the central place of regional survey in her Oxford diploma course and the demand from teachers for good regional data. Recognising the mismatch between geographical region and county, MacMunn was pragmatic in using county boundaries to frame her study because they were so widely used. Her key argument was that geography affects and controls human behaviour, especially economic activity, which in turn explains the history of counties and countries. The paper is quite novel for its comparative nature and MacMunn concludes with comparing the continuities and changes in the historical economic practices and status of the two areas. 'If any proof is wanted of the geographical isolation of Cumberland beyond the historical one, all that is necessary is to go and stay in one of the valleys tourists do not frequent, and feel it ... One gets a good idea of what it must have been like all over the north-west of England before the rush for its mineral wealth began, and some realisation of the great change of equilibrium which has taken place in this country since that time' (MacMunn 1907: 38). MacMunn was also co-author with A.J. Herbertson and F.J. Unstead of the important GA publication *Guide to Geographical Books and Appliances* (1909), which became an indispensable reference for teachers. The book recommended good texts, the use of sand models and object lessons; it also promoted the use of popular fiction with geographical settings including mainstream masculine imperial literature such as Kipling and Haggard, but more surprisingly included the more counter-hegemonic novel *The Story of an African Farm* by Olive Schreiner (1883), with its exploration of gender roles, feminism and socialism (Maddrell 1998).

Maps and historical analysis are a strong feature of both MacMunn's teaching and publications and she summarised the best historical maps and atlases in a brief report for teachers; her belief in the *practice* of cartography and active learning shines through in the advice: 'it is generally better for pupils to trace their own maps. It takes little time to do, they remember them better, and only such features need be put in as are specially important for the time or occasion which they are to illustrate, and then the effect will be much clearer' (Herbertson, MacMunn & Unstead 1909: 37). In 1915–16 *The Geographical Teacher* responded to public demand for information and insight relating to the theatre of war in Europe, producing symposia on French and German Borderlands and Geography and the War. MacMunn's contribution was entitled 'Geographical aspects of the political history', and drawing from her lectures on the region she demonstrated how the area was categorised topographically as an intermediate area, characterised by its 'between-ness' historically and in the present. Later, she and Geraldine Coster used the late Andrew Herbertson's outline to write *Europe. A Regional Geography* (1922), a project MacMunn was to have assisted him with. This was intended as a comprehensive book for older school pupils and students in teacher training colleges, a text which would be readable and engaging rather than packed with overcompressed detail. Thinking very much in terms of accessibility and learning strategies they argued 'a fuller treatment, though it looks more formidable in bulk, provides something in reality easier to assimilate' (1922: ii). The authors changed little of Herbertson's draft introduction (pp. 1–12), included a brief discussion of 'political divisions and peoples' by O.J.R. Howarth in the appendix and acknowledged their use of George Adam Smith's *The Historical Geography of the Holy Land* and Marion Newbigin's *Geographical Aspects of*

Balkan Problems. They succeeded in producing a clear text, which applied the regional method to each subdivision of Europe, including the general character of a broad area, for example, the Balkan Peninsula (including climate, soils, vegetation, routes and occupations), followed by detailed accounts of the smaller natural regions therein. The text is amply illustrated with clear sketch maps and monthly figures for climate (an innovation the authors felt was particularly helpful to understanding an area's climate, as well as providing data for comparative graphs).

Shortly after MacMunn was appointed at Oxford, women were appointed to teach geography at Edinburgh and Liverpool universities. In Scotland **Alice Lennie** (MA, BSc) won the first RSGS University medal jointly with fellow Edinburgh student James Cossar in 1909,[6] both being part of G.G. Chisholm's first 1908 cohort of geography students. Lennie became assistant to Chisholm in 1912 and seven years later was promoted to the post of Junior Lecturer.[7] **Edith Marjorie Ward** was employed as tutor in geography at Liverpool, working with Percy Roxby, making her 'one of the first women in this country to hold a university appointment as a geographer' (Allison 1955: 279). Born in Leigh, Kent, c. 1888, part of Ward's childhood was spent in Odessa before the family returned to the UK, settling in West Kirby in Liverpool (ibid.). She attained a BSc at Liverpool University in 1908, followed by a Special Certificate in Geography in 1910, and according to Allison was appointed as tutor on the course the same year (Steel & Lawton (1967) have her as in post 1911–13). Ward continued to study alongside lecturing and was awarded an MA in geography in 1915. Allison (1955) describes Ward as having 'retired to private life' in Grasmere in the Lake District during the First World War and it was during this period after leaving the university that Ward's career as an author flourished. She died in Grasmere, spring 1955.

Ward's publications focused on agricultural (1913–14) and coastal geographies (1915–16, 1920, 1922), with three papers appearing in *The Geographical Teacher* and one in *The Geographical Journal* (1920) on the evolution of the Hastings coastline, based a paper she presented to an afternoon meeting of the RGS. Her coastal work culminated in her book *English Coastal Evolution* (1922); other books include *Days in Lakeland, Past and Present* (1929, reprinted in 1948). She also wrote a series of novels, described by Allison as 'slight but pleasantly written novels, each with an obvious geographical background', for example, *Deborah in Lakeland, Far Easedale* and *Sea Wind.* The latter was set on the Lancashire coast and featured a geomorphologist (something Allison thought to be unique in fiction) and a female botanist: 'For six months she had been wandering about Europe botanizing, and writing learned articles about her discoveries' (Ward 1938: 27), illustrating Ward's use of her knowledge of the locality and its geomorphology, as well as her views on women's education and public roles.

Catherine Raisin

Although a botanist and geologist, Dr Catherine Raisin taught and administered geography for periods of her career at Bedford College and came to influence the development of geography there. Her career progression was also significant in paving the way for other women to follow. Raisin was born on 24 April 1855, daughter of Sarah Catherine (née Woodgate) and Francis Raisin, who was a pannierman at the Inner Temple (Creese 2004b). Raisin was educated at the North London Collegiate School, known for its academic curriculum, and went on to study at UCL from 1873, alongside teaching at her old school until 1875 (Creese 2004c). She studied botany, geology and mineralogy, taking the honours examinations as a private student and continued to study when it became possible for women to take degrees at London University from 1878 onwards. She studied geology with T.G. Bonney and zoology with T.H. Huxley. Raisin stayed at UCL in a voluntary capacity as research assistant to Bonney after gaining her BSc, and went on to become only the second woman geologist to receive a DSc from London University in 1898 (the same year Newbigin gained her DSc as an external candidate of the university). Raisin continued to achieve a number of 'firsts' as a woman scientist. Notably she became the first woman to head a department at Bedford, when, only four years after being appointed demonstrator in botany, she was promoted to head of geology in 1890. In so doing she became the first woman to head a university geology department in Britain. She was also head of botany 1891–1908, vice principal 1898–1901, and responsible for the sub-department of geography 1916–20 where she taught both geology and geography. In 1893 Raisin was the first woman to be awarded the Lyell Fund by the Geological Society – even if Bonney had to receive it on her behalf as women were not admitted to the Society until 1919 (when she became one of the first women fellows) (Creese 2004c).

Raisin established a reputation for microscopic petrology and mineralogy, publishing 24 scientific papers 1887–1905; she was one of the first female professional geologists in Britain, took a keen interest in women's educational and political progress and formed the 1000 strong Somerville Club[8] in 1880, which provided a forum for women's discussion (Creese 2004b). Clearly Raisin embodied many of her school's aspirations for women and in both her professional life and retirement she sought to extend these values and opportunities to other women. Remembered as a charming woman, solicitous of students, Raisin was equally known for her strength of character, for ruling her department and being a formidable opponent (Creese 2004b). It was to the benefit of geography that she adopted the subject as one of her causes. Coinciding with Raisin's promotion to head of geography, **Miss M. Heath** was appointed to teach economic geography in 1916, but was released

Figure 6.1   Blanche Hosgood.
Source: courtesy of Royal Holloway University of London Archives

from this contract to work for the Admiralty in relation to Intelligence Handbooks (Hilling 1994: 133), as was the case for a number of geographers from other institutions. Raisin served on the university-wide Board of Studies in Geography and Hilling (1994) credits Raisin and the geology department with generating the pressure to enhance the status of geography at Bedford. Raisin seems to have negotiated a guaranteed three-year salary for Miss Heath from an anonymous donor and the teaching was undertaken by Blanche Hosgood and Professor Lyde of UCL (Hilling 1994).

Blanche Hosgood

Blanche Hosgood, in common with many other early twentieth-century geographers, had taken the University of Oxford Diploma in Geography (Darby 1983b), where she studied with A.J. Herbertson and gained a distinction (Smee 1953). On appointment to Bedford, she became head of the now separate Department of Geography, a post she held for 23 years until retirement in 1948, having been promoted to Reader in geography in 1923 (Hilling 1994). The fact that Dr Raisin left a bequest of £100 in 1944, to endow a new prize for fieldwork equipment for a second-year geography student (Smee 1953) suggests Raisin's respect for Hosgood's work in the geography department.

Despite this longevity in post and subsequent promotion, relatively little is known of Hosgood's work and Hilling (1994) describes her as a 'shadowy figure ... [she] did not write for publication and her views, however they may have been expressed, are now lost'; 'she published little, and was never conspicuous at meetings of geographers' (ibid.). Despite the fact that Hilling describes Hosgood as remembered by her students as 'remote', 'strange', 'forbidding' and that her lectures were mostly remembered for their chalkboard illustrations prepared by a technician, the development of the geography department seems testimony to some abilities in addition to the attraction of geography for women students in the early twentieth century. By the mid-1930s, the 50 or so women students at the Bedford geography department constituted one of the largest cohorts of geography honours students in the country (Smee 1953), with some students going on to complete PhDs (Hilling 1994). In contrast to Hilling's account, colleague Dora Smee's (1953: 320) obituary for Hosgood describes her as 'a stimulating and exacting teacher with much originality in her methods'. No doubt the opportunity to acquire the validation of diplomas and degrees played a part in recruitment too: a formal programme for a University of London Diploma in Geography was followed by the BA General degree in 1919 and the BA Honours and BSc Special degree from 1924 (Hilling 1994: 133). Hosgood lectured on economic geography, France and other regional geographies from around the world; she gave weekly tutorials and led biennial field trips (Hilling 1994). This reflected the curriculum to be found at London University geography departments in the interwar years, as all the London colleges worked to the same University Board of Studies syllabus, on which Hosgood served (Darby 1983b). Hosgood also played a part in early discussions leading to the foundation of the Institute of British Geographers in the early 1930s. Hosgood retired in 1948 and her successor, Gordon Manley, was appointed as the first professor of geography at Bedford College.

Hosgood's short obituary appears in *Geography*, the journal of the Geographical Association, perhaps because the majority of Bedford College students went on to become teachers (Kirk 2003) and therefore its readership was considered the most appropriate constituency. Smee recorded of Hosgood: 'She never tolerated poor work yet she gave to each of her students, of whatever ability, her full interest and attention. She was never too busy to discuss with a student any problem, academic or otherwise. Among the earlier pioneers of geography in this country she was in many ways unique, and her personality and her industry will long be remembered by all who came in contact with her' (Smee 1953: 320).

Hilda Ormsby (née Rodwell Jones)

Hilda Rodwell Jones was born on 1 November 1877 at the family home in Hanley, daughter of Sarah Ann (née Cuthbertson) and the Reverend

Figure 6.2   Hilda Ormsby DSc c. 1931.
Source: LSE Archives

William R. Jones, a Wesleyan minister.[9] She was part of a large family and another sibling, her brother Llewellyn, was to share her interest in geography. The family moved regularly with the short-term postings characteristic of the Methodist Church; by the time Hilda was 21 years old, the family had lived in Hanley (Staffordshire), Bristol, Sheffield, Sunderland, Huddersfield, Canterbury, Maidstone and St John's Wood (London). Her parents finally retired to Ealing in 1904 (Maddrell 2004e). At the age of 31 Hilda Rodwell Jones started the Diploma in Geography at the London School of Economics and Political Science (LSE), 1908–9.[10] As a student she is recorded as heading a deputation to the Secretary of the LSE petitioning for more advanced courses in geography and better standards in the Diploma (Harrison Church 1981), and by 1912 a Higher Certificate had been instituted including extra courses in surveying and physical geography at University College under Professor E.J. Garwood's supervision.[11] The exact dates of all her studies and appointments are not recorded, but it is known that Rodwell Jones was a teacher in England and France and studied at the École Normale de Melun in France (probably 1909–11). There she studied mostly French and German languages, fitting a three-year course into two years, and she later recounted how impressed she was with the centrality of maps to French education.[12]

Hilda Rodwell Jones initially came into contact with Halford Mackinder's work through his Saturday lectures for school teachers at the LSE, which she attended with her brother Llewellyn in 1911. Hilda was appointed as demonstrator in geography in 1912, succeeding Alice Thistle Robinson who left to marry, becoming Lady Bottomley. The department was now able to provide the compulsory first and special optional geography courses (three out of nine final papers) for the BSc (Econ), the Higher Certificate and the Academic Diploma in Geography established by Mackinder during 1915–6 along Oxford lines. Having been the main teacher of geography in the department during the First World War (Harrison Church 1981), as well as working on large-scale terrain maps for the Western Front for the Naval Intelligence based at the RGS, Ormsby attained the BSc (Econ) in 1918. She was appointed lecturer in geography at LSE in the same year[13] and took on responsibility as organiser for the Diploma and Honours BSc (Econ) under Professor J. Sargent (Maddrell 2004e, 2006). In 1919 Llewellyn Rodwell Jones joined his sister in the LSE department (initially as Mackinder's assistant), having previously been lecturer in railway geography at the University of Leeds and having completed distinguished war service (Wise 1980). This was to be a sister–brother partnership unique in all the UK geography departments (Harrison Church 1981), but although they shared responsibility for some courses, for example, General Regional Geography and Detailed Geography of Europe (LSE Calendar 1927–8), Ormsby implied that their work was a division of labour rather than a joint endeavour, Rodwell Jones being 'a lone man and went his own way – asking help from no man, but always ready to give it, if asked for'.[14] The year 1920 was pivotal in Hilda's life on both personal and professional fronts: she became a Recognised Teacher at the LSE, University of London, and in the same period ('five glorious weeks' according to Harrison Church), at 43 years of age, she was promoted to lecturer in commerce with special reference to commercial geography, appointed assistant to Mackinder (replacing her brother) and married George V. Ormsby (26 June 1920), a Reuters journalist who went on to become chief of the London bureau of the *Wall Street Journal* (Maddrell 2004e, 2006). The couple lived in a flat on Fetter Lane, where they gave regular soirées as well as occasional graduate supervisions by Hilda – one graduate student recalled her smoking in public and once greeting him at the door in her black silk pyjamas,[15] both indicative of a relaxed somewhat avant-garde style.

Ormsby's key period of research and lecturing occurred between 1920 and the mid-1930s, culminating with the publication of *France* (1931) and she was awarded a University of London DSc in 1931, largely for her work on *France* (being one of the few geographers to achieve this award at this time (Harrison Church 1981)). As a result she was promoted to Reader in the following year (1932). Her brother had been appointed to the departmental chair in geography in 1925. She retired from the LSE in 1940, but continued some part-time lectures for LSE students evacuated to Cambridge, where she also undertook intelligence work. Ormsby contributed to work on the Naval Intelligence Division 5 Handbooks, writing part of the handbook on France as well as completing some mapping work (Harrison Church 1981). Both Hilda and George Ormsby were also air raid wardens after the Battle of Britain,[16] which gives a sense of the multiple roles many civilians fulfilled during the war years, academics included (Maddrell 2008).

Records of the events leading to the founding of the Institute of British Geographers are a little vague, but Steel (1984) notes Ormsby, along with her brother, as one of the 13 founding members. Blurred memories and varying lists aside, Ormsby was one of four women associated with different stages of the Institute's beginnings, including Blanche Hosgood, Dora Smee and Alice Garnett; and Ormsby has particular significance as the first woman to serve on the IBG Council (1936–8). In 1962, at the age of 85, Ormsby was honoured within the academic community when she was created an honorary fellow of both the LSE and the RGS.

As a teacher Ormsby was described as attaining 'a high degree of technical efficiency' as well as someone who could 'handle with marked success every class from raw beginner to the finished honours product' and could 'control a department independently'.[17] Whilst not remembered for clarity in her lectures, students remembered her enlivening classes for her zest and sense of fun, such as the time she modelled a Romanian skirt. She gave memorable courses on Europe 'instilling in many a student a deep appreciation of France in particular' (*Geography* 1967: 423) and ran a celebrated seminar on London in the late 1920s and early 1930s. She was at her best in tutorials and map interpretation classes and provided students with insightful feedback on their work (ibid.).

This litany of qualifications, posts and publications (see below) makes clear that Ormsby was undoubtedly a significant producer and disseminator of geographical knowledge. R.O. Buchanan wrote of the expansion of university geography in the 1920s: 'most of the appointments, even to responsible posts, were relatively young men, men with ideas and energy, men who were conscious of their opportunities and their responsibilities, even though they were heavily overburdened with teaching and administrative duties (Buchanan 1954, 1–14). We might ask whether Ormsby was an honorary 'young man' at this time (she was a significant figure in the geographical world and was promoted to Reader) or as a woman in her forties was she side-lined? Archive correspondence relating to the proposal of the LSE/King's College Joint School in Geography established in 1919 make no reference to Ormsby, which is surprising given her major teaching role;[18] this was prior to her receiving recognised teacher status within the university, but could also reflect the proponents' strategic or otherwise representation of geography as a 'manly' endeavour within the university. An archived 1923 testimonial from Sargent suggests she may have been applying for promotion at that point, bur her brother was promoted over her in 1925.

Ormsby and her husband retired to 'Water End', an Elizabethan cottage in Hertfordshire. Always a lively conversationalist and host, Ormsby continued to entertain former students and colleagues there; she was 'warm-hearted and vivacious, frank in her opinions, a good listener as well as a good talker' (*The Geographical Journal* 1974: 177). George Ormsby died in 1950 and Hilda lived on to within a few days of her 96th birthday, dying on 23 October 1973 at the Guest House, Althorne, Essex, where she was then living. At the time of her death it was noted that 'she probably claims the record for longevity of any British geographer' (*The Geographical Magazine* 1974: 218). A student bursary at LSE was endowed by Hilda and George Ormsby.

Publications

Best known for her work on London and France, the regional method of early twentieth-century British geography defines Ormsby's work (Maddrell 2006). Whilst heavily influenced by Mackinder whom she found inspirational, she was equally influenced by the work of the leader of the French regional school, Vidal de la Blache. She was involved with the Field Study Group and the Le Play Society (Harrison Church 1981), which not only facilitated collective field trips, but also stressed the importance of linking empirical findings to social policy in both urban and rural settings (Dickinson 1969; Walford 2001). This characteristic can be identified in Ormsby's regional writings in several places, notably articulated in 'Regional Survey in a Large City', in which she identifies the unique and dynamic character of a region: 'The geographer with his [sic.] training in regional studies can perhaps best present to those experts in social welfare the picture of a unit area made up of a multitude of interacting forces, but having a history and a personality and even a dignity of its own, which must be recognised and respected by those who desire truly to administer to the welfare of its inhabitants' (1927: 45).

Ormsby's first book was an historical geography, *London on the Thames* (1924), which she dedicated to her husband. The book was innovative in treating the London area as a whole and was aimed at teachers of geography as well as antiquarians. It included 34 maps, reflecting her belief stated in the preface that 'The map is the chief reference library of the geographer, and it is also his [sic.] natural means of expression' (Ormsby 1924: vii). The preface acknowledges Mackinder's influence and Duncan Montgomorie for 'his attempts to restrain my rashness in matters antiquarian' (1924: x). This reflects her general tone of modesty in the preface which she opens in a most self-deprecatory way: 'one feels that the publication of yet another book on old London demands an excuse, if not an apology' (vii). She goes on to justify the book as none others have treated London as a geographical whole, but concludes 'even if greater learning and future research prove the theories I have been bold enough to put forward to be utterly wrong, the book may be of some value in suggesting lines of investigation' (viii).

True to regional style, she began with the physical factors before proceeding to discuss settlement, defence and trade and how these related to the river and its hinterland, but this was all framed within the context of Vidalian possibilism. She also placed London within the broader European context, both in geological and economic terms. The maps, which were always central to Ormsby's work, gained particular acclaim, especially her contour map of London. Her analysis of the area's topography and her fresh style were also welcomed (*The Geographical Teacher* 1924); and one reviewer only regretted the limitations of her subject: 'Mrs Ormsby writes with such enthusiasm, and illustrates her points so clearly with sketch-maps and diagrams that one almost regrets that in the course of time human ingenuity triumphed over, and almost obliterated, London's topographical characteristics, and so set a limit to her book' (*The Geographical Journal* 1924: 534). Another reviewer thought it a 'pleasant and instructive book', if too heavy on the geology as topographical influence (Thompson 1924: 98). Reviews by historians were more critical, acknowledging Ormsby's reasoning but questioning the evidential base for her challenge to existing theories on the early Celtic settlement of London. As noted above, Ormsby used London as a local case study in her teaching and wrote her paper on applying regional survey to a large city (Ormsby 1927). She also contributed a chapter on the Thames area to *Great Britain: Essays in Regional Geography* (1928) (the only female author in the volume). This volume, edited by Alan Ogilvie, grew out of papers given by British geographers at the IGU meeting held in Cambridge in 1928 and was intended to provide a resource to fellow lecturers and teachers in a 'synthetic study of nationwide regions' (Ogilvie 1928, introduction). Darby (1983b) described the book as consciously intending to show the state of the discipline in Britain and equated the contributors with the intellectual core of the IBG. Ormsby's place was very much part of this core, as one reviewer noted earlier of her book on London, 'it gives her a serious claim to a place in our front ranks' (*The Geographical Teacher* 1924: 324). Each author was given a free hand to approach their region in their own way in the 1928 volume, resulting in a methodological diversity which would later be criticised as 'a bewildering variety of points of view' (Steers & Woodbridge 1976: 136). The volume was also criticised for too much emphasis on the natural environment as a starting point which never moved beyond the determinist versus possibilist debate in theoretical terms (ibid.). Consequently some considered Albert Demangeon's *Les Iles Britanniques* (1927, English translation published 1939) to be far superior (Clout 2003). Putting these critiques of the wider volume on one side, it is apparent in all of Ormsby's writing on the Thames area that she was very conscious of the ways in which physical factors were rapidly being obscured or even obliterated by the concentration of human activity in and modification of the environment. More than that, in terms of policy, she articulated the growing influence of London's exceptional economy on the whole of the southeast of England): 'the ever-tightening grip of London on the whole area ... gradually welding, not only the London Basin, but the whole of South East England into a geographic whole' (1928: 68). This prefigured the concerns of the Barlow Commission investigating inequalities in employment distribution a decade later and conceptual developments in British regional thinking which would give precedence to human factors (Maddrell 2006).

Ormsby travelled widely in Europe with the Le Play Society and other colleagues in the interwar years (Harrison Church 1981) and the fruits of that work can be seen in her publications on the 'Danube as a waterway' (1923), 'The limestones of France and their influence on human geography' (1932) and her most political piece 'The definition of *Mitteleuropa* and its relation to the concept of Deutschland' (1935). The latter was initially read to Section E of the BAAS in 1935 and subsequently published in the *SGM*. It focused on the writings of modern German geographers and concluded that Hettner's idea of Central Europe, initially accepted by non-German geographers as a convenient description of an area, had come to take on political significance. Ormsby concluded that her study of a wide range of German work 'leaves me with the impression that their authors are frequently more concerned to find support for the pre-conceived idea of a definite Central European entity, than merely to make a convenient division for descriptive purposes' (1935: 337). The paper would be widely read as the implications of her comments became ever more pertinent in the lead up to and duration of the 1939–45 war (Sinnhuber 1954).

Ormsby's major text appeared in 1931 and earned her a DSc. *France. A Regional and Economic Geography* was made up of 11 regional studies following a brief introduction focusing on climate, and concluded with three chapters on the economic geography of the country with particular reference to the regions previously discussed. Her regional divisions were based on watersheds, partly for convenience, partly due to the significance of water to trade and communications, although she is at pains to point out that she had '[not] allowed myself to be hidebound by the system I have adopted' as this would lead to absurdities. As a consequence, Brittany for example is defined by its coastal waters rather than its drainage system.

In the preface to *France*, Ormsby intimates some of the factors influencing her production of the text. Firstly that her method has been to work 'on the map' (and the book is intended to be read alongside a 1:200000 topographical map), combining that map work with field observation and a good use of French geographical literature. Conceptually, her use of watersheds to define regions indicates the importance given to physical factors in the book. She argues: 'There can be few countries where the interrelation of physical and human elements is more striking than in France and I have been at pains throughout to emphasise the importance of the physical basis of the human geography and of the continuity of geographical influences and controls, though exhibited under varying forms and with changing incidence' (1964 second edition: v). On the opening page she cites Vidal de la Blache's (1911: 8) *Tableau*, where he states his possibilist creed that a country is 'a medal struck in the effigy of its people'.

Ormsby also gives generous acknowledgement to the people who assisted and influenced her writing. In terms of the subject, she acknowledged the special influence of Professor Sargent who first got her interested in France and to Halford Mackinder for his inspirational teaching. She also acknowledged London colleagues who read sections of her manuscript: Stamp, Wooldridge, Sargent, Benham, Beaver and O'Dell. Her brother was also on this list and she added: 'I take this opportunity to acknowledge, with affectionate gratitude, a debt of long standing, constantly renewed, to my sister, Miss Dora Jones, for her never-failing encouragement and ready help' (1964: vii). In keeping with the tone of her preface to *London on the Thames*, Ormsby justifies writing the book in the interim whilst awaiting a more authoritative economic geography of France: 'From the view of an economic geographer … a major work has still to make its appearance. While awaiting the publication of such a work from the French geographers, and because no detailed geographical study of France exists in the English language, I have been emboldened to publish the present book in which I have embodied the results of close study of the subject, covering a number of years' (1950: v). In one deft move Ormsby simultaneously deflects any criticism from French writers for straying onto their territory and intimates the extent of her research and therefore authority to write the book (Maddrell 2006). Her reference to the influences of Vidal, Mackinder and Sargent also lend their own weight to her authority, as do the list of colleagues who acted as critical readers.

France received generally very positive reviews from H.J. Fleure and Raoul Blanchard, but both were critical of her emphasis on drainage basins. Whilst Fleure (1932) felt more attention could have been given to roads in France, the nation's demographics and the continuity of local social relations, his main criticism (shared by Blanchard) was the limitation of using watersheds as the defining characteristic for regions. Fleure also quibbles over the use of 'economic' to describe what he sees as essentially 'commercial' geography, and gives a hint of his view of Ormsby's character: 'the authoress has adopted a position of her own which she maintains with characteristic firmness' (1932: 511). The treatment of the wine industry and the stamp of Ormsby's field observations were praised, as were the original maps and useful bibliographies accompanying each section. Whilst making it clear he would have preferred a different approach to the book, Fleure concludes: 'within the chosen limits it is to be warmly welcomed and, one hopes, will be widely used' (1932: 512).

Blanchard, writing in the *Geographical Review*, confessed pleasant surprise that a book written by a non-native was so faithful to the spirit of French geography and that it was such a substantial text. He noted that 'the regions which Mrs. Ormsby has evidently visited she has seen with the eye of a geographer; her accounts are precise, informed and touched with life' (1933: 156). However, he too criticised the use of drainage basins as the organising principle for the work and disagreed with some consequent demarcations (e.g. Perigord in the Central Massif rather than Aquitaine); but his deeper criticism was that the concept of the region was applied in an old-fashioned descriptive way rather than an explanatory framework. Despite these criticisms he concludes 'One must nevertheless thank the author for having produced on a foreign country so full, careful and impartial a study' (ibid.: 157). Indeed, the study is very even-handed and although Fleure commented on the relative lack of attention to Paris, given its economic dominance, this was clearly intentional on Ormsby's part, a reflection of her view of industrial complexes acting as regions of economic 'radiation' rather than 'concentration' and her commitment as a geographer to study and understand peripheral as well as core economic regions (Maddrell 2006). Economic as well as environmental possibilism is at the heart of Ormsby's work, alongside a precise analysis of specific *genres de vie*, based on detailed study and observation of localities.

It is in this detailed study that it is possible to discern a sensitivity to gender difference not necessarily found in generic figures and surveys. For example, her study of the Breton crofters includes reference to the whole family's participation in the economy, notably women's role in the seasonal work of fish processing and canning, as well as more traditional practices such as seaweed gathering. In this she demonstrates women's as well as men's participation in waged labour, part of the wider blurring between peasant and capitalist economies emerging in that region. She also notes the specificity of gender roles in textile production areas and the impact of male conscription and war mortality on the agricultural workforce and the consequent changes to production patterns in upland areas. Ormsby's authorial voice was masculine and her reader was represented as the 'neutral' masculine, but this was the literary convention of the day and it may be presentist to imply she should have written differently (Maddrell 2006). In her London book she makes the comparison between the everyday occurrence of Samuel Pepys taking his wife by Thames barge to shop in London 'as it would be to-day for anyone [assumed masculine] to take a taxi to Westminster and drop his wife in Knightsbridge with a little present to buy a frock' (1924: 130). For today's readers this is redolent of masculine and classed tones, and Ormsby may have been absorbed in and unconsciously reinforcing of this discourse; alternatively, in the light of her tendency to flout at least some social norms, it may have been written with humorous irony by the 47 year old.

To Ormsby's surprise and gratification,[19] *France* was still in demand in the late 1940s after being out of print for 10 years, and a second edition of the book was published in 1950 with updated statistics and an appendix on economic conditions (1939–47) provided by F.J. Monkhouse (Ormsby having been retired for 10 years). The second edition remained in print into the 1960s and overall *France* was a standard university text for three decades. Ormsby's *France* should have been followed by a book on Germany, but the long-prepared manuscript was lost in German bomb damage during the Second World War (Harrison Church 1981). She obviously decided against spending her retirement in rewriting the volume, no doubt at least in part influenced by her husband's illness and then death.

Eva Taylor

Eva Taylor was born on 22 June 1879, daughter of Emily Jane (née Nelson) and Charles Taylor, a solicitor. She had a difficult childhood – her mother left her husband and children when Eva was three years old. Eva and her older brother and sister were subsequently forbidden toys or pets, a response Eila Campbell described as 'the "sin" of the mother … visited upon her children'. However, Campbell suggests that it was in this formative period that Taylor turned to nature and the garden as a source of entertainment and information (Campbell/Baigent 2004). After initial home education, Taylor went to Camden School for Girls and then to the North London Collegiate School for Girls. From there she won a scholarship to Royal Holloway College for women where she achieved a University of London first-class honours degree in chemistry in 1903. As was common for educated women of this period, Taylor became a school teacher, teaching science at the Burton-on-Trent School for Girls (1903–5) and the Convent School in Oxford (1905–6) where she combined teaching with her own further study at the university. The headmistress at Burton-on-Trent, in a glowing reference, described Taylor as giving 'such interesting lessons which are the delight of all our girls including the quite little ones' and that she was 'most painstaking in the preparation of her work and imparts her informa-tion with the clearness and ease of a good class-teacher or lecturer'.[20]

In 1906 Taylor enrolled as a student for the Certificate in Regional Geography at the University of Oxford School of Geography, followed by the Diploma of Geography (1907–8), passing both courses with distinction. Freeman (1976) describes it as a 'happy chance' which brought her to

Figure 6.3   Eva Taylor.
Source: RGS-IBG Picture Library

Oxford and her subsequent disciplinary shift from chemistry to geography. However, Taylor's obituary in *The Times* records that she specifically went to Oxford to improve her qualifications, intending to take the Diploma in Education, but being persuaded by the women's tutor, a Miss Cooper, to consider the geography courses on offer.[21] Taylor later told a news reporter that she moved into geography as a discipline with career opportunities for her as a woman: 'she told me with the utmost frankness, she could see no opportunity for a woman in chemistry from the financial point of view and so she turned to geography in its broadest and truest sense as a field of study that was not overcrowded'.[22] Clearly push factors were just as strong as pull factors in this process. There are interesting parallels with Marion Newbigin here, who similarly moved from the biological sciences into geography, influenced by the need to earn a living (Maddrell 1997; see Chapter Three). No doubt the same is true of men who saw the new discipline as a field of opportunity; as Freeman (1976b) points out, Taylor and Newbigin were in no way unique in coming into geography 'by chance'. Taylor's understanding of geography 'in its broadest and truest sense' may have been related to her lifelong passion for nature study and natural history and, as with Newbigin's biology, meshed science and geography within the study of physical processes and regional studies.

Taylor became private research assistant to A.J. Herbertson (1908–10), Herbertson having succeeded Halford Mackinder as Reader at the Oxford University School of Geography in 1905. At the same time H.O. Beckit was appointed official assistant to the Reader (Scargill 1999). Beckit wrote of

Taylor in a 1916 testimonial that Herbertson had considered her 'the most able and brilliant of the many women students trained under him at Oxford'.[23] Taylor was inevitably influenced by first being taught by and then working so closely with Herbertson, but she also experienced being a student on a geography course in which the presence of women was the norm. This was in part because of the close association of the emerging geography department with teacher education, especially the biennial Oxford Summer Schools for teachers, and, it might be argued, a more politically expedient need to demonstrate student numbers to the university and the RGS which subsidised geography at Oxford until 1926. Whether strategic or not, Mackinder actively recruited women students (Scargill 1999) and Herbertson continued this policy. The Diploma course had women students from the outset and Nora MacMunn had been appointed in 1906, so Taylor would have been taught map and practical classes by a woman (see above). She moved to London in 1910 and she had three sons in 1912, 1915 and 1919; the second sadly dying in infancy.

In his account of her life Gordon East described Taylor as having married, but that she had not allowed the children to interfere with her career.[24] Neither statement was accurate. Described as a devoted mother (Campbell/Baigent 2004), Taylor combined her children's early years with drafting maps and writing texts. She took part-time posts from 1916 to 1918 (after the death of her second child) at the Clapham Training College for Girls and the Froebel Institute and then at the East London College in 1920 (when her third child was in his second year). It was not until 1921, at the age of 42 and some 13 years after receiving her Diploma that she achieved her first post as university lecturer in geography, when she was appointed by Unstead as a part-time lecturer at Birkbeck College. Birkbeck had originated in the 1823 Mechanics Institute and provided part-time evening and weekend courses for those in employment by day; the college became a formal college of the University of London the year before Taylor was appointed (de Clerq 2007). Taylor's appointment was also one year after Hilda Ormsby gained recognised teacher status at the University of London and her full lectureship at LSE (see above).

Taylor had an affair for some years with Herbert Edward Dunhill (of the wealthy tobacco family); the couple shared a house on Oakley Street in Chelsea and Taylor's first son Spencer was recognised by the Dunhill family. Anthony, the youngest, was given the Dunhill name, but there seems doubt about his paternity; Eila Campbell searched unsuccessfully for a birth certificate, but his marriage certificate located by her names Dunhill 'a gentleman' as his father. In correspondence relating to a proposed biography of Taylor, Campbell reported that Taylor had refused to discuss the matter and became distressed whenever the subject was raised. Campbell records an anecdote that when Anthony known as Tan was born, a story spread around the East London College that he had been rescued from the *Titanic* in order to explain the unmarried Taylor's child, and she speculated that the child may have been born in Canada. Campbell also noted that Tan had been brought up by the Latchfords, an elderly working-class couple in Hertfordshire, presumably while Taylor was working full time.[25] East (ibid.) described Taylor as 'a feminist in her early years and one letter in the British Library would seem to support this (as well as her comments about career opportunities for women and rates of pay – see below). The undated letter was written by Taylor to birth control pioneer, Marie Stopes, recalling when the two had been at school together. 'Seeing your address in today's "News" I thought I would write as I am so interested in your work. I remember still your paper on "tadpoles" when you were on the "Gallery floor" at the North London & I was on the "Hall Floor", & have followed you in the daily press for years'. After asking about details of any forthcoming lectures, Taylor continues, perhaps seeking approval for her choice: 'I have taken (from conviction!) the rather drastic step of having children without being legally married to their father, so that I can still sign myself, Yours Sincerely, Eva G.R. Taylor'.[26] Whilst this choice does not seem to have affected her career in the long term, her father would not recognise her sons and income she received from her father's estate was stopped after her death. According to Campbell, Taylor had to adopt her own sons in order to become their guardian.[27] Taylor kept a personal diary which after her death went with her other books to Birkbeck only to be lost when the college librarian died, much to Campbell's frustration as a hopeful biographer. Consequently we are not privy to the details of her personal motivations and choices, but her public statements combined with what personal information is extant suggest that Taylor could be both strategic and socially brave in her choices (even if these might seem contradictory) and that in later life she relished in her achievements as a woman, her nonconformity and challenging views different from her own. Walter Freeman (1976) likened Taylor to George Eliot – certainly the commitment to having their work taken seriously and unconventional domestic arrangements were common to both. As a capital investor and shareholder,[28] Taylor does not seem to have been an economic radical, beyond promoting women's access to high paying jobs. However, when writing to Kierwan at the RGS in 1946 she did suggest that her political views were at least considerably more progressive than those of some other members of the RGS: 'Anything I might have suggested regarding political geography ... would not have commended itself to a Society whose officers genuinely believe that:

> The rich man in his castle
> The poor man at his gate

God made them high and lowly
And ordered their estate ...[29]

Taylor taught geomorphology and historical geography at Birkbeck,[30] a combination which would be unlikely if not unheard of today, but was more common in the first half of the twentieth century when it was considered that any geographer worth their salt could address and combine both physical and human geography (Crone 1966). In 1929 Taylor achieved her DSc in Geography from the University of London and in the year following this academic validation of her work was appointed professor of geography (1930). She succeeded J.F. Unstead and Campbell stresses that Taylor was awarded the chair in open competition. East (*op. cit.*) in turn underscored the fact that she achieved the post despite being the mother of two children; that the first female professor of geography, appointed in 1930, was an unmarried mother of two is remarkable in the context of social mores of the day. She also leapfrogged several normal stages in the progression to a chair, being promoted from demonstrator to professor, but it was felt to be 'fully justified'.[31] There was obviously some confusion about her marital status, however, as some colleagues (such as East) referred to her as married and that she retained the title of 'Miss' simply out of principle. Nonetheless, such an appointment would have been unlikely outside London University with its nonconformist principles. Notably the university had to its credit both the first professor of *geography* when L.W. Lyde was appointed at UCL in 1903 and the first *woman* professor in the UK when Caroline Spurgeon became professor of English at Bedford College in 1913. Taylor held the post until her retirement in 1944 and it was during these 14 years as professor and head of department that Taylor's academic reputation was established at a national level, culminating in the popular appellation 'Queen' of the BAAS, echoing Mary Somerville's title some 100 years earlier (see Chapter Two).

A 1935 news report on women at the forthcoming BAAS meeting, although containing some biographical inaccuracies, sheds some light on Taylor's storying of her career, including combining work and motherhood, salary and the role of mathematics in female education and employment. The article reported that:

> During their [her sons'] early childhood Professor Taylor decided that domestic claims came first and that her studies – and the fruits there of – must be a part-time affair. At the same time she decided that she would not even later take a whole-time job unless it meant at least 'four figures a year'. Her children are now almost grown-up and she has achieved her object, for her salary and income from her books are well past the £1,000 a year mark. ... Women are supposed never to know north from south nor east from west. To Professor Taylor, triangulation, survey maps and longitudes are child's play. Yet she made me one confession with regard to her sex in science. 'It is

in mathematics that women students are weakest' she said. 'And as mathematics play an important, if not the most important, part in practical science – engineering, constructional work, architecture, land survey, actuarial work – it will not be until girls' schools in England have a system of teaching higher mathematics properly and thoroughly that women will have an equal chance with men in those branches of applied science where there is big money to be made.[32]

During the war years Taylor kept the Birkbeck geography department going, largely through weekend classes, spending weekdays out of London. This was when Eila Campbell first started to assist Taylor. From 1939 to 1940 Taylor instructed officers of the Eastern Command in map interpretation (causing her to query the accuracy of the War Office maps).[33] She also chaired the RGS committee preparing a report to the Royal Commission on the distribution of industrial population in the UK from 1938. The committee was initially to be chaired by Dudley Stamp of the Land Utilisation Survey fame, but he was absent for the most part in India, so by default the role fell to Taylor, resulting in some tussling for status and recognition as Taylor thought she should be the official chair having done the work in practice. The RGS secretary wrote to her in diplomatic ease: 'I do not think the [Research] Committee made any formal resolution on the subject of the Chairmanship, but they clearly desired that you should be principally responsible for preparing the evidence for the Royal Commission and for laying it before them. They also resolved that you should open the discussion and be followed by Fleure and Daysh: so that there is little scope left for the nominal Chairman, though I do not think he was ever actually replaced'.[34] Taylor was appeased but Stamp ultimately co-presented the Committee's findings to the Royal Commission. The RGS submission constituted a memorandum and over 40 maps largely prepared and assembled by Taylor, who was great believer in the power of maps. Both the geographical interpretation of the issue and the visual images of the maps influenced the Commission, notably Taylor's masking of the maps to show an axial belt of industry running diagonally from London to Liverpool. This axial belt became known as 'Professor Taylor's Coffin' (see Figure 6.4; Willatts 1971) and drew attention to the marginalised peripheries where only important nodal towns or resource sites thrived. Subsequently Taylor was appointed to Lord Reith's Panel of Reconstruction (1940) and the Consultative Panel of the Ministry of Works and Buildings (1941); she also supported the work of the Association of Planning and Reconstruction (*TIBG* 1968). The axial belt as a doctrine or theory was challenged by Baker and Gilbert (1944), but Taylor asserted it was simply a model rather than a theory and in the long term the model was found useful in planning circles for decades (Willatts 1971).

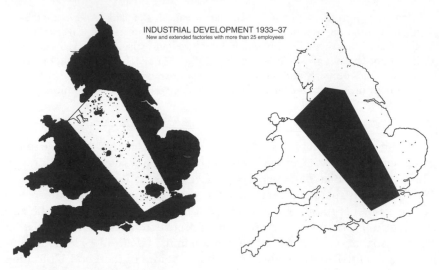

Figure 6.4   The axial belt of industry.
Source: Willatts 1971: 312 GJ

During her career Taylor served on numerous committees of geographical and related bodies; she was also honoured with a number of awards. She joined the RGS in 1922 and served on the Society's Council 1931–5, soon after being promoted to professor at Birkbeck, and again 1937–41; she was only the second woman to be elected to Council, swift on the heels of Elizabeth Ness elected in 1930. Taylor was also a founder member of the IBG. Elected president of Section E of the BAAS in 1939, the meeting was dissolved after the outbreak of war, so she was re-elected in 1947, when she spoke on 'Geography in war and peace'. During this time she also chaired the BAAS Committee on a proposed National Atlas, with S.W. Wooldridge as secretary, but the war hampered progress on this matter. Taylor was awarded the RGS Victoria medal in 1947 and an honorary fellowship in 1965. She was also honoured with an LLD from the University of Aberdeen in 1948, which she told Kirwan of the RGS was 'a compliment to Geography rather than me personally'.[35] In the post-war years Taylor was interested in contemporary issues such as civil aviation as well as the historical means of travel which were the main focus of her research. Reflecting those interests she was a vice president of the Hakluyt Society ('where she must have felt like a fish in water' (de Clerq 2007)) from 1945 until her death. She was also honorary vice president of the Society for Nautical Research in 1961 and honorary member of both the Town Planning Institute and the Institute of Navigation. The Institute of Navigation was a lynchpin in her post-retirement writing career (see her list of publications in TIBG 1968) publishing over 20 of her papers and commentaries and supporting her later books; she gave the Institute's first Duke of Edinburgh Lecture in 1959 at the age of 80.

Her character has been described as 'masterful' (Campbell/ Baigent 2004) and Bill Mead likened her to a Tudor chatelaine or Elizabethan matriarch.[36] The notion of a 'masterful matriarch' contains obvious contradictions, perhaps 'magisterial' would be a better adjective for her commanding personality. Her correspondence to those working for the RGS and British Library are full of requests for information, copies of pictures etc. that give the impression that whatever *she* was undertaking was of the utmost importance and no one should refuse requests which served her cause, whether it was tracing a rare manuscript or reference, reproducing diagrams, or borrowing maps for the Reith enquiry. No doubt this was partly her character but perhaps also a necessary part of being taken seriously as a leading academic. She was known for being formidable (Freeman 1976) and she did not brook criticism. Taylor's correspondence files in the RGS/IBG archive are full of characteristically forthright communications, illustrating her willingness to compromise on a point, robust in her critique of weak geographical work which she reviewed, but also a brittle response when a paper of her own on William Leybourne was criticised by referees in 1943. Forwarding a publication from another author, Taylor wrote to Crone at the RGS 'the writer of the enclosed keeps pressing me to review it in the Journal. As I did not wish to tell her that my writing is not considered good enough for the Journal I put her off saying you were short of space. However, it appears that she asked you directly and you denied this. Will you please see that the [work] is reviewed by someone who is *persona grata* with your publications' committee. I can do nothing about it'. There were soon conciliatory approaches from the RGS, including an invitation to speak to the Society and she was asked to draft a memorandum on post-war needs in geography (Kirwan wrote 'I know of no other geographer who is likely to produce a better one').[37] Taylor continued to be robust in retirement, criticising the RGS' choice of a non-British geographer for the first post-war research monograph in the absence of British material: 'since for the majority of us there has been a six-year stop on research ... and one would have thought that the officers of the Society would have been ashamed to use that stick to flog the professional geographers with' (ibid.).

Although deeply interested in the history of exploration, she clearly felt she needed to protect the interests of academic research in the face of the competing lobbies of contemporary travel and exploration within the RGS. Arguing in defence of academic interests, she wrote in no uncertain terms to Kirwan the secretary and director of the RGS:

> Herewith your card. Surely you did not expect us to be duped by the Machiavellian device of splitting the vote on scientific geography into 9 and

thus ensuring that Travel and Exploration should win hands over fist? When my colleagues [hold forth] on your charm and friendliness, I remind them that judgement must be based on deeds – and when these are examined (e.g. the choice of the new Honorary Secretary) the red light shows plainly. Are you off to Abyssinia yet?

Within a few months Taylor was writing to the director asking him to second her nomination of a former student to membership of the Society[38] and the following year she was awarded the society's Victoria medal for her 'many and varied contributions to geographical knowledge' (*The Geographical Journal* 1947: 291–2). Candid as ever, Taylor made her position clear in her acceptance speech: 'I am greatly honoured by the award of this Victoria medal and yet a little surprised because my role in the Society has often been the ungracious one of critic' (ibid.). She paid tribute to other Victoria medal holders Hinks and Heawood and especially to Halford Mackinder who died that year, describing him as 'my teacher and inspirer' (ibid.). Gerald Crone, RGS librarian and frequent correspondent with Taylor wrote on her death:

> With her strongly held opinions often provocatively expressed, and certain prejudices, she was sometimes a difficult conversationalist, but her evident determination to get to the root of the matter and her practical common sense usually carried the day … Her contempt for lazy thinking, her persistence in working for the accurate answer, and her wide knowledge are not likely to be forgotten (1966: 596).

Taylor retired in 1944. Initially living at her home 'Ralph's Ride' where she continued her research and writing and being in demand both as a speaker and author. Despite having told a reporter at 70, 'I have now qualified for bananas and when one reaches that age one should keep quiet',[39] she was ever active during her last years despite severe arthritis and other health problems, even working on her last book in hospital. She spent her last years at St Anne's Nursing Home in Wokingham, Berkshire. A stroke in 1964 limited her movement and vision and she needed much help in completing her final book. Michael Richey assisted her with the *Haven-Finding Art* (1956) and co-wrote *The Geometrical Seaman: a Book of Early Nautical Instruments* (1962), but her unofficial assistant of decades was former student and colleague Eila Campbell, who in Taylor's later years acted as amanuensis, secretary, research assistant, errand runner and made practical arrangements for Taylor's care and finances in the absence of her sons working and living elsewhere. Some have suggested that Taylor was exploitative in the case of Eila Campbell, but Taylor was also warm and supportive which can be seen in correspondence to Campbell (see Chapter Seven).

Taylor is remembered for enjoying the company of young men (according to Gordon East) rather than mentoring young women. Nonetheless, her correspondence files suggest that in addition to the lifeline that Campbell represented in later years, Helen Wallis became a friend as well as a source of information at the British Museum and Library. Taylor's *Times* obituary also noted that she was 'a stalwart champion of young university workers whom she both bullied and inspired'.[40] At Taylor's 80th birthday celebration on 22 June 1959, organised largely by Campbell, with representatives from the RGS, the Hakluyt Society, the Society for Nautical Research, the IBG (colleague Gordon East), the Institute of Navigation (Richey) and the Society for the History of Science, a rousing speech paid tribute to her (as yet unfinished) lifetime's work. '*De Juvetate*: You have carried the burden of Atlas manfully [crossed out and replaced by "bravely"] for four decades, though happily enthralled the while by Clio the muse'; the speech went on to list Taylor's work on the history of geographical ideas, discovery and navigation and continued: 'We are confident that in these fields you have ploughed a straight furrow and have shown to youth the way it should go, even though it might well have been of you that the poet George Barker wrote: She is a procession no one can follow after'.[41] Barker's popular poem 'To My Mother' continues 'But be like a little dog following a brass brand' which had an unfortunate connotation for Eila Campbell's career, who was considered by some to be Taylor's lackey rather than successor. Eva Taylor died of heart failure on 5 July 1966, not long after the publication of her last book.

The dinner on the occasion of her 80th birthday was perhaps the most significant honour Taylor was given, when those representing all her interests came together to recognise her remarkable achievements and initiated an appeal for donations to fund an annual lecture in her honour. By June 1960 £600 had been raised and a biennial lecture was decided upon. The list of subscribers in the Taylor papers at the British Museum include numerous male geography colleagues such as Balchin, Freeman, Steel, Steers, Stamp, Wooldridge, Baker and East, but one-quarter of the 200 donors were women, including geographers: Mary Marshall, Alice F.A. Mutton, Hilda Ormsby, Ethel Patterson, Dora Smee, Alice Coleman, Jean Mitchell and Helen Wallis.[42] Clearly the desire to honour Taylor was not determined by gender, but that there was widespread support from so many fellow women geographers and other academics, suggests a degree of solidarity and recognition. Donors were drawn from across the disciplines (e.g. Dame Lillian Penson and Professor Dame Helen Gwynne-Vaughan), within and beyond the academic world and from countries as diverse as France, the USA, Australia, Ghana and Portugal. The biennial lectures continue to the present day and address academic developments in work close to Taylor's own interests.

Analysing Taylor's work

With over 150 publications to her name ranging from school texts and popular articles to research monographs and academic papers in over 20 journals, Taylor was a prolific producer of geographical knowledge (see *TIBG* 1968: 182–6 for a comprehensive list). Always using a lively prose style she could write in many registers, addressing readers in *The Geographical Teacher*, *Geography*, *The Geographical Journal*, *The Geographical Magazine*, *The Scottish Geographical Magazine*, *The Pipe Book* (Dunhill 1924), *History*, *The Mariners' Mirror*, *The Journal of the Institute of Navigation*, *Imago Mundi*, *The Listener* and *The Times*. Her humour can be seen in her letter to the *Sunday Times* concerning the forgotten 'Yeti of the Alps' (1960)[43] and her (1947) post-retirement article on Elizabethans at sea entitled 'They rejoiced in things stark naughty'. It was the same wry humour and candid comment which provided good copy for journalists reporting on the BAAS meetings. It is not possible to do justice to the full range and number of Taylor's works here, but a selection of that range are addressed and evaluated, principally her research monographs.

Like most of her contemporaries in the first generation of professional university geographers, Taylor started writing school texts. As Herbertson's assistant, Taylor's first published work as the statistical appendices to his *Junior Geography* (1905) and *Senior Geography* (1907) and she then co-produced the *Oxford Wall Map of North America* with Herbertson (1909), these were 'wall maps familiar to generations of students' (Freeman 1976: 208). This was followed by a series of co-authored texts, principally with J.F. Unstead (1910–12) whom she met at an Oxford Summer School. Unstead and Taylor texts, designed carefully with curriculum and examination requirements in mind (see 1910/1926: vi), were popular in schools and sold widely. Their global scale *General and Regional Geography for Students*, first published in 1910, was in its ninth edition and 17th print run by February 1926. The book was only discontinued due to falling sales in 1965 after 55 years in print![44] The regional book started from the global scale as the only sensible way to consider phenomena such as seasons and wind systems, then moved to regions largely based on Herbertson's (1905) 'Major natural regions'; Britain as home country receives most attention, followed by the other continents reflecting 'their relative economic and political importance' (see preface), with particular reference to areas belonging to the British Empire. The preface reassured teachers and students that the material would be accessible to non-science students but diagrams such as Figure II (p. 18) illustrating the relationship between latitude and the altitude of the Pole Star belie Taylor's developing research interests. A review of Unstead and Taylor's *Commercial Geography* in *The Geographical Journal*, which criticised some commentary and statistics, resulted in a robust trading of facts and figures between authors and reviewer in the pages of the journal in 1912. School texts at this time often included bibliographies or lists for 'further reading' but few sources for data were referenced. This issue was to follow much of Taylor's writing where she drew from wide-ranging sources but was far from unique to her as it reflected writing practice at the time. There is a relative gap in Taylor's list of new publications (1912–17) when her children were born, but she continued to work on school texts. She wrote numerous entries for encyclopaedias, compendiums and collated atlases, but one of her most influential school texts was *A Sketch-map Geography* (1921). She also wrote *A Business Man's Geography* in 1923, updated to *Production and Trade. A Geographical Survey of all the Countries of the World* (1930a), and should be remembered for her 'notable contribution' to social and economic geography (Skelton 1968) as well as the historical work most associated with her name.

Taylor's first publication on historical work, entitled 'The Earliest Account of Triangulation' appeared in the *SGM* (1927) and was followed by several papers in *The Geographical Journal*, for example, on William Bourne (1928b), Roger Barlow (1929a) and Jean Rotz (1929b). These publications were crucial to Taylor's doctorate as she explained to RGS secretary Hinks: 'I am proposing to present my studies of Tudor geography to the university as a thesis for a higher Degree and as it is a condition that the work must be published, I hope you will find space for Jean Rotz: he is only 3000 words long'.[45] In response to Hinks' query regarding the timing, she wrote again explaining that she did not expect to be able to publish the whole thesis, but thought she must get at least three or four papers in print to represent a 'substantial portion'. In less than a week Hinks wrote to accept the paper. Despite bemoaning to Hinks of the difficulties of publishing work on the boundaries of geography and history, *Tudor Geography, 1485–1583*, was published in 1930 and was followed by *Late Tudor and Early Stuart Geography, 1583–1650* in 1934. These books represented her two key scholarly texts written at Birkbeck and drew together and developed themes from her numerous papers on geographical discovery and related writings and techniques.

Both volumes address not only the journeys undertaken in the periods studied, but also the scientific and mathematical developments which made them possible. In these themes we see much of her life's work in the history of geographical thought and related mathematical and navigational techniques. Both volumes are grounded on extensive bibliographies cataloguing all pertinent English sources, with commentaries on those sources and thematic chapters addressing developments in particular sub-fields and some chapters devoted to the work of individuals. The overview chapters are particularly accessible to general readers and they have been described as 'written with the absolute mastery of language and in the terse style that characterized all of her writings' (de Clerq 2007). *Tudor Geography* was considered 'by far the better of the two volumes' (Johnson 1935: 290), with its study of the developments in British geographical knowledge after 1550, focusing on the pivotal work of John Dee whom she discusses in detail over three chapters, considered to be 'the first adequate appraisal of Dee's place in the history of English science ... Miss Taylor is to be congratulated on restoring John Dee to his rightful place as the central figure in the beginning of modern science in England' (*op. cit.* 291). In addition to Dee's rehabilitation, Crone (1931) also notes that Jean Rotz's *Differential Quadrant* and Roger Barlow's *Geographia* received their first adequate treatment in Taylor's work, justifying Taylor's claim that Barlow was England's first systematic geographer, although Crone wished for more on physical geography and a general chapter to tie individual chapters together. Whilst noting a couple of errors, Johnson (*op. cit.*), of the Huntingdon Library and writing in *Isis*, the journal of the US History of Science Society, concluded the book was 'a work of sound and thorough scholarship, and by far the most important book yet published dealing with Elizabethan science'.

By comparison, *Late Tudor and Early Stuart Geography* is described as 'decidedly inferior' to the first volume (Johnson 1935: 292). Whilst Johnson welcomed the chapters on Hakluyt and Purchas, mathematical geography, navigation and surveying, and economic geography, he thought the book was prepared too hastily and had numerous inaccuracies. His main fault with both books was their bibliography, which at first sight was impressive, but on examination failed to distinguish between extant books available to other researchers and those referred to historically but no longer locatable. Parks writing for *The Geographical Review* described *Tudor Geography*: 'The book is authoritative. Surveying the geographical literature of a century, the geographer explains many matters which had puzzled the mere historian. She has made a special contribution by bringing in the mathematicians. And with much skill in historical research she has frequently gone behind the documents and resolved moot points in maritime history. The work is good history as well as geography' (1931: 686). He also stressed Taylor's achievement in the follow-up volume: 'the compiler's industry was prodigious, and the result is admirable' but like Johnson finds fault in the structure of the bibliography which dominated the text, notably the fact that manuscript items are 'left unlocated and thus in a manner re-interred' (1934: 693). However, J.N.L. Baker, who lectured on the history of exploration at the Oxford University School of Geography, wrote 'Professor Taylor is to be warmly congratulated on the production of a book which will be read with attention and profit by all geographers' (1934: 171). He went on to argue that the second book 'is likely to appeal more widely, and be more generally useful, because it is both better written and deals with the subject in a more systematic and geographical manner' (172). Like other reviewers Baker (reluctantly) bemoans the shortcomings in the bibliography which was so central to the work, but he also acknowledges the context of the production of the book: 'But Professor Taylor has reason to complain that the work was produced at considerable financial cost to herself, and that there is a limit to the amount of money which an academic geographer can afford to spend on a non-popular work of this kind. By extending the sale of this admirable volume, geographers would not only repay her, but would indirectly make the publication of similar works less difficult in the future' (ibid.).

In addition to her papers and books, Taylor also contributed two chapters on 'Leland's England' and 'Cambden's England' to H.C. Darby's (1936) *An Historical Geography of England before A.D. 1800*. Darby had studied with Taylor whilst a student at London University and was to become the UK's leading historical geographer in the post-war period. Darby's (1960) review of historical geography 20 years after the publication of this collection was only to a make a fleeting reference to one of Taylor's works: the sub-discipline was moving on to a more analytical phase and increasingly concerned with either Domesday or post-1800 material. Taylor had a 'relative lull' in her publications output during the war years (de Clerq 2007), writing mostly shorter pieces, but her output was to be prolific in her retirement years. She described her later writing as her 'hobby'[46] but it was an all-consuming occupation.

Between the ages of 75 and 87 she published three volumes considered to be 'probably her three greatest books' (de Clerq 2007): *The Mathematical Practitioners of Tudor and Stuart England* (1954); *The Haven-Finding Art: a History of Navigation from Odysseus to Captain Cook* (1956); and *The Mathematical Practitioners of Hanoverian England* (1966). *The Mathematical Practitioners of Tudor and Stuart England* was organised in three parts: a narrative account; a biographical list of practitioners and their achievements; and an annotated bibliography. The book focused on the application of mathematics to navigation and built on Taylor's earlier related work. Crone (1955: 105) welcomed this as a 'culmination of a quarter century's research' and considered that her study of over 600 practitioners and texts would mean 'this volume will remain unchallenged as an authority' (ibid.). Whilst Johnson (1957) credits Taylor's earlier work as having 'marked a significant change in the approach of modern scholarship to the study of the history of science in the Renaissance' he continued to be frustrated by discrepancies in Taylor's account and her failure to indicate the whereabouts of some of the sources cited. Despite this, the book was recognised universally as pioneering work which became an important source for specialists and general readers (Bedini 1967). More recently, *The Mathematical Practitioners of Tudor and Stuart England* has been described as 'The classic text ... in which the historical terrain of mathematical practice was first identified and its coherence demonstrated at length. ... Many of the assembled characters would previously have been considered too obscure or insignificant to merit serious study. But when gathered together, Taylor was able to reconstruct a didactic, urban and vernacular tradition, in which instrument makers and ordinary textbook writers had a place alongside better known names from the history of science' (Johnson 1957: 378). As present-day historians of the discipline we must acknowledge the ways in which she placed her subject in its wider context, looked for hidden histories and included many minor figures.

Taylor continued her work even as her health failed, writing to Eila Campbell in 1957: 'My lecture to hist. of Navg'n. went well and they are anxious to publish the "Math'l Practitioners" if monetary assistance is forthcoming. I wrote to East about dates etc. if applying to Birkbeck and the University – I hope the existence of the Athlone Press doesn't mean there are no longer any pub'n. grants' [Taylor's abbreviations].[47] In her last years she relied heavily upon addressing queries by correspondence and Eila Campbell's efforts to track sources and later to transcribe material, especially after Taylor experienced a stroke which damaged her eye muscles so she couldn't read or write. Campbell explained to a journal editor: 'I now act as her scribe ... [but] ... She is still mentally alert and is drafting a new paper on a Portuguese map of about 1503. She dictates her thoughts to me. It is, of course, a slow business.'[48]

The Mathematical Practitioners of Hanoverian England, published shortly before Taylor died, was partner volume to the 1954 book and replicated its format. The book takes a very contextual approach, setting over 2000 short biographies and developments within the wider setting of both geographical and scientific events (Calvert 1966). Calvert, writing for *The Geographical Journal* obviously relished Taylor's lively prose: 'the graphic style of the narrative gives the impression that Professor Taylor lived amongst the people she writes about', recognising the degree of study and immersion in the literature required to achieve such a book. Whilst feeling the book had too many errors and fell between the needs of students of the history of geography and those seeking a detailed reference text, he acknowledged that for one person to complete the text was an 'almost superhuman task', not least in failing health. The book was produced with financial assistance from the Institute of Navigation, the Society of Authors, Birkbeck College, University of London, and the RGS (Calvert 1966: 521) and was welcomed by White (1967: 271) writing for *The Mathematical Gazette* and Bedini (1967) writing for *Isis*, who considered it to be a major achievement despite its referencing issues. Inevitably these last reviews, written posthumously, were coloured by Taylor's death and sought to acknowledge her life's work. Indeed, Bedini included a brief obituary at the end of his review and concluded that the combined volumes on *Mathematical Practitioners* 'will become and remain a permanent and useful monument to the author' (Bedini 1967: 121).

Published between the two mathematical books, *The Haven-Finding Art* (1956) is Taylor's account of the history of navigation from the work of Homer through to the explorations of Captain Cook. In the preface Taylor admits to not being a sailor, but also explains this 'outsider' view is precisely what had roused her 'passionate curiosity to learn just how a ship is steered with such confidence across the unpathed waters of the sea' (1956: xii). As ever, Taylor drew on the expertise of a wide range of specialists as sources of information and clarification of technical puzzles; in this case particular credit was given to Michael Richey, Executive Secretary of the Institute of Navigation, who she described as 'inspirer and editor of the book'. Others included Helen Wallis and R.A. Skelton at the British Library, Gerald Crone at the RGS, and representatives of the National Institute of Oceanography, the National Maritime Museum and HM Nautical Office. The book also included a foreword by Commodore K. St. B. Collins, the hydrographer of the Navy, who noted 'In these pages you will find disproved once and for all that persistent myth that the first sailors navigated by "hugging the shore".' (1955: x). Writing for *The Geographical Journal* Naish wrote: '"The haven-finding art" is a model of packed information, narrated clearly and amusingly, with just the right amount of illustration and reference to the original sources. This is a book with wider appeal than had her "Mathematical Practitioners" but is no less important ... Professor Taylor has told her story freshly and lucidly; indeed magnificently. Geographers whether they be seamen or students, will find her book both a delight and a necessity; for she has written a story which both tickles the fancy and feeds the understanding' (1957: 86). Morison, writing in *Isis*, regarded the book as 'an excellent short history of navigation' (Morison 1958: 352). Whilst pointing out that main-braces were confused with sheets in one diagram and regretting the absence of early American navigators and relative neglect of Magellan, Morison praises Taylor's work: 'her chapters on the navigation practices of the early Irish and Norsemen, of the development of the wind rose and discovery of the compass, and the invention of the chronometer, are the best that this reviewer has read outside specialized monographs' (ibid.: 353). The volume was welcomed by author and sailor Arthur Ransome.[49] Taylor's publications for the Hakluyt Society were also appreciated by the nautical community, for example, world-class sailor Francis Chichester, who when reading her biography of William Bourne was compelled to put pen to paper: '... I must write and tell you how much I am enjoying it. Not only is all the technical navigational description so lucid and interesting but you make the reader feel that he is right on the spot at the time, taking part in what is going on down the Thames'.[50]

Some of Taylor's books which were criticised by reviewers were later described as classics, albeit flawed, but core texts nonetheless (de Clerq 2007). *The Mathematical Practitioners of Tudor and Stuart England* was reprinted three times in the 1960s and both it and its partner volume were reproduced in 1989 by American publishers (de Clerq 2007), an updated version of the bibliography of the Hanoverian volume was produced in 1980. Eva Taylor made the history of the pre-nineteenth-century sciences and instruments her own field; the history of navigation was a surprisingly recent topic and 'many people would give Professor Taylor the credit of inventing this course of study, if she did not carry her learning so lightly' (Naish 1957: 85). The history of pre-nineteenth-century exploration has become unfashionable in post-1970s' British geography and within the discipline much of Taylor's work and reputation has become obscured despite the Taylor Lectures. Ironically this fall from celebrated disciplinary figure into relative obscurity was assisted by Campbell's failure to complete a piece on Taylor for the Geographers' Biobibliographical Series commissioned by Walter Freeman, as well as the longer biography she intended to write;[51] but we do have Campbell to thank for collating Taylor's papers and their eventual deposition in the British Library. However, it is hard to understand why Taylor's work is reduced to fleeting footnotes and passing comments in Livingstone's (1992) chapter on exploration and navigation in the 'Age of Reconnaissance', considering the acknowledged impact her work had on the study of British exploration, navigation and mathematics in this period.

As the first woman professor of geography in Britain Taylor met the geographical establishment head on – and used it for her own purposes. In many ways through the subject of exploration and navigation Taylor married the interests of core constituencies within the RGS (namely the very masculine Admiralty and international trade interest groups), with cutting-edge interdisciplinary historical research which caused the Society to describe her work as 'brilliant' (*The Geographical Journal* 1959: 279) and claim her as one of their own, even if she was often a thorn in its side. In many ways Taylor's work is revisited today by historians of science and navigation rather than geographers – would that hers was 'a household name' within the history of geography and her reputation defined by those *within* as well as beyond the geographical community.

Conclusion

Women students played an important role in the demand for geography courses in the 1900s. It is also significant that the first women lecturers in geography in higher education can be traced to the foundation of the first geography departments and the first formal qualifications in geography. This can be seen as representing something of a strategic alliance between women gaining access to higher education (including teaching posts) and the needs of an emerging discipline within the university sector. Although women were able to gain entry to new geography departments, there is evidence to suggest that attaining full lectureships and subsequent promotion could be a difficult process, reflecting the notion of a 'stone floor' holding women down rather than pushing against a 'glass ceiling'. However, accreditation played an important part in Ormsby's geographical career, even if only achieved in her forties. Taylor gained popular and academic recognition, no doubt assisted by Unstead's support at Birkbeck in the first instance and by dint of her intellect and publications in due course. Once open to women fellows, the RGS played an important role as a publishing outlet for women, but crucial gatekeepers in the Society could still limit women's recognition and power within, whilst simultaneously co-opting them into the institutional hierarchy. The women in the 'first generation' of university geographers were recognised as important producers of geographical knowledge in their fields, but most have become obscured in disciplinary histories as their subjects (especially British regional studies) have become unfashionable; even Taylor and Ormsby had only a few postgraduate students to continue their disciplinary legacy.

Chapter Seven

Fieldwork and War Work: Interwar University Geographers

Introduction

Whilst the precedent had been set by the 'first generation' of academic women geographers, many of the 'second generation' were pioneers in their own departments. One department where there was continuity of women appointees in the early to mid-twentieth century was at Bedford College and this merits particular attention, as does the Edinburgh University geography department. However, it is equally important to understand the culture of those departments where there was only one woman appointed at any given time: was this a positive or negative experience for the women in question? How did male-dominated departments respond to a sole female colleague?

The 'second generation' of geographers discussed in this chapter are those women who had the opportunity to study geography as a university degree after 1917, and principally applies to those studying and beginning their university careers in the interwar period, 1919 to 1938.[1] The chapter divisions are largely organisational, but there are clear watershed periods in which socioeconomic, cultural and political shifts took place including economic depression in the 1920s, the universal franchise in 1928 and the Second World War, 1939–45. Within the university sector the foundation of redbrick university colleges in the period 1920–40 was also significant in increasing both university places and appointments. The women whose work is discussed in this chapter include those who played prominent national roles within the discipline and those whose influence was more localised: Alice Garnett, Catherine Snodgrass, Jean Mitchell, Harriet Steers, Margaret Shackleton, Dora Smee, Eunice Timberlake, Florence Miller, Katy Boswell, Alice Mutton, Dorothy Sylvester and Eila Campbell. The number reflects the growth in women's appointments to university posts, as well as exemplifying the breadth of the geographical work they undertook. Within that work there are three broad themes: the commitment to field-work; clusters of women specialising in physical and historical geography; and the contribution of many of the women discussed to geographical war work. This also applies to Ormsy and Taylor (Chapter Six) and King, MacIver, Cole and Marshall in Chapter Eight, which deals with those undertaking war service and/or being appointed to university posts during or shortly after 1939–45. War service is a recurring theme in British wom-en's geographical work for those completing degrees and working over the 30-year period 1914–45 (Maddrell 2008), so it inevitably spreads over the different chapters addressing those periods. Whilst the notion of permanent change to women's socioeconomic status as a result of participation in war work seems generally to be unsupported (especially as a high proportion of women were made redundant at the end of the war), there is evidence that many women viewed their experiences as subjectively significant and this theme will be returned to in the conclusion.[2]

Alice Garnett

Alice Garnett was born on 15 May 1903 in Wandsworth, daughter of Alice (née Brooks) and George Garnett. At the time of her birth her father was a publisher's clerk and later a company secretary. She was one of three chil-dren (sister Olive and brother Leslie). Garnett read geography at University College London (UCL) studying with Professor L.W. Lyde. She graduated in 1923 and completed a teacher training diploma at the University of London. By the autumn of 1924 was she appointed to an assistant lecture-ship in the department of geography at Sheffield University, where she would spend the rest of her career, but her teacher training would inform her work and interests throughout her working life (Maddrell 2004f).

Initially appointed as assistant to R.N. Rudmose Brown, her arrival ena-bled an Honours School in Geography to be established; the two lecturers covered the whole syllabus, which necessitated a minimum of 20 hours of teaching a week (Garnett 1983). It was only with further departmental appointments in the 1930s that Garnett was able to give more time to pub-lication and research. Continuing to lecture full time, Garnett gained her PhD from the University of London in 1937 and was subsequently pro-moted to senior lecturer after the war. At the age of 45 Garnett married steelworks manager Colin Crow in 1948, but continued to be known pro-fessionally by her maiden name.[3] Amongst her many commitments, Garnett, as one of the very few women lecturers at Sheffield, made time to support women students in the university. She initiated a Women's Day Students'

Figure 7.1 Alice Garnett c. 1968.
Source: Courtesy of the University of Sheffield geography department

Union (1935–40) prior to the existence of a wider Students' Union; became chair of the Committee of Tutors for Non-resident Women Students (1947–58), as well as being senior tutor for women students.[4] During her career she also took on the roles of sub-warden of a hall of residence, president of the Non-professorial Staff Association, president of Staff House, and organised numerous social events for 'home' students (Potter 1968; Ellis & Hunt 1989). Throughout the war Garnett taught the intense condensed 'war degree' in geography at Sheffield and worked part time on the Admiralty's Naval Intelligence Handbooks prepared at Cambridge, under the direction of H.C. Darby, where she wrote the bulk of two volumes on Yugoslavia 1941–5 (Maddrell 2004f, 2008).

Early in her career Garnett joined the RGS (1925), having been proposed by Rudmose Brown and A.M. Dell,[5] and she joined the GA soon after in 1926. She had been part of discussions at the IGU meeting in Cambridge in 1928 concerning a new society for geography academics and was co-opted onto the working committee which led to the foundation of the IBG in 1933 (Garnett 1987; Wise 1990). Her gender does not seem to have been a factor, but her geographical location was as she helped counteract suspicion that the proposed group was London dominated (ibid.). Between 1935 and 1940, Garnett was a member of the organising committee of the Student Group of the Le Play Society, which provided opportunities for geographical fieldwork at home and abroad, and she became vice-president of the renamed Geographical Field Group in 1948.[6] She was a member of the Royal Society's British National Committee for Geography 1947–53 and 1958–64; and was nominated and funded to represent the group at the IGU meetings in Lisbon, 1949 and Washington, 1952.[7] Having been elected to the Geographical Association Council 1930–40, Garnett became honorary secretary in 1947, taking over from H.J. Fleure, who had held the post for nearly 30 years. From experiencing both the best and worst of geography teaching at school (Garnett 1969), and inspired by her teacher training, Garnett was committed to supporting the discipline in schools. Early on she secured accommodation for the GA in Sheffield; these offices were located on Garnett's route home and she would complete a day's work at the geography department, then go on to the GA.[8] Garnett was an important link between the GA, IBG and RGS and filled major posts in each society. The GA thrived under her leadership, membership doubled, activities diversified, and the Association's national standing also increased (Waters 1991). Membership grew in post-war years despite increased subscription rates, not least because Garnett wrote to lapsing members, stressing the value of an unbroken run of the journal and the tax deductions available to members, a strategy described as 'one of the many quotable indications of our Honorary Secretary's effective combination of scientific insight and business enterprise' (Fleure & Fleure 1968).[9] With her team, including colleague David Linton as honorary editor and the assistance of Edith Coulthard in the early 1950s, Garnett reorganised committees, got the journal on a near-self-funding footing and stressed the importance of the Association's work in primary schools as well as the need to constantly update geographical information and methods. She also galvanised petitions on behalf of geography in the new universities as the higher education sector underwent expansion in the 1960s.[10] On her retirement as honorary secretary of the GA she was offered a 'tribute of gratitude and admiration and affection' (ibid.: 95). As Ellis and Hunt (1989: 275) commented after her death: 'She always regarded the GA as one great family and her passing marks the end of an era of parental care under which the Association flourished'. She became president of the GA in 1968 and chaired its Council from 1970 to 1973, the first woman to hold any of these offices. Likewise, Garnett, having been a founder-member of the IBG, was a member of Council 1956–8 and became its first female president a decade later in 1966. At the RGS she served on its Education Committee (1948–58 and 1961–3) and Research Committee (1953); she also sat on Council 1952–5, 1961–3 and ultimately became vice-president 1969–71.

Garnett was appointed professor of physical geography in 1962, and was only the second woman in Britain to hold a chair in geography. She took on the role of head of department in 1964 and became dean of faculty in 1967. During Garnett's career there were no other women lecturers in geography at Sheffield until 1968 when Jennifer Bray was appointed as assistant lecturer and promoted to lecturer in 1969; Margaret Wilkes was also a research assistant in the department during the 1960s (see Chapter Nine). When Garnett retired in 1968 after 44 years of teaching and research in the geography department, two prizes for geography students at Sheffield were funded by subscription in her honour. In 1970 the new geography building, which had been started while she was head of department, was completed, 'its acquisition was yet another achievement of a forward-looking and very persuasive advocate of geography' (Waters 1991: 117). This ability to stay up to date and to look to the future in practical and intellectual terms characterised much of Garnett's varied work. In subject terms this can be seen in her openness to appropriate new methods such as quantitative techniques and remote sensing (e.g. Garnett, Pead & Finch 1976) and in her call to teachers for the critical application of quantitative methods (Garnett 1969: 390). In retirement she nursed her terminally ill husband at their home in Ferndown, Dorset, but also kept up her research through the Air Pollution Unit she had set up in Sheffield, where she continued to work for nine years. In 1977, her contribution to geography, the university and the city were recognised when she was awarded the honorary DSc at Sheffield.

Promotion to a Chair and academic honours had come late in her career, 'sed longo intervallo' (Potter 1968), but Garnett was gracious in her acceptance of the acknowledgement and felt herself privileged. She was a person of indomitable energy, incisive argument, good cheer, organisation and willingness to help others and in all of this was influenced by her humanist ethics (Ellis & Hunt 1989). She achieved an enormous amount during her career, principally as a result of being goal oriented, disciplined and committed, as well as having lots of energy and an encouraging partner (Wise 1990). For David Linton (1968: 67) it was Garnett's personal characteristics which were the root of her success: 'First, a good-humoured cheerfulness that met everyone and everything with a characteristic smile. Second, an opportunism in the definition of means and ends allied to an astonishing resilience that was proof against discouragement. Third, a never failing supply of energy; and fourth, an abiding interest in students and loyalty to her subject'. She also knew how to make friends and supporters for the discipline (ibid.), something she perhaps learned from Fleure (see Garnett 1987). Some did find her intimidating (Ellis & Hunt 1989), but she certainly made lots of friends and was held in wide affection. Although she disliked the feminised and possibly spinsterish appellation 'Auntie Alice' given to her by some in the Sheffield department, she was happy for everyone to address her as Alice[11] and was remembered for organising facilities and parties for home students in the years predating the Students' Union (Potter 1968). Two pieces reflecting on her career made reference to her gender, both of which identify her success in male-dominated academia whilst fulfilling roles and/or retaining feminine characteristics. The first noted 'She has done for the Department a man's job and done it well, and for much of the time has done for the university a Woman's work also' (Linton 1968: 67), a view which, although complimentary, perpetuates the representation of a gendered division of labour within the university in which academic work is 'men's work' and pastoral and administrative work is 'women's work'. The second recorded 'In a male-dominated academic world Alice was rightly accepted as an equal, combining in a surprising way logical and intuitive faculties which enabled her to present an intellectual argument, or a spirited defence of a matter of principle, with devastating clarity, while respecting the feelings of others and thereby retaining their friendship' (Ellis & Hunt 1989: 274–5). This implies that being 'accepted as an equal' was something of note and a status which a woman had to earn. It is also revealing to read the note of surprise at her combination of 'logical and intuitive faculties'. Alice Garnett died at Poole General Hospital in Dorset on 5 March 1989.

Publications

Garnett's first paper was an historical geography of 'The Capitals of Morocco' which appeared in the *SGM* (1928). As Garnett (1983) herself noted, many of the early British geographers wrote textbooks, compelled in part by practical educational needs and part by personal financial needs;[12] her own first book, *The Geographical Interpretation of Topographical Maps* (1930), was based on her practical classes (published in Harrap's New Geographical Series edited by Rudmose Brown). The book demonstrated the 'originality and thought that lay behind her teaching' (Wise 1990: 113); it was innovative in its approach and highly valued by students and lecturers for more than two decades (Waters 1991). Reviewing the book in 1931 Alan Ogilvie (A.G.O.) praised Garnett's clear style and the ways in which she engaged the reader, encouraging analysis and interpretation. He thought Section III on interpreting human relationship to the topography of an area to be of particular novelty and demonstrated 'that serious research has contributed to the general excellence of the text'. He did, however, query Garnett's emphasis on geology as the only feature 'open to serious criticism' and bemoaned the loss of the full colour original maps in the inferior reproductions: 'perhaps it proved impossible to obtain beauty at the price' (A.G.O. 1931: 188–9). By 1953 the book was in its third edition and 'already known to a wide public as a model of its kind' (H.R.W. 1953: 491–2). Wilkinson noted that for Garnett map work was the next best thing to fieldwork and that she 'is always conscious that the mind must ever "read and not spell the map" and that "facts are not only read from it but into it"' (H.R.W., 1953).

Garnett's research work was to follow this model, being renowned for its meticulous representation of data and analysis on map, table and diagram. Her (1937) monograph 'Insolation and relief: their bearing on the human geography of the Alpine regions' was the fifth publication by the new Institute of British Geographers (IBG). It was based on her PhD research and established her academic reputation (Wise 1990): '... both Miss Garnett and the Institute may be congratulated upon it. The monograph has obviously meant a vast amount of detailed and laborious work, both carto-graphical and climatic, but the labour involved has resulted in some very useful new and precise information' (J.A.S. 1938: 63). Previously the influ-ence of insolation in mountain areas was often treated in a very generalised way in geographical literature and her work was welcomed for its insight and transferable methodology: '*Adret* and *Ubac* now come to have a mean-ing which can be measured and not left to rough approximation' (ibid.). Alfred Steers stressed that the detailed exposition required careful reading, but that any effort would be repaid, concluding it was an 'instructive and careful piece of work of real value to the geographer and also the agricul-tural economist' (ibid.: 64). Subsequently Garnett's IBG piece was described by David Linton as 'one of the most scholarly and important monographs the Institute of British Geographers ever published' (cited by Waters 1991).

Publications on sunlight and relief in high latitude settlements (1939) and ice crystals (1940) followed, but her writing during the war years was largely confined to the Naval Intelligence Handbooks. A major paper on the geographical factors of air pollution and two shorter papers appeared in 1945 (see Garnett 1945a, 1945b, 1945d). After the war she continued work on the local Sheffield climate (see Garnett 1956), for which she gained international recognition and received the 1956 Murchison award from the RGS (Waters 1991). Garnett's presidential address to the IBG focused on this work: 'Some climatological problems in urban geography with reference to air pollution' (1967) and she published several single and co-authored papers on climate-related material in the 1960s and 1970s. She continued this work on retirement, having gained external funding to establish an Air Pollution Unit and is credited with having 'anticipated by more than quarter of a century the preoccupations of today's environmental custodians' as well as influencing planning policy in Sheffield (Waters 1991: 117). Throughout her career, Garnett received funding from a number of sources, including the universities of Sheffield and London, the Leverhulme Trust, the Air Ministry and Section E of the BAAS.[13]

Climate and pollution were the recurring themes in Garnett's research but her work was grounded in an appreciation of the interconnection between physical and human geography developed in her early years with Lyde, and she is credited with avoiding simple geographical determinism, taking a more possibilist view of the relation between the two (Wise 1990). As she wrote on the Pennines in *Great Britain*:

> Few regions of comparable size in Great Britain can portray better than this the changing values that the environment may present in relation to man's changing technical skills and social and traditional needs, and, from Pennines to Ouse, the diverse physical landscapes reflect very clearly a changing but dominating pattern of the works of his hand (1962b: 353).

The list of Garnett's publications compiled by Wise (1990) includes her presidential addresses (1967, 1969), obituaries and appreciations for colleagues, overviews of the discipline, reflections on teaching geography, contributions to historical retrospectives, and publications outside the discipline. All of these combine to give a sense of Garnett's central place in the production, promotion, organisation and dissemination of geographical knowledge nationally and internationally. In the light of her long and much-valued research and institutional roles it is quite remarkable that Garnett should be reduced to three references in Johnston and Williams' (eds) (2003) *A Century of British Geography* – and not mentioned at all in other major books on the history of geographical thought (e.g. Gregory (1985)).

Catherine Snodgrass

Catherine Snodgrass was appointed to the University of Edinburgh geography department in the 1930s and after this time, the department consistently employed women until the 1960s, with as many as three in a staff of six in post-war years. Catherine Park Snodgrass was born on a farm at Cockpen, Bonnyrigg, Midlothian in July 1917, daughter of Annie Scott (née McHarrie) and Peter Lennox Snodgrass, both of whom haled from farming families. She attended Esbank Girls' School before going on to George Watson's Ladies' College and St George's School for Girls in Edinburgh. She won several class medals and graduated from Edinburgh University with an MA in mathematics and natural philosophy in 1924. After completing a one-year teacher training qualification at St George's Training College, she taught mathematics and geography at St Oran's School in Edinburgh part time while studying the First Ordinary Geography course. This was followed by the two-year Diploma in Geography, studying under Professor Alan Ogilvie.[14] She was awarded the class medal and RSGS Silver medal for university students in 1925, followed by a Carnegie Research Scholarship in 1929–31 and again in 1935. Snodgrass completed her PhD thesis on 'The influence of physical environment on agricultural practice in Scotland' in

1931, and was appointed as temporary assistant lecturer in 1931–4 and assistant lecturer in 1936, later being promoted to lecturer in 1945. Snodgrass continued in this post at Edinburgh until her early retirement in 1957 (MacGreggor 1975; Withers 2005, having spent two years on secondment (1947–9) to work on the *Third Statistical Account of Scotland* when she was replaced by agricultural specialist C.J. Robertson.[15] After leaving the geography department she became an assistant lecturer in natural philosophy.[16] Snodgrass had to give up her editor's post at *SGM* in 1967 due to ill health and she died on 13 December 1974.

As a university lecturer, Snodgrass, known as 'Snoddy', is remembered as 'a thorough, exacting and most devoted teacher', who lectured mostly on regional geography in which she combined 'a strict insistence on a sound factual basis with a clear recognition of the need for social problems to be tackled with care, sympathy and understanding' (MacGreggor 1975: 129). She may also have taught political geography during the war years, and according to colleague Swanzie Agnew, 'She was good at that sort of thing'. Archive papers suggest that students wrote essays for her on 'Where and how big are the burghs of Scotland?', 'Scotland's agricultural future' and 'Where do the Scottish people live?'[17] Agnew remembered Snodgrass for her work on agricultural geography and for being 'forthright, very Scottish'. These characteristics shine through in her review of the *Land Utilisation Survey Report on the Lothians*, written by P.M. Scola, and give some idea of what she may have been like as a teacher. She acknowledges the strengths of the report, but also holds no punches in where it is weak, e.g. its inadequate reference to altitude and slope and its dated conclusions about the demise of agriculture in the area: 'this section is disappointing, old-fashioned in outlook, and cynical – the writer still assessing economic activities primarily by financial and not by human values' (1946: 32). She reported concerns that this report was compiled at a distance in the libraries of Cambridge and acknowledged the evidence of fieldwork, but she could not help adding 'no doubt certain misconceptions would have been avoided by a worker more thoroughly steeped in the past and present of the region' (op. cit.: 28).

Snodgrass' 1931 doctoral research drew on her family background in farming to which she was deeply attached (MacGreggor 1975). Her thesis, which focused on the influence of the physical environment on Scottish agriculture, is considered the 'outstanding success' of this period of her career, as well as vindicating her risk-taking move to the new discipline of geography after her academic success in mathematics. Her obituarist's comments are also telling about women's involvement in research, some 30 years after Newbigin attained her doctorate. 'Not only did Dr. Snodgrass demonstrate that Geography could be studied at a post-graduate intellectual level, but she did so at a time when it was unusual for women to be involved in such work at all' (MacGreggor 1975: 128). Snodgrass' research reflected the causal approach to the subject promoted by Ogilvie and she established agriculture as a long-standing research interest in the Edinburgh department (ibid.).

Withers (2005 characterises Snodgrass' work as falling into three main phases and categories: land use planning; the *Statistical Account of Scotland*; and geographical and political issues relating to Scottish nationalism. Her work on land use was concentrated in the late 1930s and 1940s, reflecting both the widespread involvement of British geographers (especially human geographers) with the Land Utilisation Survey (LUS) headed by Dudley Stamp and the utility of this and related information in wartime Britain. Snodgrass also served on the wartime Scottish Reconstruction Committee. However, her work in the 1940s was more varied and papers published in the *SGM* reflect not only land utilisation but much broader concerns with economic and related geographies. Snodgrass' first paper appeared in the *SGM* in 1935; it was related to her doctoral work on agriculture and was followed by a comparative study of the Scottish agricultural population with those in England and Europe (*GJ* 1941). In the latter she described her topic as of 'intrinsic interest and practical value' and of crucial importance to national plans to redistribute the population in the face of concern regarding dangerous levels of urbanisation. She also published 'A map of economic regions in Scotland' (1943) and two papers which analysed the geographical distribution of employment and recent population changes in Scotland (1944a,b).[18] MacGreggor (1975: 128) identifies several prescient themes in Snodgrass' work at this time: 'In her publications of the 1940s it is revealing to find her appealing for action to check the flow of Scottish emigration 25 years before it reached the excessive proportions of the 1960s, drawing attention to the merits of decanting underprivileged city populations to selected outlying burghs long before the term "overspill" had been heard, and that she was deeply concerned for the conservation of the Scottish heritage when few would recognise it was in any way at risk'. Drawing on pre-war statistics, Snodgrass identified the vulnerability of the lower Clyde area because of the dependence of most forms of employment on ship-building and related industries which had been badly hit during the Depression, causing her to recommend industrial diversification as a matter of urgency. She makes similar recommendations for coal-mining areas where over 30 percent of male employment was dependent on the mines. The other areas of concern were those dependent upon textiles (Dundee and the Tweed Basin), firstly because of that singular dependence and secondly because of their high rates of female employment. Her article reflects pre- and wartime concern for planning and is interesting from a historiographical point of view because it begins with employment by gender as one of its main themes, addressing this from the outset: 'It is generally accepted by qualified town and country planners that regions and

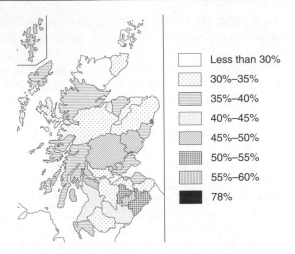

Legend:
- Less than 30%
- 30%–35%
- 35%–40%
- 40%–45%
- 45%–50%
- 50%–55%
- 55%–60%
- 78%

Figure 7.2   Scotland: number of women workers as a percentage of men workers.
Source: Snodgrass 1944:71, Scottish Geographical Magazine

large towns should provide a variety of occupations for their inhabitants and a proper balance of employment for men and women and for older and younger people' (Snodgrass 1944b: 69). Snodgrass, conscious that it was unusual to focus on female employment, identified it as crucial to understanding economic and social well-being. She also drew attention to the lack of recognition given to unpaid home workers. However, whilst supporting women's rights to work she clearly saw the employment of married women as an economically necessary evil where there is insufficient male employment and something which should be remedied by effective regional planning. Using census data from 1921 and 1931 Snodgrass compiled a map of Scotland showing the percentage of women workers in relation to men (see Figure 7.2), summarising her views as follow:

> Fig. 2 shows the number of women workers in each county and city as a percentage of the men workers. (It should be noted that unsalaried home workers are classed as unoccupied.) This very important factor in contributing to human and social well-being has had little, if any, regard paid to it in the past. If we are to build healthy and emotionally satisfied, and socially wholesome communities, a proper proportion of work should be available for women, and for young people of both sexes in every district and large town. There should not be large numbers of single women living in lodgings and hostels (except for educational purposes), nor married women working out of the home unless they have a strong personal urge to do so, nor numbers of young people in the vulnerable and impressionable teens forced to work hundreds of miles from their homes, and other communities left with an undue proportion of older people (Snodgrass 1944b: 69).

She goes on to make a comparison between the relatively high level of female employment in Scotland at 43 percent (and England 42 percent) with the lower levels found in the USA, Canada and Holland (28 percent, 20 percent and 32 percent, respectively), implicitly indicating these as targets Scotland should aspire to. Whilst feminists today may not agree with her social interpretations, the priority Snodgrass gave to highlighting and analysing female employment is to be credited. Her handwritten notes also contain an interesting paragraph not included in the final paper: 'The problem of greater sharing of responsibility in economic life and in affairs generally is beyond the scope of the present article but is one of the most urgent and vital in Scotland and Western civilisation generally'.[19] We can only speculate whether this was excised by the editor or by her own hand and what 'solutions' she might have counselled.

Snodgrass' (1944a) paper on population changes in Scotland was packed with detail, including her characteristic choropleth maps and tables of statistics. The paper followed on from Chisholm's 1911 study of Scottish population and highlights the key problems in terms of emigration from Scotland (especially in the 1920s) and migration to overcrowded industrial areas within Scotland. She asked 'What are we going to do about the vital question of loss by migration?' and asserted of the trend to conurbation formation: 'It is our duty, both as geographers and citizens, to ask ourselves whether this is necessary and whether it is in the real interests of the country or of the people directly concerned' (1944a: 37). She suggested Scotland should move, in Geddes' terms, from the 'Paleotechnic' to the 'Neotechnic' age, using the Swedish model of dispersed industrial towns built on existing nodal communities. She doesn't use the term 'new town' but this is clearly what she was suggesting. Subsequently she wrote a chapter on Scottish conurbations in T.W. Freeman's (1959) *The Conurbations of Great Britain*. Gender is not a dominant theme here but Snodgrass argues that male migration not only threatens the rate of biological reproduction, but also 'On the human [side] it means, especially for women, loneliness and the frustration which results from the limitation of expression of fundamental human instincts, and also a failure to give the family that place in society which doctors, psychologists, and sociologists believe, and the experience of centuries suggests, is necessary for a healthy and abiding society' (1944a: 36).

Deeply involved with the *Third Statistical Account of Scotland*, Snodgrass published a report on East Lothian in 1953 (also Peeblesshire and Selkirkshire). As a trained mathematician, statistics were Snodgrass' métier, for her quantitative techniques were an expected part of geographical enquiry and MacGreggor (1975: 128) notes that 'the preoccupation of the younger generation of geographers with techniques of quantification cut little ice with one who had long regarded numerical analysis as a normal feature of geographical investigation'. The *Third Statistical Account* followed earlier national surveys c. 1790–5 and 1835–40, each area study representing a major piece of research.

Finally, Snodgrass, increasingly politicised, became involved in issues around Scottish nationalism and with the work of the Scottish National Party (SNP). She addressed the SNP in 1945 on 'Scotland's great opportunity' and wrote numerous articles in the *Scots Independent* in the 1940s to 1970s including 'Scotland and the modern world: a plea for freedom, self-government and full participation' (1959) and 'A choice for Scots. Runaway growth or quality in life?' (1974). She was co-author of Scottish Secretariat publication 'A constitution for Scotland' (1964) with M.F. Somerville and J.D. MacLean. Snodgrass was also a member of numerous societies, all reflecting her interests in the university, geography, Scottish or women's interests, e.g. the RSGS and GA, the Highland Fund Association, The National Trust for Scotland, the Scottish Committee for War on Want, the Edinburgh Association of University Women, the Edinburgh University Women's Union and the Women's International League for Peace and Freedom.[20] Whether discussing the rights of young people, women or the Scottish nation, her concern was for enabling choice and facilitating full and democratic participation in society by all of Scotland's citizens.

The reception of Snodgrass' work

A review of recent literature relevant to planning, published in the *Geographical Review* (1946) included Snodgrass' map of economic regions which was credited for its clear reduction of larger scale maps used by planners and its 'reasonably reliable estimates' of agricultural workforce data. A.E. Moodie of Northwestern University reviewed *Conurbations …* (1959) in *Economic Geography* without mentioning Snodgrass by name. Naturally he would focus on Freeman's writing of the majority of the text, but beyond noting the mere 50 pages devoted to the chapter on Scottish conurbations, he does not engage with the content of that chapter and only refers to Freeman's 'co-author' and 'associate author', which served to erase Snodgrass' significant contribution to the volume. Praise for the book was forthcoming, 'There can be no doubt that Mr. Freeman and his associate author have made an outstanding contribution to the literature on the geography of Britain', but one has to go back to the book's subtitle to know who that co-author was. E.E. Evans, writing in the *Geographical Review*, acknowledges Snodgrass as author: 'Dr. Catherine P. Snodgrass of the University of Edinburgh covers Scotland in a single chapter devoted to Clydeside and six smaller areas and pays close attention to urban expansion in relations to topography', but only pays tribute to 'the [lead] author for his great labours' of value to scholars of urbanisation.

In 1944 Snodgrass was awarded the RSGS Newbigin prize for her contribution to geographical research, her article on 'Recent population changes in Scotland', although short by modern standards, was deemed the best paper published in the journal that year. She was the third person and one of only two women to receive the award before 1970.[21] Her involvement with the RSGS continued when she was appointed to the Society's Council on which she served for 20 years (1945–65), and she became editor of the *SGM* in retirement (1964–7), which must have been a role close to her own heart (see Chapter Three). Described by colleague Swanzie Agnew as 'hardly able to step over the border for [fear of] contamination',[22] nonetheless Snodgrass did travel beyond the Scottish borders on occasion. For example, she presented a paper on her postgraduate research at the 1930 Bristol meeting of the BAAS[23] and was the second woman to serve on the IBG Council (1949–50). But ultimately her concerns were Scottish to the core, and to this end her brief editorship of the *SGM* at the end of her career was a significant contribution.

At the time of her death her *Third Statistical Account of Scotland* volume on East Lothian was considered her 'most notable' publication (MacGreggor 1975). Snodgrass' work was quickly forgotten within the university sector perhaps because of her early retirement (ibid.); or perhaps it was marginalised in academic memory because of her move into political advocacy. MacGreggor suggests that it was her concern for the welfare of Scotland and its people which caused the switch from her academic success in mathematics to geography and certainly she used the geographical mode of enquiry to further knowledge and understanding of Scotland's population, landscape and agricultural economy, making political points about each in the tradition of all human geography being 'political geography' rather than undertaking what we might recognise today as the sub-discipline of political geography. Described as 'undeservedly neglected' (Withers 2005), Snodgrass' work and papers would be of interest today to those researching the history of Scottish national identity and independence, as well as the economic and related political geographies of Scotland, including women's employment.

Jean Brown Mitchell

Jean Mitchell was born on 26 July 1904 at Manchester, daughter of veterinary surgeon James Mitchell and Margaret Cowan. She was brought up by an uncle in Cheshire and educated at Macclesfield High School, returning to the family farm in Galloway in the summer holidays. Mitchell recounted how she would read books in the hayloft and go to market with the farm workers and Adrian (1990) attributes her love of the countryside and interest in the lived experience of rural economies to this background. Mitchell studied at Newnham College, Cambridge 1923–6 and 'was one of the vigorous group of young geographers, initially under the leadership of Frank Debenham, who helped to establish geography as a discipline in Cambridge University' (ibid.). She attained double first class marks in the newly formed Geography Tripos Part I in 1924 and Part II in 1926 and was awarded a College Scholarship 1924–6. She completed her MA in 1931 and was appointed as junior lecturer at Bedford College 1926–30, the only years in her academic life which she would spend away from Newnham College (Coppock 1990). Mitchell returned to Newnham in 1931 as research fellow, a post she held until 1934, when she became college lecturer in geography 1934–68 and director of geographical studies 1933–68 (as well as archaeology and anthropology 1954–7 and architecture 1953–68).[24] Marjorie Sweeting, a student at Cambridge during the 1930s, captures something of Jean Mitchell in her recollection: 'I well remember my first meeting with her, an elegant, well-dressed woman in her thirties, driving an open-top car and full of vigour and fun', going on to note that Mitchell already had a formidable reputation in geography and was widely noted for her intellect: 'her mind was as sharp as a needle' (1991: 120). During the war Mitchell contributed to the Naval Intelligence Handbooks prepared at Cambridge, writing three of the chapters in the volume on Greece, as well as teaching army and navy personnel. After the war she was appointed university lecturer in geography at Cambridge (i.e. a member of faculty), a post she held from 1945 until 1968. She played an important part in the development of geography at Cambridge: 'She nurtured with intellectual missionary activity this newly emerging discipline … Historical geography flourished and her lectures were commanding and full of detail' (Sweeting 1991: 122). Despite long illness in the 1940s and 1950s, when she often taught from a day bed in her college rooms, she is remembered for her vigour and stress on fieldwork, indeed Sweeting thought she was essentially a physical geographer at heart, but turned to historical geography as a more manageable specialism, given her health limitations. Mitchell took Cambridge geography and education students and international groups (including those from the IGU) on geographical excursions to the villages and towns of East Anglia and Leicestershire, where she made their 'communities come alive' and demonstrated 'unparalleled' knowledge of mediaeval church architecture (ibid.).

Mitchell joined the RGS in 1926 and served on its Council 1952–5 and 1958–61, as well as sitting on a number of other national committees. She was an active member of the RSGS and served on its Council from 1945 and was made a fellow of the Society in 1983 for her contributions to geography. She served on the Newnham Council for several periods and became an emeritus fellow in 1974. The Jean Mitchell award was funded by monies collected in her honour on her retirement and is 'given to encourage independent geographical work'; she took great pleasure in the award and an interest in its recipients (Sweeting 1991). Mitchell retired to Edinburgh where she had many years enjoying an active retirement, including as external examiner for geography at Oxford University for three years. She also learnt to cook well (having dined in college for the most part) and pursued hobbies and interests, such as the Silver Society, Scottish silver being a keen interest (Adrian 1990). In her last years she was increasingly debilitated by blindness and Parkinson's disease and she died at the Royal Victoria Hospital in Edinburgh on 5 January 1990. She was remembered as intellectually sharp but convivial, direct but self-effacing, a person of deep but unobtrusive faith, possessing much fortitude and supportive of others, 'a serious woman, but one who believed that we should all enjoy the pleasures of life' (Sweeting 1991).

Publications

Jean Mitchell's publications were centred on historical and regional geography. An early paper on 'The Matthew Paris maps' (1933) was followed by 'The growth of Cambridge', in H.C. Darby's (ed.) *The Cambridge Region* (1938). Mitchell's single most influential publication was *Historical Geography* (1954), which was one of five volumes in a 'Teach Yourself Geography' series published by the English Universities Press. In her introductory chapter Mitchell discussed the still-contested nature and boundaries of geographical knowledge. She identifies 'place' in all its physical and human complexity as the object of geographical study, arguing that geographical study must begin at home and on the ground (i.e. in the field), asking two key questions: 'where?' and 'why there?' in order to address distribution and issues around location, although she was quick to acknowledge that clear answers to the latter are rarely found. She went on to argue that historical geographers spend more of their time establishing the first question of 'where?' than other geographers might, because it is often not immediately apparent in the contemporary landscape, which she described from the point of view of the masculine subject in the gendered language of the day: 'The field he has to till is almost virgin' (1954: 12). In keeping with the pre-war unified epistemology of Vidal and Mackinder, she argued for the physical basis to historical geography: 'The historical geographer then must be a good physical and biological geographer. Physical and biological geography form the basis of all the rest of the study, for if the relationships between man-made features and the natural environment are to be assessed, the natural environment must be fully understood'. This reflects her own expertise across the geographical divide, an ability which allowed her to teach both physical and human geography to Newnham students for a number of years (see Sweeting 1991). For Mitchell, geography was analogous with biography of place and historical geography the study of a place's 'childhood and youth' relative to present-day maturity, something essential to the synthetic understanding of the geography of place. While the book was 'readable', Coppock (1990) considered it 'some pale reflection of that dedication' Mitchell gave to teaching her students, but Lucy Adrian observed that 'Although geography has changed much in the thirty-six years since that book was published, the volume can still be read with advantage. It retains many of the characteristics of its author: small in size, vigorous, erudite, precisely argued with exemplary simplicity and clarity, and imbued with enthusiasm and humanity' (1990: 243).

Mitchell went on to edit *Great Britain: Geographical Essays* (1962), published by Cambridge University Press. This volume brought together the work of many of the leading geographers of the day and became a valued geographical text. In her preface Mitchell stresses two points: the book is not a study of *the* regions of Britain, but rather a collection of regional studies, which may overlap or fail to insect. Equally it was not a new edition of the 1928 collection *Great Britain: Essays in Regional Geography*: 'That collection stands enduring, a testimony to the Britain and to the geographers of its period' (Mitchell 1962: xi). Mitchell contributed a chapter on population to the overview section of the volume and co-authored the chapter on East Anglia with Alfred Steers, drawing on her years of study, fieldwork and teaching in the region. Her chapter on population gave a vivid picture of the historical changes to Britain's population, including the impact of invasions, the industrial revolution and other geographical, political and economic contextual factors. She provided a detailed contemporary picture based on the 1961 census, tracing possible future trends (such as economic decline around the industrial areas in the Pennines) and Britain's geographical and political position relative to the Cold War superpowers (p. 39). Mitchell's publication list is not a lengthy one, but as Lucy Adrian explains 'Jean Mitchell was not a prolific writer; it is by the distinction of her intellect and the superb quality of her teaching that she has most influenced generation of geographers' (Adrian 1990: 243). Sweeting (1991) similarly stressed Mitchell's gift as a 'Superb supervisor ... [her] sharp and incisive mind taught us how to think clearly and precisely' (1991: 121–2); Jean Grove (see Chapter Nine) and Alan Baker (2003) also testified to her influence. J.T. Coppock's (1990: 64) personal memory of her was 'tiny, bird-like figure, with silver hair, perched on a stool in a vast lecture theatre, capturing the attention of a hundred and sixty ex-servicemen with her infectious enthusiasm, a characteristic she retained to the end'. Whilst Mitchell wrote and edited two books which were nationally influential within the university discipline, her contribution to the geographical community as a teacher clearly exceeds this textual legacy for her former students at Cambridge.

Harriet Wanklyn/Steers

Harriet Wanklyn[25] was born on 2 January 1906 at Turner Heath, Bollington near Macclesfield. Her parents were Grace (née Debenham) and John Alfred Wanklyn; her father was employed as a mill manager and in the 1930s became director of the Fine Cotton Spinner's Association in Manchester. Wanklyn was educated at St Margaret's School, Bushey, Hertfordshire 1920–4 and she studied history at St Hilda's College, Oxford (1924–7), attaining a second class honours degree. Whilst at St Hilda's she coxed the Oxford Ladies first eight to success, despite the disapproval of the rowing fraternity (coxing being considered a male preserve) (Mead 1990). Between 1928 and 1933 Wanklyn undertook various language studies, including a period at School of International Studies in Geneva, three study tours of Eastern Europe (researching a book) and voluntary work for the Overseas Department of the League of Nations Union, the Royal Institute for International Affairs and the Fellowship of Reconciliation, her work varying from translation (French, German and Italian) to journalism. From April 1933 onwards she worked on her *March Lands of Europe*, which she reported was 'carried out with the approval of Professor E.G.R. Taylor' and accepted by a publisher, subject to revision; and assisted in the Cambridge department of geography, giving special lectures on the economic and political geography of Eastern Europe and student supervisions.[26]

Frank Debenham was a distant cousin of her mothers and after helping Wanklyn gain access to the Cambridge and RGS libraries, was soon persuaded of her talents, asking her to lecture at the Cambridge geography department following her successful talks to clubs and societies. She was accepted as a graduate member of Girton College and 'narrowly missed' a Rockefeller fellowship. Her enthusiasm, keen support for the League of Nations Union, language skills and extensive travel in Eastern Europe and Russia featured strongly in Debenham's recommendation to the RSGS, as did his reassurance of her geographical credentials, writing 'If she has not an honours degree in geography, her career for the last three years has been very definitely geographical'. Debenham acknowledged her weakness in physical geography, but added 'Her personality and address is all that could be desired, and indeed is not unlike that of Dr. Newbigin herself. Keenly interested in all geographical and political topics, she has an energy and critical faculty which will, I think, take her far in whatever career her she finally settles down to.' He also noted she had sufficient private means not to have to seek further part-time work to supplement the editor's 'inadequate salary'.[27] H.C. Darby also wrote to Professor Ogilive recommending Wanklyn, describing her lectures and supervision as an 'unqualified success'.[28]

Wanklyn was appointed as assistant editor of the *SGM* and her 1933 paper on the 'Niemen River: a neglected waterway' to the BAAS was followed by contributions in 1936 and 1938 when she spoke on 'The Turkish ports' and was a discussant on the regional concept with S.W. Wooldridge, R.E. Dickinson, R.O. Buchanan and C. Daryll Forde. After three years at the *SGM* (1934–6), Wanklyn returned to Cambridge. During the war she wrote a volume on Czechoslovakia for the Admiralty Intelligence Handbooks produced at Cambridge, drawing on her experience in the area. It is unclear why, but the manuscript was returned to her and not published in the series (Clout & Gosme 2003). She married colleague Alfred Steers in 1942, taking his family name, but continued to publish as Harriet Wanklyn. Harriet Steers was offered a fellowship at Girton, but declined it in order to focus on her young family, however, she did take on the directorship of studies and became a fellow of Wolfson College, Cambridge, after her children had grown up. She retired to Edinburgh where she died in 1990; her funeral was held in St Boltoph's Cambridge where her husband had been church warden. In his obituary Mead (1990: 353) recorded her pleasure in seeing Eastern Europe released from the Iron Curtain in the months prior to her death, as well as the couple's 'successful academic partnership and … for their particularly happy family life'; by contrast Stoddart (1988: 112) made no reference to Alfred Steers' family life in his obituary, but commented that Harriet gave her husband 'enormous support over the years'.

In both teaching and research, Eastern Europe was Harriet Steers' forte, an expertise refreshed by her regular travel in the region. *March Lands* (Wanklyn 1941) 'remained the standard statement in English for many years' (Mead 1990: 353) and her wartime study formed the basis of *Czechoslovakia* (Wanklyn 1954) brought up to date with post-war changes. She published a biographical memoir of Friedrich Ratzel (Wanklyn 1961), written with the express purpose of redressing what she considered a hysterical rejection of his work in post-war years, after Nazi appropriation of some of his ideas on *Lebensraum*. She called for a detailed contextual reading which recognised the shortcomings and strengths of his work. 'Too often, unfortunately, modern geographers have seized upon passages and even sentences from Ratzel's writings without appreciating their context … [but stressed] this is not to whitewash Ratzel….' (1961: 1–2). In many ways this was still polemical, appearing only 16 years after the end of the war, but her measured biographical account, which drew on many German and other international sources, has remained an authoritative source in English, one which did much to rehabilitate Ratzel in late twentieth-century histories of geography in the English language. For this reason the book was welcomed by Crone (1962), and Clarence Glacken (1962: 468), reiterating Ratzel's importance in nineteenth-century scholarship, considered 'this gracefully written memoir' and bibliography to be one of the two best guides for the student of Ratzel. In 1962 Wanklyn was one of several Cambridge colleagues contributing to *Great Britain: Geographical Essays*, edited by Jean Mitchell. Her chapter, on this occasion written under her married name, was on the Grampians, and its discussion of the geological foundations of the area showed little evidence of early concerns about her lack of training in physical geography. However, Harriet Steers is best known for her work on Eastern Europe and her substantial collection of books on the region was donated to the Cambridge University's School of Slavonic and East European Studies (Mead 1990). Harriet Steers served on the Council of the RGS 1943–6 and she and her husband were jointly made honorary fellows of the RGS in 1977 for their 'exceptional services' to geography (*GJ* 1987: 438).

Margaret Shackleton/Mann

Margaret Reid Shackleton was educated at Mayfield, later known as Putney High School, she went on to take a BA general degree at UCL prior to the validation of the geography degree at London University in 1921. Once the degree in geography was available she began to study for the examination, where she 'came under the notice of Professor L.W. Lyde, [and] combined private study with helping him' (W.G.E. (East) 1950). She sat the exam as an external student and attained a first class honours degree – 'a recognition of her critical and scientific mind and also of her enthusiasm for geography' (ibid.). As senior assistant at UCL, she lectured mostly on European political and applied geography (Clout 2003b). She left UCL in September 1929 (Clout 2003b) and married Dr F.G. Mann of Cambridge University in 1930 thereby becoming part of both the London and Cambridge geography networks.

Shackleton's book, *Europe. A Regional Geography*, was published in 1934 under her maiden name, part of a five-volume University Geographical Series edited by Dudley Stamp.[29] The book was considered her 'chief contribution to geography' and was described as 'well-known and lucidly written' (W.G.E. 1950). Shackleton's regional approach was deeply embedded in the physical geography of the region (East attributes this to the influence of Professor Garwood) and drew on her wide travels in Europe, especially in Germany, Hungary and Romania. The introduction to the book makes it clear that she was a Europhile, believing Europe not only to occupy a physically but also a culturally and spiritually advantageous position in the world. Her text and references show the influence of Wegener, Steers, East, Moodie, Demangeon, Cvijic, Ripley, Coon, Fleure and Newbigin among others. At the time of publication, *Europe* was reviewed favourably, described as 'well planned, ably written, and admirably illustrated' and was expected to 'be in wide demand' (E.E.E. 1935). At the time of her death in 1950 the book continued 'to be deservedly popular as the best survey in English of this broad field' (W.G.E. 1950: 268). Shortly before her fatal illness she had completed wide-ranging revisions for the post-war fourth edition, which included new chapters on Russia. Writing in Cambridge and Honolulu (where her husband was working and where she was herself a Visiting Professor) she expressed her thanks to Gordon East, Norman Pounds, Alice Mutton and Mrs. F. Anderson for their assistance. Later Alice Mutton and Gordon East would update the fifth (1954) and sixth (1957) editions, both of which had new impressions made and which my two second-hand ex-library copies show were still in use in the 1960s, giving some indication of the book's wide and continuous use over 30 years.

During the war, Shackleton taught Queen Mary College, London students evacuated to Cambridge and used her expertise on Europe to write most of the volume on the Netherlands, volume two and part of volume three on Yugoslavia. She was a fellow of the RGS for 30 years. Still engaged in research (on the Danube area) she died on 26 January 1950 in Cambridge after being ill for some time. East concluded his obituary 'Mrs Mann's lively and cheerful personality, her carefully prepared lectures and her unfailing and kindly attention to her students have endeared her to a succession of geography graduates of University College, as well as to her colleagues' (1950: 268).

Dora Smee

Dora Smee was born on 10 September 1898[30] at Carson Road, West Dulwich, London, named Dora Kate for her mother, Kate (née Mandry). Her father, Frederick Smee, was an auctioneer and both parents emigrated to New Zealand before the First World War, leaving Dora at 15 years of age 'almost on her own' (Timberlake 1983) to complete her education. After qualifying as a teacher she taught for a few years, before becoming a student at Bedford College. Whilst teaching Smee took a holiday job as companion to Constance Ismay, American wife of the White Star shipping magnate Charles Bower Ismay. The Ismay family supported Smee during her degree course and gave her the opportunity of extensive international travel during vacations and a leave of absence in the spring term of 1938. This provided Smee with first-hand experience of the world beyond northern Europe, which was rare amongst British geographers in the interwar years (Timberlake 1983). What started as a seemingly formal relationship became a close and lasting friendship.

Smee graduated in 1925, and went on to complete her MA in 1927 and PhD in 1929 and taught evening classes at Hillcroft College for

Working Women. Subsequently she was appointed as junior lecturer in geography at Bedford College for the Lent and Easter term in 1930. She was reappointed for the academic year 1930–1, then appointed as lecturer in 1931, and promoted to senior lecturer in October 1949. After her retirement in 1965 she became a warden for a student residence for several years (Timberlake 1983).[31] As colleague and friend Eunice Timberlake wrote, such a brief outline of her career 'can give little idea of whole-hearted and invaluable involvement in the affairs of the College and of the University Board of Studies in Geography' as well as the Institute of British Geographers of which she was a founder member (Timberlake 1983: 120).

During the Second World War, Hosgood, Smee and Timberlake accompanied their students to Cambridge where their college was evacuated in the autumn of 1939–44. A long, mostly handwritten list of geography and related books found in her papers suggests she may have been one of the volunteers clearing the Bedford library for evacuation to Cambridge during the Second World War. The London colleges had access to some of the Cambridge University lectures, but largely followed their own curriculum, combining forces and resources (Bedford and Queen Mary colleges shared a room in the Cambridge geography department (Hilling 1994)). Smee led student fieldwork in Huntingdonshire and Fenland, surveying land for the war effort 'plough up' campaign (Hilling 1994), which targeted marginal agricultural land. As Smee had already been deeply involved with the Land Utilisation Survey with Dudley Stamp, this type of work was very much her métier. Smee also undertook fire prevention duties at the Downing Street department (exempting her from civil defence duties under Fire Prevention Order 1941).[32] At the same time Smee regularly travelled between Cambridge and Haselbech, where Constance Ismay was removed to the lodge cottage while the main house was requisitioned.[33] In the absence of her own family in New Zealand, Constance Ismay's house at Haselbech became Dora Smee's home, and the locality the focus for her research.

Smee (along with Hosgood and Timberlake) were all founder members of the newly formed Institute of British Geographers (1933) (Steel 1984). Smee is recorded as having attended the first formal meeting of the London geographers, which went on to launch the IBG (ibid.); she later served on the Council of the IBG for two years 1952–3. Her commitment to the IBG was acknowledged in 1974 when Smee received honorary membership of the Institute (along with four former presidents and Jean Eames, the first administrative assistant of the Institute); 'throughout her career [she had] been a stalwart and generous supporter of the Institute' (Steel 1984: 119).

Smee was known as 'a forceful character who did not suffer fools lightly, but as an inspiring teacher and demanding tutor she was respected and rewarded with great loyalty by her students' (Hilling 1994: 124). This suggests that Smee was in the mould of her tutor and then head of department Blanche Hosgood. Hilling (1994: 124) reports that 'Her main interest was historical geography but she travelled widely and was able to teach a wide range of regional courses from first-hand knowledge'. Academically Smee was known for her high standards and for her pioneer championing of fieldwork (Hilling 1994; Timberlake 1983), enlisting students in the first national Land Use Survey in the 1930s and taking students on day and longer field trips to Ireland, the Lake District and the Midlands she was 'instrumental in making field work ... an essential and memorable part of the students experience for those who accompanied her .... Her recipe for good fieldwork was to go, observe, record, and tap local knowledge – still sound advice' (Hilling 1994: 124–5). Timberlake records Smee's contribution more strongly: 'She was a pioneer in promoting fieldwork and was partially responsible for the inclusion of field studies as a fundamental part of undergraduate training' (Timberlake 1983: 120). Eunice Timberlake was herself another important element in that promotion of fieldwork (see below). Smee enlisted students to assist her in her long research project on the historical geography and soil survey of agricultural land around Haselbech, Northamptonshire and the lists of names found in the back of her address book of women and men visiting for weeks or fortnights at a time, typically in groups of 8–10,[34] are most likely these field study groups. Timberlake describes Smee's field parties (some of which she assisted with) as 'conducted with great enthusiasm and energy over many years, [they] were memorable for their combination of hard work and happy adventure' (ibid.), hence her description of Dora 'entertaining' these groups at Haslbech.

Smee's pioneering of fieldwork was grounded in her own experience, having both studied with Hosgood who included fieldwork in the curriculum (see above) and been actively involved with the Le Play Society, the BAAS and the Council for the Promotion of Field Studies (CPFS). She travelled with the CPFS to the Haverfordwest centre (the photographs can be found in the Northamptonshire archive)[35] and co-led a successful five-day field meeting in Galway with A. Farrington for the Belfast meeting of the BAAS in 1952 (Steel 1984: 102–4), drawing on her earlier research on ports and trade in Ireland. Even in her leisure she was gathering course material, gaining knowledge of the UK during walking holidays. She clearly loved not only travelling but also engaging with the places she visited and recording them in photographs and cine film, all of which bore fruit in her field and lecture-based classes: 'She was an inspiring lecturer and tutor in many aspects of her subject, but her courses in regional geography were outstanding – that on the British Isles was unique' (Timberlake 1983: 120). No doubt some of this expertise was utilised in her chapter on Ireland's ports and commerce in volume *The British Isles. A Geographical and Economic Survey* (1933), principally written by fellow London geographers Dudley

Stamp and Stanley Beaver. The book went to six editions, however, Smee's name was removed from the revised fourth edition (1954) when her chapter was expanded – but still included much of her original material including tables and sketch maps. This represents another example of how women as minority, but not minor, contributors, have been excised from major geographical texts (see Snodgrass above).

Undoubtedly, Smee's greatest published legacy was the soil and topography survey in her home parish in *Haselbech Soil Mapping: a Case Study at Haselbech, Northamptonshire, 1954–1974*. The surveys on which this work was based were largely carried out by groups of students over several decades, and included 75 000 hand-auger soil samples, which 'must be something of a record for a small area survey of this type' (Hilling 1994: 125). A record from Smee's diary for Friday 15 July 1945 gives a sense of the collective effort: 'Tim, Vollans and Mountjoy arrive late afternoon and Leslie later, to stay at Gate Cottage – for Survey and Ridge and Furrow Research'. In her introduction to *Soil Mapping* Smee stresses the contribution of colleagues and students to the study and underscores the fact that, until the last couple of years, all the soil auguring had been done by women. Thank you notes from friends and neighbours who received a copy of her book on publication, suggest that in addition to a whole generation of Bedford College geography students, numerous local people were involved in helping with this monumental survey. One correspondent recorded that receipt of her book 'brings back very happy memories of the period when I helped with the soil augering.[36] A local farmer recorded the practical benefit of her survey: 'Dear Miss Smee, It seems inadequate to say "thanks for the books"! This monumental compilation which must be one of the first classifications of life's most basic need, will be a reference for years to come. Used in conjunction with crop performance I am sure many problems can be solved and money saved. In expressing my gratitude I am very aware of the advance of scholarship you have made, and for which I am thankfully a beneficiary by your use of Haselbech as the example'.[37] Smee intended to write a companion study on ridge and furrow but never completed it, which is not surprising given she was 78 years old when her *Soil Mapping* volumes were published. Dora Smee died six years later on 5 April 1982.

While the Smee archive at Northamptonshire Record Office provides more detailed information about her life than is available for many other women geographers, the information is fragmentary and we can only speculate as to some of the connections. Her papers include bills from Harrods and Burberry. Her reasons for requesting details of any heraldry associated with the Smee family from the College of Arms in 1935 can only be guessed at.[38] Smee was a beneficiary of Constance Ismay's will[39] and having struggled financially in her early adulthood, she was generous in her wealthier later years. Timberlake wrote of her friend: 'her wisdom was enriched by the experience of both hardship and good fortune' (Timberlake 1983). There seems to be evidence of her endowing a travelling grant at Cambridge, known as the 'Bedford Travelling Grant in Geography'.[40] A letter from 'Pat' thanking her for signing her RGS application form, combined with Smee's address book with details of former students and colleagues suggests she maintained a healthy friendship network within the discipline; as a local scholar Smee's opinion was often sought, for example, as an expert on landscape by the local history group.[41]

Eunice Timberlake

Eunice Timberlake started her lifelong attachment at Bedford College when she studied as an undergraduate 1923–6. She was to return as junior lecturer in the department of geography 1931–4, and promoted to lecturer 1934–65 (having been given a life appointment in October 1945).[42] It was only on Dora Smee's retirement that Timberlake, at the age of 60, was promoted to senior lecturer in 1965, which illustrates the extent to which promoted posts were rationed within academia prior to 1970.

Timberlake had studied general science before undertaking special studies in geography and was one of only a few academics in London University to hold a BSc special degree, and it was this combined grounding in the natural sciences and geography which 'formed the basis of her distinctive contribution to the teaching of geography', notably in leading the teaching of mathematical geography and surveying for the London University colleges (not just Bedford College) (Vollans 1992: 247). It was this expertise that enabled her to contribute to various wartime research projects undertaken at Cambridge University geography department, which included secret work on map projections for air navigation charts and tables to assist in air rescue for naval convoys. She went on to publish the associated astronavigation tables with E.E. Benest at the end of the war in 1945 (Benest & Timberlake 1945). In addition to normal teaching, Timberlake also lectured air force and navy cadets on astro-navigation and did the cartographic work underpinning models made for the different theatres of war (Vollans 1992; Maddrell 2008).

In post-war years, through to retirement, Timberlake taught a wide range of courses on the undergraduate syllabus at Bedford, but also played an important role as a key teacher of advanced mathematical geography and surveying on an intercollegiate course, which led to professional accreditation by the Royal Institute of Chartered Surveyors (Hilling 1994; Vollans 1992). Although she was obviously gifted in things mathematical, she was sympathetic to those who were not, and is remembered by students for 'her clarity of explanation, cheerfulness and infinite patience' (Hilling 1994: 135).

Her teaching included field astronomy and map projections as well as being part of the team leading the associated annual three weeks of surveying in the field. As noted above, Timberlake was instrumental with her close friend Dora Smee in both developing fieldwork at Bedford College (surveying the Abbey National building in Baker Street was an annual event (see Hilling 1994)) and assisting Smee in her study of Haselbech: 'it was she who planned the survey and preparation of a large-scale contoured map of Haselbech parish (now in the British Library)' – part of the 'remarkable partnership' between the two women (Vollans 1992: 248).

'Tim' as she was known in undergraduate days is remembered by Eleanor Vollans (a colleague from 1948) as someone who dedicated her life to the study of geography at Bedford College and within the wider London University.[43] But she also had varied outside interests, such as her family's cattle farm near Tring and widespread travel in Europe. She appears to have taken a back seat in the wider world of geographical affairs compared to Smee, but as Vollans points out, Timberlake very much played the support-ive role in their friendship. Already dedicated to her profession, her work load grew with her appointment as departmental academic adviser in 1965, which coincided with Smee's retirement and an expansion of student num-bers and courses under new head of department Monica Cole. This was a difficult period for many academics and particularly those in Bedford College's geography department (see Chapter Eight) and much can be read between the lines in Vollans' (1992) appreciation of Timberlake: 'It fell to Tim, as Departmental Academic Adviser guiding students in their choice of courses, to counter such difficulties, removing uncertainties, allaying anxie-ties, and generally devising expedients to smooth transition from old to new. It was a task which Tim performed admirably, sometimes taking pleasure in her own ingenuity, but above all seeking always to be accessible to students in need of help. One colleague, recalling her good sense and resolution in the face of difficulty, spoke of her "as a tower of strength". She had passed her sixtieth birthday when she became Adviser, and she retired at the age of sixty-four. For those four years she was owed a profound debt of gratitude'. Tim retired in September 1969 and spent her last years in Berkhampsted, Hampshire,[44] where she died in February 1991, aged 86 (Vollans 1992).

Florence Miller

Florence Clark Miller was born in 1889 and graduated from the University of Leeds with a BA honours history degree. In common with many other women qualified in the humanities, she went into the teaching profession, working in Northumberland and Montgomeryshire (Bird 1968). She was appointed lecturer in geography at the University College of Southampton in 1921 and stayed there until retirement. Head of department W.H. Barker, a keen regional geographer and educationalist who left in 1922 to become Reader at Manchester University, was succeeded by O.H.T. Risbeth. After Professor Risbeth's retirement due to ill health in 1937, Miller was appointed 'lecturer in charge' under the direction of the supportive economics professor. She was only formally acknowledged as head of department 12 years later in 1949 when the department gained independent status and began a period of growth. Miller retired in 1954 and her successor, F.J. Monkhouse, was appointed to the re-established chair in geography. Florence Miller died in 1967.

Known as 'Flo' to staff and students, 'Miss Miller was a most original teacher who caught and easily held the interest of students. Her pioneer work for the subject will always be remembered by those who were fortunate enough to have studied under her' (Bird 1968: 328 – Bird had been a student of the department going on to become professor). Students included members of the RAF during the war years, to whom Miller taught map reading. Former students remembered her as a 'lady in the traditional sense' but a demanding teacher: 'It was inviting a sharp rebuke to make a loose statement, let alone an inaccurate one' (Frank Bevington, student 1951–4, in Wagstaff 1996: 9). Wagstaff's account of the department at Southampton includes students' caricatures of her physique with her large round glasses and handbag under her arm: 'She cut a rather comic figure: short and dumpy, with thin bobbed hair, a round face, sky-blue eyes', she had a disarming smile and 'reminded me of a perky robin' (Arthur Hunt, student 1934–8, ibid.). Several students recorded by Wagstaff, including Mary Jeffries, commented not only on the small interactive tutorial discussions, but also how Miller made them *think*. On her retirement past students belonging to the 'Geog-Soc' (student society) contributed to a retirement gift for Miss Miller, who chose a refrigerator, something she has 'always wanted' (Wagstaff 1996). This substantial gift must be taken as indicative of the esteem in which Miller was held. The Geog-Soc had been founded in the late 1920s, financially supported by Miller, and organised field trips and lectures as well as social occasions. Mary Grant (later Jeffries) was the first president (ibid.).

Whilst noting that she 'did much to shape the character of the department and its approach to teaching', Wagstaff records that she was not 'research active' in today's terms. However, she did publish a paper in *The Geographical Teacher* on 'Early maps of China and the Mediterranean' (1921–2), which stressed the contrasting interest in land-based and coastal measurements and features depicted on Chinese and European maps, respectively. In the same year she also wrote a report on 'The [Geographical] Association's tour to the natural regions of North England and South and Central Scotland'. She was also local secretary for the geographical section of the BAAS in 1924 when the annual meeting of the Association was held in Southampton in 1925. In contrast, head of department Risbeth gave a paper at the session but was not on the organising committee.[45] Whilst leading male geographers were the local secretaries for these meetings in other institutions, it is hard not to read this particular division of labour in gendered terms, but we do not know if Miller was satisfied with the arrangement or not.

Katy Boswell

Katherine C. Boswell, born in 1889, was a Southampton graduate in biological sciences (1906) and after working as a teacher returned in 1923 to take the College Diploma in Geography – but in fact graduated with an honours degree in geography. She was subsequently appointed in 1926 as demonstrator and then assistant lecturer in 1932 (Wagstaff 1996), followed by lecturer until she left in 1944. Boswell completed her MA in 1936 on the economic geography of Southampton, but essentially she was a physical geographer and was responsible for teaching the physical courses. She devoted a lot of time to translating Walter Penck's (1924) *Die Morphologische Analyse*, only completing this with the help of Hella Czech shortly before her death (Wagstaff 1996). She was remembered as an enthusiastic teacher and an 'outdoors type' who relished detailed research and fieldwork. As her Quaker Society obituarist recorded: 'It is in keeping with her character and her faith that while others were theorising about river profiles, Katherine was out with dumpy level and target staff, measuring heights up and down the 37 miles of the Test, finding out for herself what the profile of a particular river was like' (Veysey 1953, cited by Wagstaff 1996: 13). One student recalled her as the 'fount of all wisdom' (Mary Jeffery, ibid.).

Boswell's faith informed her social conscience: she housed European refugees in the early war years, helped evacuees and was clerk of the local War Victims Relief Committee; she was then inspired to become co-principal of the Friend's College, Highgate, in Jamaica 1945–9, where she had particular responsibility for the Rural Development Centre from 1945. At the end of what had been a demanding four years of service, and finally completing the translation of Penck's work, Katy Boswell crossed the Atlantic to attend the International Geological Congress at Algiers in September 1952. Apparently motivated by her desire to see 'real desert', she died of heat exhaustion at 63 years of age. Recorded as the first to scale a hill during a stop on the desert excursion, she never quite recovered from the exertion and died later that day at Beni-Abbes, where she was buried with a simple service reflecting her Quaker beliefs. Florence Miller summarised Boswell's life in *Geography*: 'A most able geographer and teacher, she produced in her students a lasting enthusiasm for the subject, and an enduring affection for herself. Her friends are everywhere, in every station of life, of every colour and creed. Her unselfish goodness was her gift to all of us' (1952: 236).

Before discussing Alice Mutton's work, a brief outline of her predecessor is given. **Bettie Tunstall** (born 1902) lectured in regional and economic geography, map work and historical studies 1924–33. Tunstall was a Bedford College graduate (1922, first class) and one of the founding members of the IBG (see Steel 1984: 146; Sheppard 1994). She left on her marriage to John Pedoe in 1933, initially intending to research on social and economic geography. Her letter of resignation gives an idea of her workload: 'During the nine years of my service here, while the geography department was being built up from very small beginnings, I have necessarily been obliged to spend a disproportionate amount of time and energy on teaching and direct preparation of courses, as compared with the little research and independent study my leisure or my salary permitted me'.[46] Tunstall was replaced by John Lebon (full time) and Alice Mutton (part time). She returned to teach at Maria Grey College in the 1940s, and her sons Hugh and Daniel recounted that she completed a thesis on soil erosion for a higher degree around this time, but that it was 'turned down as non-geographical by a formidable woman professor' (presumably Eva Taylor), and the manuscript was subsequently destroyed during the blitz. In post-war years Bettie Pedoe accompanied her husband on his academic postings around the world. She died in 1965.[47]

Alice Mutton

Alice Florence Adelaide Mutton was the daughter of two teachers, she was born on 4 March 1908 in Lee, Kent and brought up in west London. Educated at the Haberdashers' Aske School in Acton, she went on to read geography at Bedford College under the tutelage of Blanche Hosgood, and graduated with a first class honours degree in 1929. She completed a teaching diploma (1930) and taught briefly before returning to Bedford, as holder of a Gilchrist studentship, to undertake her MA on the historical cartography of Sussex. This was no doubt in part inspired by her family's links to Sussex and she was awarded her master's degree in 1932. This was followed by a University of London Postgraduate Studentship (1931–2) and an appointment as part-time lecturer at East London College (to become Queen Mary College (QMC) in 1935), University of London in 1933.[48] Mutton was also an active member of the IBG, RGS and GA. During the 1930s Mutton, a fluent speaker of French and German, travelled widely in western Europe and led three Le Play Society field study trips in Germany and Czechoslovakia. She spent some time studying at the University of Freiburg-im-Breisgau and was awarded her PhD in 1937 for her work on the historical geography of the Black Forest region, presenting a paper on this research at the 1938 IBG conference.[49] Attaining a PhD, still quite rare for British geographers, resulted in her being 'recognised' by the University of London, but she only succeeded in gaining a full-time post in 1939 (Sheppard & Rawstron 1980). The summer of 1939 was spent in the USA and Canada, guest of the American Federation of University Women.[50] With the outbreak of the Second World War Mutton was evacuated with QMC and other London University students to Cambridge and her expertise on Germany was called upon for the Naval Intelligence Handbooks prepared there. In addition to her teaching, when she 'held the Department together and participated in the teaching of most of the branches of the subject' (GJ 1980), she made a major contribution to the Admiralty volumes on Germany based on her recent research there (Maddrell 2008). While in Cambridge Mutton was also president of the local branch of the GA 1944–5 and was chief examiner for the Cambridge University Higher School Certificate Examination 1940–5, as well as examining for several other boards.[51]

In post-war years Mutton still carried a heavy teaching and examination load but was able to continue her vacation travels abroad, including organising field excursions for students at a time when travel was difficult. Despite her workload she was remembered as an accomplished and devoted teacher, who drew on her extensive travels and put her own photographic slides to skilful use in lectures. 'She gave unstintingly of her time and talents to students … The large and enthusiastic attendance at her retirement reunion in 1973 was a demonstration of the high regard in which she was held by those she had taught. Queen Mary College was fortunate to have been served by such a devoted and versatile teacher during the formative decades of the geography department' (Sheppard & Rawstron 1980: 124). That devotion for almost 40 years was all the more remarkable given her dim view of the university's tardiness in assigning her a permanent appointment. This experience coloured her personal politics, especially within the university, and her own early battles 'left their mark in a readiness to take a firm stance on matters of principle' (ibid.) and 'she sometimes spoke from bitter experience of the severe difficulties that faced budding academics in the interwar period and was always anxious to help younger colleagues gain the security and additional remuneration which she had previously lacked' (GJ 1980: 160).[52]

Mutton failed to gain a Readership in 1946 but was promoted to senior lecturer in 1955, with Percy Crowe, Arthur Smailes and Eric Rawstron being promoted over her during her career. She was successful in gaining promotion to a Readership in 1972 shortly before her retirement, supported by Professors Smailes, Edwards and Mead.[53] She took early retirement in 1973 after a period of serious illness and died in her home at Wilmington in Sussex on 1 March 1979. Her last article 'The Weald remembered' was published in 1978 and she had almost completed a book on the historic towns of Europe when she died (Sheppard & Rawstron 1980). QMC established the Alice Mutton award in her memory, which true to Mutton's own interests, is given for a student travel project in the summer vacation.

Publications

Mutton worked principally on the historical and regional geographies of Europe and North America, with her publications focusing on the former. *Central Europe. A Regional and Human Geography* (1961) was her key publication, and was warmly welcomed by Harriet Steers (1962a) as a much-needed text in English. Steers appreciated Mutton's long experience in the region and knowledge of its geographical literature, as well as her dedication to such a large task in covering Central Europe. Whilst tactfully pointing to the difficulties of defining 'Central' Europe and gaining equal access to up-to-date information in the bifurcated Europe of the Cold War, Steers generally applauds the volume, despite the fact that sometimes the 'writing at times seems to tire under its own load' of research. 'Students who, up to now, have had to search here and there for the factual material about Central Europe will be entirely grateful for so much that is now easily accessible and coherently written up' (Steers 1962a: 79). Mutton was also one of the consortium of five London University authors who published *An Advanced Geography of Northern and Western Europe* (Harrison Church *et al.* 1967) as a core text for the London degree. Each wrote on their specialist country or sub-region, which in Mutton's case was Switzerland. Fellow London University geographer, Gordon East, praised the book for its pithy style and focus on the human environment. For East good regional geography demanded the best work from geographers and he credited these authors with achieving this, concluding the book 'should arouse, and not merely inform, the student who wisely selects it for study' (1968: 90). Mutton's scholarly papers were in the field of economic regional geographies. Her paper on the human geography of the Black Forest (1938), based on her doctoral work, was typical of pre-war regional studies, begins with a long overview of the physical characteristics of the region, before proceeding to discuss factors such as agriculture, industry and population. This work was clearly pertinent to her contribution to the Admiralty Intelligence Handbooks on Germany. In post-war years her attention turned to the rapid development of hydroelectricity in Western Europe (1951) and Norway (1953). These papers were based on her own field research conducted over three summers and supported by London University's Central Research Fund; the paper on Norway was presented at the 1953 IBG annual conference. Like her teaching, Mutton's writing was illustrated with useful maps, charts and photographs. She also wrote useful reviews of French and German literature on energy and water.

Dorothy Sylvester

Several women contributed to the teaching of geography at Manchester University on short or part-time junior contracts from 1914 to 1947: **Gladys Marten** 1914–22; **Margaret Willis** 1926; **Marjorie Kendrick** assistant lecturer 1928–30; **Mabel Crawford Wright** (married to Frank Osborne) assistant lecturer 1930–2, special lecturer in geography 1932–9 and assistant lecturer once more 1939–40; **Margaret Davies** (married to Elwyn Davies) assistant lecturer 1941–4; and **Kathleen Holmes** (married to Thomas Kenyon) assistant lecturer 1944–7 before going on to UCL.[54] While this shows junior/temporary posts going to women geographers who were the spouses of male lecturers in the department, it can be interpreted in a number of ways ranging from the women being a highly qualified pool of cheap reserve labour called upon and dropped according to institutional needs, to women benefiting from appointments by nepotism; alternatively they were qualified women undertaking the most appropriate local war work, or a reflection of individual and institutional reluctance to give permanent posts to suitably qualified partners and/or married women.

In 1944 Dorothy Sylvester was the first woman to gain permanent employment as a senior lecturer in geography at Manchester. Born 1906, Dorothy Sylvester grew up in Cheshire and was taught geography by Miss D.M. Preece at Crewe County Grammar School. Sylvester wrote of Preece: 'No one could have been more fortunate in their Geography teacher'[55] and it was a combination of her gifted teaching and the experience of fieldwork which inspired Sylvester to make geography her life's study. Climbing Cardingmill Valley to Bodbury Ring on a 'landmark day', a 'day of wonders', she was struck by the landscape, vegetation, livestock and geomorphological features, as well as the Prehistoric ring and ancient rocks, concluding that 'if history had any rival claims to my interest, this alone would have disposed of them'.[56] Sylvester went on to study geography at Liverpool University 'under Roxby's inspired and King's sound teaching'. At a time when she felt herself maturing into adulthood during her second year, Sylvester found her commitment to the subject solidifying, helped once more by a field course when a group of geographers from Liverpool, Aberystwyth, Manchester and Leeds combined at Marlborough College during Easter. She was inspired by Fleure's talk on the Marlborough Downs, which became 'living hills to me, as during the coming week they were to become part of me'.[57] Sylvester graduated in 1927 and this was followed by completion of a teaching diploma in 1928 (she also attended the IGC meeting in Cambridge that summer (Stoddart 1983)). Sylvester taught geography in girls' public schools for two years before being appointed as lecturer in geography at the University of Durham in 1930, the same year she was awarded an MA from Durham for her thesis on 'Aspects of the historical geography of Shropshire'. During the war years from 1940 to 1944 Sylvester was acting head of department at Durham. In 1944 (around the time she married William Taylor) she was appointed as senior lecturer at Manchester geography department; almost 20 years later and near retirement she was promoted to Reader in 1965. She resigned in October 1966 at the age of 60.[58] In retirement she continued as president of the Crewe and Nantwich GA, pursued her historical work, and was an active member of the Offa's Dyke Association and her local Methodist church at Wells Green, where she was on the church's social responsibility committee. Her non-geographical writing included unpublished short stories in her youth and published poems and hymns in later years ('We live not to ourselves, O Lord' particularly embodies her social responsibility ethos).[59] Dorothy Sylvester died in 1989 and her papers were deposited in the National Library of Wales in 1992.

Sylvester was a specialist in historical geography, especially of the northwest of England, whose work culminated in the *Atlas of Cheshire* (1958, co-edited with G. Nulty). She was a fellow of the RGS and an original member of the IBG (Steel 1984: 145), serving on the IBG Council during 1951 and 1952. She worked with other leaders in the field such as H.J. Fleure and Clifford Darby, as well as local colleagues such as Norman Pye and Walter Freeman. She is remembered as the 'kingpin' of the Crewe and Nantwich branch of the GA where she made a huge input of time and energy as president for over 20 years from 1957 to 1980.[60] Programme cards for the GA 1958–70 show that the number of women on the local committee increased from two in 1958–9 to typically five or six for the rest of the period; the number of vice-presidents was also increased to six and typically included two women in their number. Most of the annual programmes of approximately nine speakers included women (e.g. Sylvester herself in 1965 and 1969; Monica Cole from nearby Keele in 1961 and 1964; and Miss P.A. Nicklin in 1959 and 1963).[61]

Sylvester published a number of articles on rural and settlement geography in England and Wales, notably on the Welsh borderlands; methodologically and stylistically she was driven by an early appreciation of facts 'to grind our teeth on'.[62] She presented a paper to Section E of the BAAS in 1939 on 'Domesday Cheshire', in which she stressed the economic importance of the River Dee and Port of Chester in addition to sources of salt in the area; published an article on creating panoramas from contour maps in *Geography* (1943); and two papers in the *Sociological Review* (1933, 1945). A number of other papers appeared in county history society journals or the

Agricultural History Review. In 1952 Sylvester published *Maps and Landscape*, drawing on years of teaching practice and refining of technique, particularly at Durham. In the preface she indicated her motivation for writing the book: 'This volume has been written in response to the requests of many teachers and students for a guide to topographical map reading and field work such as would cover the ground adequately for the higher examinations in practical geography and suggest methods by which students could set about preparation of independent regional surveys'. She covered both quite elementary and advanced material to meet the needs of students and teachers. There are clear parallels between this and Garnett's *The Geographical Interpretation of Topographical Maps* in its third edition at this time, but Sylvester considered the third section of the book, which dealt with the three-dimensional landscape to be quite innovative (it dealt with field of vision, field sketches, block diagrams, drawing from plans and relating landscape photographs to the map). The reviewer in *The Geographical Journal* considered the book to be a wonderful tool kit, principally for teachers and lecturers. She appreciated its full colour maps, well-chosen examples and the awareness of the limitations as well as all the 'potentialities' of maps, stating it would 'be to the geographer what a seed catalogue is to a keen gardener – a mine of information of a specific, comprehensive and interesting nature, systematically set out, and designed towards practical ends' (A.C. (Alice Coleman) 1955: 368–9).

The Rural Landscape of the Welsh Borderland was published in 1969 after Sylvester retired and whilst some post-retirement publications cement an academic reputation, Bowden's (1972: 215) cutting review suggests this volume only succeeds showing how old fashioned her work was:

> Sylvester's book pulls together the labour of a lifetime primarily expended in the northern Borderlands, where the book is strongest. It shares the author's valuable intuitive reflections upon a great amount of evidence and will introduce general readers to some settlement problems of the Border. It will satisfy Borderland Antiquarians and those who set at nought advances made in geography since 1950. The cautious sensitivity to historical sources of the 'Darby School', the awareness of locational analysis, and the knowledge of the contributions made to Border literature by other historical geographers since 1955 unfortunately all are absent from an overly expensive work.

He goes on to contrast what he describes as Sylvester's 'latter-day example of the work of the first generation' of historical geographers, with Baker *et al.*'s (1970) edited collection of recent papers drawn from journals published under the title *Geographical Interpretations of Historical Sources*. Clearly there are issues of evidence in some of Sylvester's claims and her style may be better described as 'geographical history' rather than 'historical geography'. However, Bowden's expressed purpose was to demonstrate how much the sub-discipline of historical geography had progressed in the previous 15 to 20 years, and to pit the single text of a retired academic against a collection of what was considered the best of recent innovative conceptual work, including Bowden's professed 'heroes', seems at best an unequal competition. In contrast E.W. Gilbert (1970: 260–1) described the book as a 'useful and interesting volume' and Sylvester 'well known as the author of many useful papers on the complex subject of rural settlement'. Gilbert only complained that the book's length and lavish production with 58 maps and eight pages of photographs made it so expensive that even libraries might balk at the price of £8. In retirement Sylvester subsequently published *A History of Cheshire* (1971, 1980) and *A History of Gwynedd* (1983) in the Darwen County History series, which were much used by local historians and regional historical geographers.

Eila Campbell

Eila Muriel Joyce Campbell was born on 31 December 1915 at Ropely, Hampshire, daughter of Lillian (née Locke) and Walter Campbell, a poultry farmer and later shopkeeper (McConnell 2004). She was educated at Bournemouth Girls' High School 1924–34 and trained as a teacher at Chichester Diocesan Training College, Brighton 1934–6, attaining a Diploma in Education from the University of Reading. From August 1936 until December 1941 Campbell taught full time at Potter Street Senior School, Northwood, for Middlesex County Council, taking numerous part-time courses including weaving, spinning and dyeing, physical education and play leadership, indicative of a culture of self-improvement. More ambitiously, she began studying geography part time at Birkbeck College, University of London, under Eva Taylor 1938–41, taking unpaid leave in order to sit her examinations.[63]

In 1941 Campbell graduated from the University of London with a BA honours geography degree, third class. The testimonials of Eva Taylor and E.C. Willatts both stress that she should have achieved a good upper second class degree had war conditions not affected the outcome.[64] In January 1942 she moved to Southall County School in Middlesex to teach geography and mathematics and attended vacation courses in mathematics at the University of Oxford during the summer of 1942 and 1943.[65] In 1941 Eva Taylor gave Campbell her 'strong personal recommendation' and Campbell was appointed as part-time demonstrator and lecturer in geography at Birkbeck College. Writing again in 1944, Taylor outlined Campbell's two and a half years' work as demonstrator and then lecturer, when she was in charge of Intermediate map work and some advanced map work, and taught regional geography. Three years later Taylor wrote in support of her promotion, underlining Campbell's MA research and teaching credentials.[66] Campbell was duly upgraded to a permanent appointment as assistant lecturer 1945–8 and was able to leave her full-time school teaching post in July 1945.[67] Undaunted by her poor results in the BA, her colleague-tutors supported Campbell in undertaking an MA at London University and she was awarded the degree with distinction in 1947. Within eight weeks she had registered as a part-time PhD student with S.W. Wooldridge to study historical geography, but the PhD was never to be completed.[68] In 1948 she was promoted to lecturer at Birkbeck, then Reader in 1963, largely on Darby's recommendation after the publication of *Domesday of the South East*. Mead (1995) maintains that working with Darby deepened her interest in historical geography and that Wooldridge may have inspired her commitment to field studies, but both these interests can be also be traced to the formative influence of Eva Taylor. Finally, Campbell was awarded a chair in 1970 and made head of department – but this represented a long haul to promotion.

Campbell's correspondence shows her place in a wide network of London and Oxbridge geographers, including other women such as Monica Cole, Alice Coleman, Marjorie Sweeting and Dora Smee. She was on friendly terms with male colleagues too, Wooldridge was fulsome in his praise of her in 1947 and described her as the best sort of 'academic woman', implicitly in contrast to unreliable ones: 'Between friends fulsome and honeyed words are out of place but I will say that although I have been indebted to you for very much in the past I have never been more relieved and satisfied to have your support than during the last few days. [Then I reflected ?] that it is as well that there are some "academic women" in the best sense on whom one can rely on such occasions …'.[69] Stamp wrote in 1953 to ask if she would be interested in coming to work with him at the University of Malaya for a year or two, teaching historical geography and replacing Paul Wheatley: 'Anyway, Eila, do let me know what your initial reactions would be, and if you are interested, I'll set the official machinery into motion. If not, well, it was worth trying anyway. As I said, if I'm still in charge, we could have some fun here together; it isn't too bad a place provided one can keep a sense of humour going!' She received another similar invitation in 1969 from colleague Michael Chisholm to spend a year as a university visitor in Ibadan, Nigeria.[70]

Campbell didn't take up these offers, perhaps in part because she cared for her mother from 1951 (McConnell 2004), or perhaps because she wanted to focus her energy on plans for promotion. In 1964 she applied for the chair at Bedford College and was supported by Dora Smee,[71] but Smee, who was near retirement, held little sway and Campbell lost out to former Bedford student Monica Cole who was younger and had a stronger research reputation. Subsequently she applied without success for a chair at Reading and a post as principal of Queen Elizabeth College in the same year.

Campbell may have been hampered by her third class degree, but when encouraging her to apply for a post at LSE in 1947, Wooldridge pointed out to her that Daysh got a Chair and Macdonald Davis a Readership with the same degree classification.[72] It was only in 1969 that Campbell finally achieved her ambition when she applied for the Chair in geography at Birkbeck on Gordon East's retirement. Using Darby, East and Smailes as referees she was offered the position by a panel which included Clifford Darby, Bill Mead and Michael Wise.[73]

One of Campbell's achievements during her career at London University was her work with John Davis to establish a Master's degree on the USA, to which she brought her historical expertise and supervised many of the dissertations undertaken by its students (Mead 1995). Eila Campbell retired in 1981, when she was conferred the title of Emeritus Professor University of London as well as honorary membership of the Senior Common Room at Birkbeck College. Whilst Campbell received these personal honours on her retirement, her field of the history of cartography simultaneously 'disappeared' from the syllabus of the University (Wallis 1994). However, she was still in demand elsewhere, for example, supervising graduate students at Brighton Polytechnic in the late 1980s[74] and gave four lectures at the Cartographic Institute of Catalonia, Barcelona in 1990 for the inauguration of a new course on the history of cartography (Wallis 1994). She was commissioned to write a number of biographical pieces, notably on Eva Taylor and Marion Newbigin for the Geographers Biobibliographical series, but only completed the Taylor entry for the *Dictionary of National Biography* (1972).

Eila Campbell was well known as a committee member within geographical and cartographical circles and this represents one of the biggest elements of her geographical work. She served on various committees of the Royal Society, including six years as chair of the sub-committee for cartography, and was a member of the National Committee for Geography. She was a member of the RGS library and maps committee 1948–9 and 1961–87, served on the RGS Council 1963–6 and 1974–5. She also served on the IBG Council 1956–8 and was president of Section E of the BAAS in 1975. She became president of the Society of University Cartographers 1974–9, was secretary-editor of the Hakluyt Society for 20 years (1962–82) and was vice-president in 1983. She was also a council member and trustee of the Society for Nautical Research and served as vice-president from 1993 until her death (Wallis 1994, 1995). Campbell was a proud member of the US Society of Woman Geographers based in Washington DC and was a staunch supporter of the International Federation of University Women. After many years as corresponding editor and conference organiser for *Imago Mundi*, she became editor and chair of the board of directors of the journal in 1974 (ibid.).

Campbell was also a great academic traveller after the war, her first major journey being a year spent as research fellow in New Zealand 1949–50, funded by New Zealand's government and the Federation of University Women, with additional travel in Australia and Canada via Hawaii. Her personal papers record the pressure of the Domesday work she took with her, and hopes for career progression on her return. The letters also reveal her rather shoestring existence with repeated money worries, her oscillation between academic insecurity and pretensions to intellectual grandeur, and strategic correspondence with colleagues (including East, Buchanan and Wooldridge) with a view to job openings. They also record her personal views and experiences including her affection for her mother, brother and former student Dorothy Peterken; disdain for her unfaithful father; the ups and downs of her outward-bound affair with the ship's purser, Don Jackson; her diet; and her often unflattering views of New Zealand society. Her correspondence with her family is very frank on all these matters, statements sometimes seeming rather bald, such as her request to her brother to file her letters 'for my biography', reflecting both her direct manner and the shorthand of family correspondence. However, it is clear that whilst missing the intellectual society of the University of London and suffering from a degree of culture shock in New Zealand, she came to appreciate the attention she had received as an honoured guest.[75]

Her correspondence also reveals her sense of gendered disadvantage specifically as a woman academic, as well as her desire for but difficulties in finding a partner. Writing to her brother about her romance with Jackson, Campbell revealed that her career constituted a barrier to long-term emotional commitment. 'Don asked me whether I would sacrifice my career for marriage and I was forced to say that I didn't know, especially as I was committed to this year's research. In the end we both decided that conditions on board a ship were not natural'.[76] However, another letter revealed underlying intellectual snobbery: 'If Don was a University man, I'd marry him to-morrow but an adoring husband might become a bore in time. Now I know why Academic men marry unacademic women'.[77] When the relationship turned sour, Campbell put it down to her own foolishness in engaging with a non-university man.[78] During her time in New Zealand Campbell worked on her Domesday material, travelled and gave speeches to the Federation of University Women, but her letters continue to speak of her isolation: 'I'm tired of mental loneliness which must be the lot of any woman unless she lives in a College community'. She counterbalanced this with planning her career: 'I've decided in my next Five Year Plan – (war permitting) to settle down if possible in a London flat and go all out to get enough published work for a Readership. I'm going to abandon my husband seeking ... As you know ... [I am] ... is independent and if she can't get one door open, she'll get another. But sometimes, I feel very tired of having to struggle in a man's world'.[79] She was clearly an affectionate person, but seemed blinkered to any options of community and love outside the academic world.

Campbell became a loyal attendee of International Geographical Union Congresses and regional meetings for over 40 years 1952–92 (Wallis 1994). Conferences and meetings took her to Budapest in 1971, Poland in 1973, Maryland in 1978, Berlin in 1979, Tokyo and China in 1980, Ulm in 1982, Portugal and the Netherlands in 1983, Perth in 1984, Ottawa in 1985, South Africa in 1985, Prague and Rhode Island (USA) in 1986 and Mexico in 1987.[80] Campbell obviously took great pleasure in these conferences, describing them as an opportunity for unhampered exchange; she was an active participant and was often asked to chair sessions, 'which she did with inimitable, but always gracious, ferocity!' (Petellier 1995: 8). Many of these conferences at home and abroad were attended with Helen Wallis, and the IBG annual conference was considered incomplete without this pair and the hilarious social evenings they organised (Mead 1995).

Campbell received several awards, notably a £1000 Parry award in 1970 from Birkbeck College to fund completing research on settlement changes in Chile; a Senior Research Fellowship at the Clark Library at UCLA 1972–3; the RGS Murchison medal in 1979 for 'contributions to the history of cartography and exploration' (GJ 1979); and an award from the Map Collectors' Society for services to the history of cartography. In turn, the Eila Campbell memorial prize is awarded to the best final year geography student at Birkbeck and another prize in her name is awarded by the British Federation of Women Graduates.

Eila Campbell died on 12 July 1994 and a commemoration was held for her at the RGS on 16 November 1994, at which the affectionate gratitude of her students was apparent and her 'infectious enthusiasm' saluted (Mead 1995). She was also memorialised in 'A Celebration of the Life and Work of Eila M.J. Campbell, 1915–1994' edited by Sarah Tyacke. Her friend, Helen Wallis, described Campbell as 'one of the outstanding geographers of London University ... particularly notable for her work in promoting the history of cartography', who left an 'impressive legacy of achievement' but whose 'greatest achievement was in bridging the gap between geography and cartography (Wallis 1994: 361).

Eva Taylor and other influences

Eva Taylor was a great influence on Campbell's geographical career, being her mentor from the outset. Taylor wrote to Campbell who needed to catch up on some fieldwork while studying part time: 'Dear Miss Campbell, the College gives you £2 for fares etc., and I should like you to be my guest at the boarding house. It is very important that you do some supervised field work – for testimonials as well as your examination. Let me have a wire'.[81] Taylor gave Campbell an entrée to teaching at Birkbeck, but was also a monitor of her work, checking the text of early lectures for example (Mead 1995). She wrote testimonials for Campbell and heavily prepped her for the MA examination: 'I want you to put up a good show before [East?], and we can bring the assignments into an oral [examination]'.[82]

Taylor also chaired the session when Campbell presented a paper to the RGS, as well as seeing the related paper through to publication. In turn Campbell gave great assistance to Professor Taylor during her working life and retirement, often acting as a personal assistant as much as a colleague. However, as Mead's memory of a 1946 GA meeting shows, acting as Taylor's assistant did not always enhance Campbell's own reputation: 'It was on "Airways of the Empire" and I recall her assistant Eila Campbell (later Professor) holding up hand-drawn map throughout the three-quarters of an hour talk while E.G.R. Taylor pointed at its features intermittently'.[83] Campbell wrote to her mother that on return from her New Zealand sabbatical, she wanted to avoid accommodation at Courthauld House in order to resist getting too tied up with looking after the now retired Taylor. However in the 1960s she became the lynchpin in Taylor's ongoing writing career, even becoming an amanuensis and personal shopper for Taylor whilst she was invalided in the nursing home during her last years. Campbell was crucial to seeing Taylor's (1966) *Mathematical Practitioners of Hanoverian England 1714–1840* through the press shortly before Taylor died. She was also central in helping to found and then organise the annual E.G.R. Taylor lectures, jointly sponsored by the RGS and other learned societies, initiated on the occasion of Taylor's 80th birthday (see Chapter Five). In some ways Campbell seems to have seen herself as Taylor's successor and certainly adopted Taylor's forthright tone at times, but she did not engender the same degree of respect; it is also suggested that she compared unfavourably with friend Helen Wallis, who achieved much through a combination of acuity and charm.[84] While some remember the retired Campbell as an 'oddball' who held forth at IBG meetings, Anne Buttimer, who met her at IGU meetings and translated conference papers for her on occasion, offers a different view: 'Eila Campbell was one of the first to try to document the extent of the participation of women in geography … I knew her in the context of the Commission on Geographic Thought. …. She was good, she was funny also, but she could be as pompous as could be … Again her male colleagues have not been kind to her at all …[in what way?] they spoke cynically about her, they just [said] "Eila" [with dismissive shrug]. She must have been very much alone you know'.[85]

Publications

Campbell co-authored several school texts with Eva Taylor and D.W. Shave. She wrote a chapter on 'The history of navigation' in Wood's (1952) *Exploration and Discovery* and numerous entries on explorers and cartographers for the 1961 edition of the *Encyclopaedia Britannica*. Campbell's MA thesis on 'The history of cartographic symbols' was the foundation for numerous publications in historical cartography, including 'An English philosophico-chronographical chart' (*Imago Mundi* 1949) and 'The early development of the atlas' (*Geography* 1949a). While in New Zealand in 1950 Campbell hoped that Gordon East would petition the university to have her made a 'recognised teacher' of London University. She wrote to him, asked her brother to send a copy of his Faber book and appears to have literally been keeping him sweet by having her post-war sugar ration sent to him while she was away! In a letter to her mother she wrote 'academically and from the point of view of my career, the year in N.Z. has been good. If East has succeeded in getting me Recognised [by the university] it will have been worth all the sugar. If you can please continue to send it until my return to London'.[86] Campbell was recognised by the university in October 1950, her publications having played an important role in this acknowledgement.[87]

Having presented a paper on Fiji at the Edinburgh meeting of the BAAS on return from her year's sabbatical in New Zealand, she forwarded it to the *GJ* with the validation of senior colleague professors Buchanan and Cumberland. Taylor who sat on the RGS research committee wrote to Campbell in 1951, revealing something of the internal politics of the review process: 'Your Fiji paper is accepted. Both Stamp and his co-[professor?] made sly cracks about your over-running your time ... but Wooldridge gave them a good smack back. The Brit. Ass. [her abbreviation] programme is very heavy and papers must be snappy'.[88] As Taylor's papers show, her relationship with Stamp was always edged with friction.

The Domesday Geography of South-East England edited with H.C. Darby (1962), the third volume in the Domesday series, was a pivotal publication for Campbell. According to Wallis (1994) this collaborative work gave Campbell valuable experience in the early years of her academic career, however, while Darby was recognised as an innovative leader in the field of historical geography and had established the pattern for the volumes, Campbell was more than a junior collaborative editor here, she was the major contributor of the content. As the reviewer (H.T.) in *Geography* (1963) noted, Campbell wrote no less than six of the 10 county chapters in this 'meticulous survey', covering Bedfordshire, Hertfordshire, Middlesex, Buckinghamshire, Berkshire and Kent. Darby only wrote the concluding overview chapter. Campbell's archived work records, correspondence and drafts show that she undertook more innovative work in the analysis of the Domesday folios than she has received credit for (Barber 1995; McConnell 2004). Barber concluded: 'It becomes clear that the reason Eila did not publish more in her own name was because she devoted so much of her time to work that has appeared under the names and to the credit of others' (1995: 11).[89] In 1962 Campbell also co-wrote a major overview 'Landmarks in British cartography' with G.R. Crone and R.A. Skelton. This appeared in the *GJ* and Campbell's piece, building on earlier work, addressed 'The beginnings of the characteristic sheet to English maps', or what she described as 'the explanation of signs' (see select bibliography in *Imago Mundi* 1995: 11–12).

Campbell's outstanding contribution to research was through her 48 years of work for *Imago Mundi*. Her involvement with the journal preceded that of Peter Skelton and she undertook a heavy load for the journal long before she became executive editor in 1975 (Barber 1995). At this time the continuation of the journal was due not only to her work but also to her secretly underwriting the costs of volume 27 (Margary 1995). It was within the history of cartography that Campbell 'became an international figure, with her meticulous editorship of *Imago Mundi* setting the seal upon her reputation' (Mead 1995: 7); 'Thus the truest monument to her scholarship is *Imago Mundi*' (Wallis 1994: 361). Her work is little cited within geography today, in part because the topics in which she specialised, cartography and the history of early modern geographical ideas, attract limited attention within the discipline.

Conclusion

The 1920s and 1930s were a period of consolidation for academic geography with the widespread establishment of degree courses in universities and the foundation of the Institute of British Geographers (1933). Women played a significant role in both. These biographical accounts of women beginning their geographical careers in the interwar years demonstrate what Patmore (1987) described as the 'Herculean' breadth of degree-level teaching undertaken in departments with a small number of staff at this time. These teaching loads were much increased by the demands of different types of war work undertaken during 1939 to 1945 as women geographers, with their male colleagues, taught armed forces cadets; drew on their previous travels, research and publications in their contributions to the Naval Intelligence Handbooks (Mutton, Garnett, Shackleton); used fieldwork skills to survey industrial and agricultural land (Smee); and formulated astro-navigation tables for the armed forces (Timberlake). There has been little appreciation of the nature or extent of women geographers' war work within histories of geography (Maddrell 2008) and the war work of 1940s' geography graduates will be examined in the following chapter and the theme addressed more fully in the Conclusion. Specialisation within geography will also be addressed in more detail in Chapters Eight and Nine, but it is important to note its early developments in the careers of the interwar geographers, especially in European regional studies (Mutton, Shackleton, Steers), physical geography (Garnett, Smee, Timberlake and Boswell), and historical geography (Mitchell and Sylvester). Fieldwork was foundational to the London degree and the majority of these women either studied, or took the external degree in geography, at the University of London. Whether a London graduate or not, the majority adopted and incorporated fieldwork at the core of their geographical epistemology and methodology, and practised this in their teaching, research and organisation of field parties. These field parties included those organised through the Le Play Society, the Council for the Promotion of Field Studies, the Geographical Field Group, the IBG and the BAAS. Women were not only active in fieldwork-based research and teaching (Garnett, Boswell, Mitchell, Shackleton and Mutton), but were leaders of international field parties (e.g. Mutton), which were often dominated by women, all of which challenges representations of fieldwork at this time as a masculine preserve. While S.W. Wooldridge is celebrated for his promotion of British fieldwork, he was parochial in his UK focus (Stoddart 1986); in contrast the women discussed were at the heart of both British and European fieldwork and the latter was particularly valuable in their war work. Several of the women had also been involved in the Land Utilisation Survey coordinated by Dudley Stamp (e.g. Snodgrass and Smee), which had great value in planning for post-war Britain.

Chapter Eight

The War Years and Immediate Post-War Period

This chapter continues the themes of fieldwork and war work developed in the previous chapter.

Margaret Dunlop/Davies

Margaret Dunlop was born on 30 September 1914 in Bury, Lancashire, the only daughter of Matthew and Alex Dunlop, who were both school teachers. She was a pupil at Bury Grammar School for Girls and went on to study geography at the University of Manchester 'where she was a favourite pupil of Professor H.J. Fleure'.[1] In 1935 she completed an undergraduate dissertation on 'The influence of climatic change on the history and development of North Central European forests with particular reference to the beech', and in 1936 contributed two chapters on soils to Marion Newbigin's posthumous *Plant and Animal Geography*, edited by Fleure. Deeply influenced by Fleure, it was he who directed Dunlop's interests to archaeology, leading to a series of studies in pre-historical geographies. She completed both her MA (1937) and PhD (1942) at Manchester, writing respective theses on 'The distribution of Bronze Age objects and implements in France, with special reference to British and more distant connections' and 'The megalith monuments of the coastlands of the Irish Sea and North Channel'. As a young graduate she was befriended by Lloyd Praeger and his wife, from Dublin; she undertook botanical fieldwork with them and considered this to be a great good fortune in her life. After taking her first job as secretary of the Manchester Literary and Philosophical Society, Dunlop married geography lecturer Elwyn Davies in 1940.[2] He was called up on war service in 1941 and she replaced him as lecturer in geography until 1944. When Fleure, now retired, visited the USA in 1947, Davies was appointed as

Figure 8.1   Le Play Society field party to Guernsey, Davies second from left front row.
Source: M. Davies Papers, National Library of Wales

acting honorary secretary of the GA with Neville Scarfe.[3] The Davies moved to Cardiff after the war when Elwyn moved into university administration[4] and Margaret became involved with the Cardiff Naturalists Club, going on to become its president in 1951.[5] So began a career of public service for civil society and the state, which drew on her portfolio of expertise and interest in geography, archaeology, botany and related conservation issues. The bodies on which she sat included the Council of the National Museum of Wales and its Archaeology, Art and Science Committees; the National Parks Council (with Alfred Steers) and its successor the Countryside Commission from 1959 – she later became the first chair of the Countryside Commission Committee for Wales. After 25 years in Cardiff, the couple then moved to Tenby in Pembrokeshire in 1970, where Margaret sat on the Pembrokeshire National Park Committee until 1980 and represented the Countryside Commission on the Milford Haven Conservancy Board from 1962 until her death. She also served on the Welsh Council; the Nature Conservancy Committee for Wales; the Department of the Environment's Review (Sandford) Committee; the Committee of the Tenby Museum; the Offa's Dyke Association (of which she was president in 1971) and the Welsh National Education Board. Both Margaret and Elwyn Davies were awarded CBEs for their public service in the New Years Honours List in 1973 and her obituary in *The Times* noted that Davies as an English woman was accepted by the Welsh 'in a way few non-natives are'.[6] She had been known affectionately as 'Dr Margaret' to those she worked with and was described at her funeral as someone who gave much in a 'quiet and unobtrusive way'.[7] Margaret Davies died on 6 October 1982 at Withybush Hospital.[8]

Publications and public work

In many ways Davies might be described as a 'public geographer', and was in great demand as a speaker. In 1960 she addressed the Section E meeting of the BAAS in Cardiff, on the subject of 'The landscape of the Cardiff region'; and between 1969 and 1970 She gave over 30 lectures, speaking to audiences ranging from the National Farmers Union to the Youth Hostel Association and the Townswomen's Guild. Educational audiences she addressed ranged from high-school prize days to the GA, university student societies and the Federation of University Women in Swansea. She also made a radio broadcast on the BBC on the 'Proposed Cambrian Mountains National Park' in July 1969.[9]

Although not pursuing a career in academic geography after her stint as a wartime lecturer, Davies continued to publish in historical geography and her related interests. As a young researcher she published two papers under her maiden name of Dunlop on her archaeological–historical–geographical work (1939) on the significance of Bronze Age trade routes in France, and with H.J. Fleure (1942) on the Glendarragh burial site in the Isle of Man, based on excavations of the site over three seasons 1935–7.[10] In 1946–7 Davies made brief notes in the *GJ* relating to correlations between flora and Prehistoric sites in Ireland; although she robustly defended herself against criticism from A. Farrington (1947) on this matter (see Davies 1947) she did not contribute to the journal again after this encounter. Her research on open field systems of Wales was published in *Geography* (1955) and the *Agricultural History Review* (1956) (it was here that Davies overlapped with Dorothy Sylvester's work on Wales, which she cited). She published *Wales in Maps* (1951) written principally for the crossover between school and university students. It was reviewed favourably in the *GJ*, although the reviewer would have preferred more analysis, and Davies's skilled line drawings were singled out for particular praise (J.G.T. 1952: 216).[11] Davies edited the first *Brecon Beacons National Park Guide* (1967), in which she stressed the primary function of the area as agricultural land. Whilst emphasising the attractions to visitors, she underscored their need to respect the working landscape: 'Upland farms and their stock are very vulnerable to the thoughtlessness of visitors. The ruined farmsteads of the Black Mountains and Mynydd Du should remind them not to add man-made hazards to the problems posed by nature' (1967: 2). In addition to the introduction, Davies wrote the chapter on geology and scenery and one on 'medieval to modern times', in which she stressed the different insights visitors would derive from the varied ways in which the Park's history was written in its landscape. Davies was on the editorial board of the 'New Naturalist' series[12] and in 1970 produced a revised version of Fleure's popular *Natural History of Man in Britain* (1951); it was printed in paperback in 1971 and reprinted in 1989. Reflecting her deep interest and involvement in Pembrokeshire, Davies wrote *The Story of Tenby* (1979) for the Tenby Museum, based on the 40 years of talks and guided walks she gave visitors to the area. Although not Welsh herself, her husband commented that Margaret was known for her 'remarkably full and sensitive knowledge of the Welsh countryside'.[13] *Pembrokeshire Children in History* (1983) was published posthumously.

In a footnote to her husband's obituary in *GJ* it was noted that 'Margaret (Meg) Davies will also be remembered as a significant contributor to geographical literature' (*GJ* 1987: 152). In contrast to this representation as a disciplinary footnote within the record of the history of British geography, Margaret Davies's death was marked by two obituaries in the Welsh press and an obituary in *The Times*. The *Western Telegraph* stated that 'Pembrokeshire, and indeed Wales, has lost one of its leading academic figures'. Stressing her contribution to the National Parks committees, the *Tenby Observer* noted her activity 'in many spheres – as a cartographer, author, lecturer and geographer', and encapsulated her life's work: 'Paramount among her many interests was the welfare of the countryside'. However, as *The Times* recorded, she brought her sharp intellect to bear in all her work: 'her standards were high and she could be devastatingly frank to pretentious and careless writers who tried to climb on the conservation bandwagon'.[14]

Lois R. Latham

Lois Rosamund Latham was one of three sisters, originally from Wakefield. She studied geography at Sheffield, graduating c. 1932, so would have been taught by Rudmose Brown and Alice Garnett.[15] She went on to become editor of the *SGM* (1936–9), taking over from Harriet Wanklyn, and knew some of the Edinburgh geographers well, including David Linton.[16] During the Second World War Latham assisted with work on the Admiralty Intelligence Handbooks, producing the maps for Alice Garnett's volume on Yugloslavia. Lois Latham spent most of her working life at the University of Hull 1946–77, but despite these 31 years there is no mention of her in the departmental history of the early years at Hull written by Harris (1978) except in an appendix listing all staff. Whilst one can understand that an overview may omit short-term appointments, it is hard to credit a departmental history that omits three decades of geographical work, and this inevitably raises the question of why?

Former Liverpool student Herbert King was head of geography at Hull 1928–58 and he created the post of assistant lecturer in 1930 to broaden the staff base required for teaching the wide-ranging London University syllabus. The college relied on part-time staff (e.g. Miss D.E. Trotter, Mr. R.J. Church (1942–5) and Miss M.S. Williamson 1944–6) to keep the courses going while King undertook war duties in London and nobody graduated in geography 1943–6 (Harris 1978). In the immediate post-war years, Harris records that 'Staffing problems were eased only in 1946 with the return to Hull of both King and Fowler and the appointment of an additional member of staff' (1978: 82). That 'additional member of staff' was Lois Latham.

For most of her career Latham was the only woman in the geography department at Hull, even at the time of her retirement in 1977 when there was a staff of over 20. Jay Appleton, whose career at Hull largely overlapped with Latham, remembered her as having a 'somewhat retiring personality' and someone who was dedicated to her teaching: 'As a teacher she was perhaps too conscientious to leave much time for research'.[17] This conscientiousness also impressed other colleagues, as well as frustrating them when it slowed down processes: 'I suspect she did more background reading when marking an essay, dissertation or exam script than the candidates themselves had done! Consequently she was rather slow and younger colleagues became frustrated waiting for scripts to come their way'.[18] Anecdotal evidence suggests that by her retirement her approach to teaching seemed old fashioned and that Professor Harry Wilkinson (head of department 1958–74) was not impressed with her work and that this may explain why he never appointed another female lecturer.[19] If such attitudes existed and were shared, this may account for Latham being expunged from the textual account of the departmental history. She was obviously 'on the margins', but whilst it was remembered that she didn't socialise much with younger members of staff and didn't go on first-year field weeks (ibid.), another colleague remembered her participation in the department's main social events, including the student Geographical Society's annual dinner dance.[20]

Latham, who was widely travelled, used her trips as opportunities to make field observations and to take slides which she catalogued meticulously and used liberally to illustrate her courses. Derek Spooner recollects Latham noting in her speech at her retirement dinner in 1977 that 'she thought she had one accomplishment that none of her colleagues could match – she had travelled in every continent except Antarctica! This revealed a sense of humour, which I had probably underrated' (ibid.). Others clearly associated her with geographical travel as I found a first edition copy of Newbigin's (1922) *Frequented Ways* in a second-hand bookshop in 2005 with her name on the inside cover; the book had been a gift, inscribed 'To L.L. with best wishes from R.B.'. Latham taught mostly regional courses including Britain,

Latin America, USSR and North America, and Spooner who took over the Britain course from her, remembered Latham's approach as very traditional, certainly without reference to the recent literature on regional development or quantitative techniques or models.[21]

In retrospect, Spooner concluded: 'Indeed when I look back I wish I had talked to her more and found out about her life; we all underrated her, and it cannot have been easy for her in a department where there were no other female academics. She was undoubtedly a much more interesting person than I realised at that time'.[22] He also indicated how primary materials are lost, as he kept some of her slides when she retired, belatedly appreciating their value, but could not find anyone who wanted them when he retired in turn.[23] A previous generation's resources tend to be seen as outdated and redundant in the light of contemporary developments and fashions. It is difficult to know whether Lois Latham was marginalised in the Hull geography department by dint of her own personality or departmental culture – or both – but the excision of her from the departmental history seems to tell its own story. Without a lineage of postgraduates or publications to mark her impact on the subject, it is easy for the geographical work of Lois Latham and others like her simply to appear old fashioned to the next generation and her contribution omitted. Latham, on the other hand, did not erase the geography department from her memory but left a bequest in her will, which in acknowledgement of her love of travel has been devoted to a student travel bursary.

The only other women on record as teaching at Hull's geography department during this period were **Doris Trotter** and **Margaret Williamson**, appointed as temporary lecturers during the war. Trotter, who held a first class special degree in geography from UCL was released from her position as geography teacher at Kingston School to take up a part-time assistant lectureship at Hull during the early war years. Margaret Williamson grew up in Ashton-under-Lyme in Lancashire, where she attended the Grammar School 1927–34. She studied at Manchester University gaining a BA (honours) (1937), a Board of Education teacher's certificate (1938), A Diploma in Education (1939) and an MA (1940), the latter completed in conjunction with part-time work. She went on to teach in several institutions (1939–44) and as part of her war service was assistant inspector in the Ministry for Health in Rochdale (1941–2). She was also an 'active worker' in the Workers' Educational Association in North Yorkshire. A student of Fleure's at Manchester, he warmly recommended her: 'You could hardly find anyone more enthusiastic or hardworking', but counselled that she should not head the department as she had no teaching experience. Subsequently Williamson was appointed to the Matlock Training College in 1947 and then at the Manchester Training College in 1963 after a long period without teaching (coinciding with her marriage when she became known as Mrs Dilke).[24]

Figure 8.2   Cuchlaine King c. 1968.
Source: Courtesy of Cuchlaine King

Cuchlaine King

Cuchlaine King, born on 26 June 1922 in Cambridge, grew up in London, daughter of Margaret (née Passingham) and William B.R. King, renowned geologist. Her initial schooling was at a tiny school, followed by Frances Holland Day School, but her sister Margaret hadn't been happy there and the two girls were sent to boarding school at Eversley, Hampshire. King reports that she would have preferred to be at home and that the school didn't offer sufficient tuition in mathematics and physics,[25] but despite this followed in the footsteps of her father and sister to Cambridge (Newnham College had been recommended by family friend, geologist Gertrude Ellis). Her father, Professor of Geology in London in the 1930s, was a huge influence on King's professional life as she grew up with fossil hunting and the geomorphology of the Yorkshire Dales from the age of two when the family took over a holiday home there. Her mother, a more creative person, was influential as a personal role model. Cuchlaine might have followed her father into geology but she didn't have sufficient background in the sciences. Whilst at the geography department at Cambridge during the war years, King experienced a variety of teachers including Jean Mitchell at Newnham and Alfred Steers in the university department.[26] Her sister Margaret was a year ahead of Cuchlaine at Cambridge and whilst she favoured historical geography, Cuchlaine pursued her interest in geomorphology. King graduated from Cambridge in 1943 and served in the Wrens as a meteorologist 1943–6.[27] After the war she returned to Cambridge and was encouraged to undertake research by Bill Williams and her father, both of whom had been involved in D-Day landings-related research during the war. Their involvement with this, Williams' possession of a large set of army-collected data observations from the Blackpool coast and the Cambridge department's possession of a wave tank combined to determine King's research topic. '[It was] both lab and field based – the whole point of the thesis was to tie the two together'.[28] She contrasted the Blackpool findings with data she collected 1947–9 on a sand bar in the Mediterranean near Narbonne. The tideless Mediterranean required special techniques for assessing the beach profile. Williams visited the site with her on two occasions and King credits him with being a very good supervisor. After the successful completion of her PhD in 1949 (examined by Brigadier Bagnold) King was appointed demonstrator in geography at Durham University in 1950, then moved to Nottingham University in 1951, where she was to spend the rest of her career. King was initially appointed as assistant lecturer, replacing Helen Allen, and was promoted to lecturer in 1953. This was followed by promotion to Reader in 1962 and Professor in 1969. This progression sounds unproblematic, but King recalls another male colleague of similar standing being promoted over her: 'I remember Professor Edwards saying that he [the male colleague] should have the professorship first although he thought I probably deserved it, but it wasn't … I can't think how to put it … he thought it would it be better if he had it first and I had a Readership at that stage [Why?] I don't know, I just don't know, I wasn't up in the politics [of the department or university] perhaps they didn't want a lady professor [wry smile]'.[29] It is a reflection of her character that King bore no resentment about being passed over in this way, indeed she preferred both the research orientation of a Readership and the avoidance of administrative burdens, 'I thought a Readership was better than a Chair because it was given on scholarship rather than on administrative ability, which I was never good at … so I think it was fair enough … I hated being Head of Department 73–76 and the sooner it was over the better! … I think there were two other lady professors [at Nottingham] when I got the Chair, amongst I don't know how many'.[30] There were also three other British geography departments headed by women in this period: Alice Garnett and Monica Cole were promoted to Chairs in 1962 and 1964, respectively; Kay MacIver, head of department at St Andrews, did not hold a professorship.

Cuchlaine King retired at 60 to live permanently in Wensleydale, North Yorkshire, in what had started as the family holiday home and continues to be her home at the time of writing. At the end of her career Cuchlaine King was recognised as both an 'eminent scholar and a warm hearted teacher' and received numerous gifts on her retirement 'subscribed to by a large number of donors as a measure of their personal esteem and in recognition of her important services to the Department' (*East Midland Geographer*

1982: 36). One of the physical geography laboratories at Nottingham department of geography was subsequently named for her.

During her working years it seems King's life was given over in its entirety to research and teaching. Her early years at Nottingham included many hours of practical classes, resulting in her carrying the heaviest teaching load in the department. Whilst she would spend vacation time with her parents in Wensleydale, she only had one foreign holiday with them to Switzerland, her long vacations usually being committed to research field-work, summer schools or British Council exchanges. These included a nine-month exchange at Canterbury, New Zealand, December 1956 to September 1957, a year at Binghamton University, New York with Don Coates and Marie Morisawa, a summer teaching a summer school in Boulder, Colorado in 1969, visits to Chile and Argentina in the early 1970s (where she opted to stay with an English-speaking family rather than a hotel and was still in touch with them 40 years later) and China in 1982. She chose to retire promptly at 60, wanting to make the most of her retirement in the Dales, joining a bell-ringing group and other local activities.

Fieldwork was central to King's work and she was a pioneer in normalis-ing women researchers in the Arctic (Sack 2004). This was achieved by her participation initially with a research team organised by Jack Ives in Iceland in 1953 (see Figure 8.3). When King approached Ives about joining the Iceland team he admitted he hadn't thought of including women in the group, but responded positively and suggested getting some other female team members as 'company' for King; the process was repeated in 1954 and again when King joined Ives to undertake work on Baffin Island for the Canadian government (1965 and 1967).[31] King was also part of the team led by Vaughan Lewis working on the Austerdalsbreen glacier in Norway 1955, 1958 and 1959. As King reported to Dorothy Sack, 'In all these expeditions to isolated and difficult areas there was some initial hesitation about taking women, but this was soon overcome and women became a regular part of many similar expeditions. Even the Canadian government overcame its ini-tial reluctance and women became regular members of expeditions to the Arctic' (King, cited by Sack 2004: 448). At the beginning of the Baffin Island fieldwork 'The men stayed at the Dewline Stations [US Air Force] whereas they wouldn't take women there so we stayed at the nursing station in Hall Beach ... they served the local people so we saw more of the local life'.[32] Once in the field proper King relished the isolation, writing in the Nottingham departmental 1968 newsletter 'It's very nice to know that one's nearest neighbour is many miles away' and Joan Fuller as editor noted 'Dr King once more holds the departmental record for the "farthest" [vacation travel]'. However, King was just as likely to be leading a field course in the British Isles, for example Westmorland in early September 1967 or working in the Isle of Man in 1964. In 1964 she attended the IGU Symposium on

Figure 8.3   Cuchlaine King surveying in Iceland 1953.
Source: Courtesy of Cuchlaine King

Glacial Geomorphology at Exeter and Cambridge and was secretary to Section IIIb on Glaciology, Oceanography and Hydrology at the London Congress.[33]

Memories of surveying classes and fieldwork with King featured strongly in student accounts collected by Cole (Cole 2000: 187–9). '[On the Scottish field trip including Skye] There was also the long trek we did with D. King to see Loch Coruisk – an ice-age remnant, I believe, in a wonderful example of a glaciated U-shaped valley. We were dropped off at some spot and set off across miles of rough terrain with a very long haul back; and at the time I thought I was pretty fit, but Dr King was obviously even fitter!' (J. Martin Shaw, student 1962–5). Whilst some students obviously found King's subject matter difficult, another student recounted her approach to geomorphology as 'a model of analytical research' (Cole 2000: 190).

It was a natural progression from King's use of mathematics and statistical techniques in her early research to the quantitative techniques of the 1960s. 'The very first time I got to use any sort of statistical technique was when I was doing my initial research and I learnt how to do regression analysis, and that was from a geologist who was at Reading University – Percy Allen – and once we realised its potential we developed it further'.[34] King wrote *Techniques in Geomorphology* (1966) and when John Cole returned from Ohio State University enthused with quantitative methods in the mid-1960s he wrote some leaflets for the department which led to the co-authored book *Quantitative Geography* (Cole & King 1968). They were assisted by a Nottingham University mathematician: 'Mike McCullagh was very good at computer programming, which was new to all of us then – there hadn't been any computers anywhere around before that – and it was with him that I developed the "Spit Sim" program, the computer analysis of the development of the spit on the south coast, so he also played a part in the development of my interest in quantitative techniques'.[35] This was followed by *Numerical Analysis* (1971) with John Doornkamp. In King's own estimation *Techniques in Geomorphology* was her one of her best books and she particularly enjoyed the collaborative process of writing *Quantitative Geography* with John Cole.[36]

Publications

King published widely throughout her career, largely relating to three themes: glacial geomorphology; coastal geomorphology; and geographical techniques. Her first paper, co-written with Bill Williams, on sand bars and wave action appeared in the *GJ* in 1949. This was followed by many single and co-authored papers including a 1969 paper on 'Moraine types on Henry Kater Peninsula, East Baffin Island, N.W.T., Canada' and another the same year co-written with Jane T. Buckley (from the Canadian Geological Survey) on 'Geomorphological investigations in West-Central Baffin Island, N.W.T., Canada'. Much of King's longitudinal study of Lincolnshire coastal processes was published in the *East Midlands Geographer* (*EMG*) with its strong empirical emphasis (see Barnes & King 1955, 1957, 1961; King 1964, 1968, 1978). A whole issue of *EMG* in 1978 was devoted to research at and around the Gibraltar point, which became iconic for Nottingham students and coastal researchers (Matless 1999).

King's books also relate to her three key research and teaching themes: techniques (as discussed above), coastal and glacial geomorphology. Her first geomorphology book *Beaches and Coasts* was published in 1959, soon to be followed by six others: *Oceanography for Geographers* (1962), *Introduction to Marine Geology and Geomorphology* (1975a), *Introduction to Physical and Biological Oceanography* (1975b) (the last two expanded from the 1962 book); *Glacial and Periglacial Geomorphology* (1968) with C. Embleton, *Northern England* (1976). She also edited two volumes of papers on

Periglacial Processes and Landforms and *Geomorphology* (both 1976) for the Benchmark Papers in Geology series.

Beaches and Coasts (1959) was well received. Contrasting it with other more descriptive publications, W.G.V. Balchin praised the volume for its emphasis on processes, as well as its 'exhaustive analysis', use of recent experimental data and combined laboratory and field data. Balchin considered the book offered 'a new and unique approach with a mathematical and quantitative treatment' (1960: 231), which was not only of interest to academic study but of relevance to coastal management. 'In the past the geomorphologist has derived a great deal of help and information from the civil engineer so far as coastal matters have been concerned. With the appearance of Dr. King's book, however, this debt has undoubtedly now become a credit balance' (ibid.). Writing principally for an American readership, Evelyn Pruitt also considered King had performed a courageous and 'valuable service' to coastal specialists, combining and 'filtering' such a huge range of data and sources. Whilst Pruitt considered King's fourfold classification of coasts and the layout of the book to have limitations, she thought these 'can be discounted in appraising the great usefulness of the work' (1961: 460). Swallow (1963: 548) described *Oceanography for Geographers* as a 'wholly admirable book' much needed to fill a gap in the literature between introductory volumes and detailed research monographs on the subject, but while commending King for her 'encyclopaedic knowledge' and up-to-date references, found fault in what was perceived as her lack of passion: 'A comment might be made that the whole treatment is too objective, giving little indication of the excitement of new discoveries and the rapid growth in recent years of a world-wide science of peculiar interest to this country with its maritime history' (ibid.). King's books on oceanography sold particularly well in the USA and were the source of her largest royalty payments.[37] Reviewing *Techniques in Geomorphology*, Sparks considered it a valuable 'high-level summary' comparable to *Beaches and Coasts* as a major contribution to geography and an invaluable reference text for advanced students and young researchers. As with her first book, it was King's ability to handle such a wide range of difficult material which generated the highest praise: 'the range of techniques covered by Dr. King leaves one in astonishment and admiration that it was possible for any one person to cover such a field' (1966: 556). Co-authored books are inevitably more difficult to read in terms of their reception in relation to individual contributors and can be written in quite different ways ranging from fully integrated co-authorship to pasting completely independent chapters together. *Glacial and Periglacial Geomorphology*, dedicated to the memory of Vaughan Lewis, was welcomed for its emphasis on explanation and as the first textbook to recognise glaciers as 'dynamic features of the present landscape ... the authors' emphasis on glacierization processes, mass balance, and flow theory is refreshing and innovative' and the material on periglaciation as 'the most comprehensive English-language summary to date' (Marcus 1969: 635–6). Marcus concluded that the book 'should become a standard reference, and we should be grateful to Embleton and King for producing it' (ibid.). George (1968: 589–90) considered it an 'encyclopaedia rather than a text book', but nonetheless considered it an 'invaluable compendium for the practical geographer'. Jack Ives (1977) thought the revised 1978 edition would 'occupy a vital place in the literature' but considered it flawed, not least in relation to simultaneous publications which challenged their interpretation of nivation in cirque formation. *Quantitative Geography and Numerical Analysis* received much more mixed reviews. The first was intended to be a handbook for all geography students, regardless of mathematical background and as such welcomed; but its very breadth was also considered its downfall: despite its 'forbidding bulk' (Court 1971) the authors 'possibly paid the penalty of neither covering any topic in sufficient depth nor satisfying the needs of any particular market' (Goddard 1970). It was also criticised for failing to distinguish between the merits of particular techniques (ibid.). Hence the reviewers recommended as 'useful but not the definitive text' and it should be used with 'careful selection' (Court 1971). *Numerical Analysis* was met with similar criticism in that it needed to provide a better indication of why a particular test should be used in a particular case, as well as needing more conceptual coherence in the introduction (Clayton 1972a). Specialists also found fault with the selection and editing of the two collections of Benchmark Papers in Geology edited by King (Caine 1976; Beckinsale 1978). Despite these criticisms, King made a huge contribution to coastal and glacial geomorphology literature throughout her career.

King was awarded the RGS Gill Memorial Award in 1961, and (like her father before her) was approved for a ScD by the University of Cambridge in 1973. She received the David Linton prize in 1991, awarded to geomorphologists making leading contributions to the discipline over a sustained period. In receiving their own distinguished career awards, in 1998 and 2000, respectively, John Andrew and Jack Ives acknowledged their debt to King. Despite King's assertion that Andrew needed little supervision for his PhD and knew more on his subject than she did, he stated 'Professor King ... taught me by her example and her kindness what research was about'.

Mary Marshall

Mary Marshall, born 5 April 1917, was the only daughter of Mr and Mrs Walter Marshall. Her father was a machine tool fitter at the time she went to university.[38] She was educated at the Municipal High School at Doncaster

and was awarded an open exhibition scholarship at Lady Margaret Hall
(LMH), University of Oxford, to read geography in 1936. She studied
under J.N.L. Baker and seems to have thrived academically, continuing her
excellent examination record. Marshall graduated with first class honours
in 1939, was awarded the H.O. Beckit Memorial prize and went on to be
appointed as a research scholar at LMH and to receive the LMH Maude
Royden travelling scholarship (Buxton 1984). Marshall started research on
the Canal du Midi in Southern France, but the outbreak of war forced her
to halt her research project and return home (Scargill 1999). She resigned
her research scholarship and worked as cartographer to the Oxfordshire
War Agricultural Committee and in 1941 was co-opted by Professor
Kenneth Mason to assist with the production of the Oxford volumes of the
Naval Intelligence Handbooks which he was commissioning and editing.
Marshall is cited as co-author with Mason for the volumes on Turkey and
Western Arabia and the Red Sea.[39] During this time Marshall was also com-
pleting her major Land Utilisation Survey report on Oxfordshire (1943), as
well as examining for the Honours School in Geography and fire-watching
by night.

Under the University of Oxford's federal-like college system, Mary
Marshall was attached to her own undergraduate college, Lady Margaret
Hall, as part-time lecturer in geography from 1944 and as the Susette Taylor
fellow from 1948. Her appointment at LMH was to the great pleasure of
geography students there and she gained a reputation for being an 'exacting
but extremely stimulating tutor' (Buxton 1984: 35). One former student
remembers Marshall's inclination to take students with potential rather
than relying on existing performance,[40] and she was known for her kindness
(including financial support) and for maintaining correspondence with
former students and friends (Buxton 1984). Marshall was appointed departmental lecturer at the University of Oxford School of Geography in 1945,
was promoted to senior lecturer in 1954, a post she held until she took early
retirement in 1982 (Buxton 1984; Scargill 1999). As part of her university
responsibilities Marshall taught a range of courses with the history of landform studies as her main focus; her papers[41] also show her role in organising
and co-running field courses in Montpellier, France in 1964 and 1970. At
her best in more informal teaching settings such as tutorial and fieldwork,
her lectures, whilst up to date and well prepared, lacked the same 'sparkle'
of her smaller group work (Buxton 1984); an evaluation echoed by a former
student who remembered Marshall's lectures as boring but appreciated the
'meticulous feedback' on essays.[42] Her papers on the history of geography
and ideas include a collection of news cuttings on geographers and travellers, such as reviews of a biography of Gertrude Bell and an old article
about women's admission to the BAAS, suggesting an interest in women's
geographical knowledge and status within the Academy. The outline of a

1965 co-taught geographical ideas seminar[43] includes quantitative methods in her section on physical geography and references ranging from Mackinder to Chorley and Bunge, showing she drew on both historical and very up-to-date sources and debates in her teaching.

Marshall suffered from ill health throughout much of her career, thought to be attributable to overwork during the war years.[44] During the 1950s severe tinnitus caused her to keep unusual hours at the School of Geography as it was easier for her to cope when ambient noise was lower in the evenings. Her part-time post at LMH was discontinued in 1961, which was considered something of a scandal at the time,[45] but she continued as honorary research fellow from 1961 (LMH Register 1879–1990) and in her post as university lecturer where colleague friends such as Freddie Martin and Paul Maggitt supported her.[46] A few years later Marshall was appointed as fellow of the new St Cross College, Oxford, a graduate college for the university, from 1965. This gave her a new college base within the university, but she continued to be plagued by ill health and in the early 1970s underwent three major operations on her sinuses in 10 days and it was over three months before she could see to read or write. She took early retirement in 1982 and died soon after in 1983.

Kay MacIver

Kay MacIver was born in Dundee on 3 February 1921, daughter of Jessie (née Reid) and Kenneth MacIver, an Edinburgh graduate and minister in the Church of Scotland.[47] She attended Birkenhead High School, a public day school for girls in what was then the county of Cheshire. She was head of school while the school was evacuated to Shrewsbury High School during the war. While there she studied for the University of Oxford entrance examinations. 'I really must have liked geography because I studied it at school under a woman who I disliked intensely and it was really a split between geography and history'. The idea of medieval Latin in second-year history was sufficiently off-putting to seal her choice and she entered Lady Margaret Hall to study geography. At LMH geography was treated as a science, so MacIver who had taken English, geography and history, with a subsidiary in German, for her Higher Schools Certificate had to compensate with extra study for the Oxford entrance examinations: 'I can remember mugging up on botany on the top back staircase [at Shrewsbury], trying to learn enough to get through. So I had to take botany, geology and physics, maths and chemistry, with the object of getting through in two terms'.[48]

At Oxford there were only one or two other geographers across the years at LMH and they were looked after in college by a science fellow, this being prior to Marshall's appointment. Being at Oxford (1941–4) during the war years had a big impact on university life. They grew potatoes in the college gardens and MacIver was one of a group selected for a government research project on feeding, when they were given a largely vegetable diet with 'two great bowls of pills' and examined at the beginning and end of the term. The geography department was a hub of activity with most of the geography lecturers involved with Admiralty Intelligence Handbooks and military despatch riders going to and fro on motorbikes. Inevitably the war made for a more feminised student body, but this began to change in the last years of the war. 'In my last year we began to get some of the men back and that of course brought a certain maturity of thought from the men who had seen action; I had a very good friend who came back in my third year'.[49] At the School of Geography MacIver was taught by J.N.L. Baker, Robert Steel, W.E. Gilbert, Sylvia Kock and Billy Martin but the staff were preoccupied with their war work and she remembers some lectures as 'a little bit curious'. She found Baker's historical geography classes on explorers very descriptive, with 'nothing you could get your teeth into', leading her to conclude that historical geography 'as it was taught then was not a real degree subject'. Indeed she found most of the geography lectures dull and in true Oxford style opted out where possible, going to other departments and colleges to hear more stimulating lecturers including C.S. Lewis and Oliver Quick. Lectures may not have been compulsory but university regulations dress codes were, and MacIver was reprimanded for attending a Saturday morning lecture without her academic gown. MacIver's account of one member of staff who made repeated advances to women students demonstrates that some lecturers exploited their position: 'the man to avoid, he was called Roly Poly, you never went out on a bicycle ride with Roly Poly alone, otherwise you regretted it. I remember sitting on the terrace of the Swan hotel in Goring [when he made some sort of advance] so having heard from everyone what he was like I said [firm tone] 'Excuse me Mr [x]!'[50]

MacIver gained a first class honours degree, and due to wartime conscription of women was immediately directed into the labour force. She went to London to be assessed and was classified as 'administrative', a category dominated by university women. Hoping to join the Wrens, MacIver was sent to join the Civil Service, something which seemed to be synonymous with an Oxford education. MacIver chose to join one of the Scottish offices and was allocated to the Housing Department in Edinburgh. There she shared a house with other women, including some geographers, who worked in the Planning Office and she was part of a team addressing housing conditions where babies with rat-bitten toes were a fact of life and she was advised to dust the hem of her skirt with flea powder when undertaking home visits. In her own words 'I grew up ... about six years in six months' but at a time when urban geography was largely unknown (especially at Oxford) she didn't see this work as in any way geographical.[51]

Knowing of MacIver's interest in Africa, a cousin contacted her when the Colonial Office began recruiting women for trial postings across the Empire. After many rounds of interviews to whittle the 1200 applicants down to 12, MacIver reached the final round. The interview was held at Burlington House and lasted an 1 hour and 20 minutes, sat around an enormous shiny table with the interview panel on one side and her on the other. She received brusque treatment from the panel: despite the social convention that required women to wear a hat, she was told 'Take off your hat. I never interview a woman with a hat on'; and in response to an opinion she had given, the ex-governor of Nigeria had responded sneeringly, '"Oh I see, Miss MacIver ... you would see yourself as a non-religious missionary". That was it. I'd had it. So I just gave him what I thought and for the remaining twenty to twenty five minutes I threw everything out ... When I came out I thought I'd be like the one before me, in tears, but I was just absolutely exhausted. Nobody could have been more surprised than I when I got a phone call from a friend in London who said "Kay, what do you know about Uganda?" Oh Uganda, around central Africa, Lake Victoria. "Well I think you should get to know it a bit better". This was offering me the job.'[52]

MacIver took the lowly assistant's job offered and sailed from Liverpool in troop ship conditions. As a single woman she acted as a nanny for children on board and made friends with one of the medical officers. She spent six months or so in the Resident's Office in Kampala where she began what was to be a long interest in land rights. It was 'something I got my teeth into and it continued the whole time I was there, the whole business of land law, I thought I could tack that on to my geographical knowledge quite usefully, because most of the civil law in most of the colonies and protectorates were based on native law and custom, whereas criminal law was based on the British system'.[53] When she was moved to Entebbe she continued working on land law and then oversaw hospital supplies and ambulance provision. In retrospect she thought she had been too inexperienced for the two and a half year posting and found Entebbe had been 'quite hard going'.

She returned to the UK in late 1948 and applied for two academic geography posts in Edinburgh and Reading in the following year, but 'couldn't take the Reading job because I would have no time for my research,' so opted for an interim job teaching at Sherbrook School for Girls for a term. This was followed by a real contrast, teaching at the Dame Harris Harper School in Bedford, where she had to be 'a few jumps ahead' of the A-level geography class and taught other more socially demanding non-examination classes. This experience helped MacIver realise she didn't want to be teaching classes where she was going to have to control the pupils as well, but still found it a good learning experience. MacIver was then appointed by Alan Ogilvie to a lectureship at Edinburgh University where she taught from 1949 to 1953. Edinburgh University required staff to gain postgraduate qualifications, so Ogilvie encouraged MacIver to use her African experience as the starting point for a PhD. She registered with Birkbeck College as an external student and began her statistical analysis of the changing population of Uganda in 1950. Trying to get accurate data was problematic:

> It was very difficult to find out anything, there just weren't any [statistics] … I was a bit cheesed off, but I did what I could. It was very hard work then, getting hold of every single piece of paper which had been published on Uganda, but I found Rhodes House and the Colonial Office library very good. I spent two summers slogging away entirely on this… in the end I thoroughly enjoyed it, but you can't do a full-time lecturing job and do a fairly solid piece of work in a very short period of time… I had an enormous amount of calculations and had to use log tables, it took hours … and I had to find the evidence, why populations always grew despite cuts in the population, like the big sleeping sickness epidemic which ripped through Uganda, which was then the most populated area and yet within five years that population was back again with the same numbers.[54]

The PhD was completed in 1954 and examined by Frank Debenham from Cambridge, whom she remembered as 'terribly interested' in her findings.

The department at Edinburgh was full of ex-service people, both staff and students: 'There was [a student] in his naval jersey, sitting with his arms behind his head looking at me and thinking 'if that woman doesn't say something interesting I'm just going to walk straight out of here', and there I was standing at what had been Lister's demonstration bench, which at least was solid which meant they couldn't see my quaking boots! I really worked extremely hard because I had a very big lecture programme and a pretty big practical programme'. MacIver went on to describe the ex-naval student above as 'the first person I recognised as being so much brighter than I was myself, that was part of growing up'.[55] She found Ogilvie to be a very good head of department, watchful without being interfering. MacIver thought that Catherine Snodgrass regarded junior colleagues as 'young useless creatures'; and Arthur Geddes 'would ring up in the morning and say "Kay dear, would you mind taking my lecture today because I think I've got a bit of a cold and I'm not coming in this morning", so I had to learn my own course and his to a certain extent, all very much on the hoof'.[56]

In 1953 MacIver took the opportunity to move to the small geography department at St Andrews. When the head of department died MacIver temporarily managed the department, becoming senior lecturer and formal head of department in 1961. It was then she felt faced with a definite choice between being an effective head of department and following a personal research career: 'I can remember very clearly ….saying to myself "You must either get out of here and do some more research or you must stay and be head of department. There was no other way". During her time as head of department there was much time-consuming political manoeuvring in which MacIver successfully gained new accommodation for the department and fought off a proposal to replace geography with geology, as well as managing a 120% increase in student intake. All her energies went into the department 'everyone of them, because it was a hard slog'.[57]

However, MacIver and her team were successful in building a strong department during the period of university expansion. She was the only female to head a geography department in the UK until Monica Cole was appointed at Bedford in 1964; at the time she didn't particularly think of herself as exceptional, but on looking back when she retired realised how unusual her position was. MacIver's first University Senate meeting at St Andrews was a moment when she was conscious that as a woman she had to make very clear points to be heard in that forum. Also while she felt her 'colleagues were really good, they just accepted me', there were occasional social issues within the department, for example one male colleague was sensitive about a woman paying on social occasions which cut across the dominant gendered roles and practices of masculinity of the time, so when they took students out for an end of year meal, MacIver discreetly gave him the money to pay the bill!

Throughout her time at St Andrews MacIver was the only woman in the geography department, despite expansion in the 1960s and was challenged about this at the time: 'Every time there was a vacancy someone would chip in and say "Oh we see, we've heard of this before, women won't appoint women".' MacIver's response was that she would appoint a woman 'if I got one good enough',[58] insisting she went over women's applications twice, but 'there just didn't seem to be any woman who were, or who appeared anyway, to be good enough'. She attributed this to the relative scarcity of women academics in the late 1950s and early 1960s, there still being 'a prevailing thought that women didn't do this'.[59]

When MacIver first went to St Andrews she taught everything from meteorology to settlement and it was only in later years that she was able to specialise. She was active in running departmental field courses ranging from local first-year fieldwork to third-year courses abroad, typically in the South of France at Montpellier or Northern Germany. 'We had some very good field classes, [with groups of about 40 students] which was a nice number, because they were third years you'd already had them for two and a half years before you went, so you *knew* them'.[60] For MacIver, the ultimate success of her career was that the department had 'educated a good lot of students'.[61]

MacIver was active in the IBG, served on Council for three years 1964–7 and hosted the 1966 annual conference at St Andrews, which was a 'huge undertaking' for her and the rest of the department at a time when the academic hosts were expected to handle bookings, transport, accommodation, food and the academic and social programmes. She was also a long-serving member of the RSGS Council, sitting from 1957 to 1970.

MacIver, faced with asking colleagues to consider redundancy during a period of financial crisis, opted to take early retirement herself. She stayed on as assistant dean of students in the faculty of arts 1978–80, and as dean 1980–4, becoming Master of St Andrews United College 1985–8.[62] She also oversaw a hall of residence and the North American exchange programme. However, the pay for a temporary appointment represented 'about a quarter' of her previous salary. Kay MacIver was awarded an honorary DDL by St Andrews University in July 1991 and still lives in St Andrews at the time of writing.

Swanzie Agnew

Swanzie[63] Erskine was born in Heidelberg Transvaal, South Africa, 9 June 1916, daughter of Elizabeth (née Reinders) and her Scottish husband Esme Nourse Erskine. Schooled in Pietermaritzburg, South Africa, Swanzie Erskine travelled to Edinburgh University to read geography in 1932. She gained her MA in 1937 with first class honours and a University Blue for tennis. In an interview late in life she recalled a year cohort of about six students which included two other women (Jean Boyd and Sheila Froud); Professor Ogilvie taught first-year classes, David Linton taught geomorphology, and Arthur Geddes, son of Patrick, gave 'a Patrick Geddes tinge to the whole department', not least through the influence of Le Play, Vidal de la Blache and Lefebvre (students had to read the originals), as well as a regional interest in India.[64] Erskine married Sir Fulque Agnew of Locknaw on 9 October 1937 (Steven 2000), after which the couple cycled from Scotland to Montpellier.[65] Having been to the College des Ecosses at Montpellier University as a final year undergraduate, Agnew returned to the area to begin doctoral research 1938–9, but this was curtailed by the onset of war and she returned to Edinburgh, her research unfinished. There she was employed as an assistant lecturer in geography 1939–47, filling wartime staff shortfall, the department having lost Arthur Geddes and David Linton to war work. Snodgrass and Agnew carried much of the teaching in the early war years and Agnew recalled taking over Geddes' course on human geography, using a combination of his text and her own student notes. Her son Crispin was born in 1944, which makes her one of the first women university geography lecturers to have a baby while in post. In 1946 she was invited to MacMaster University, Ontario, by Wreford Watson and she spent a year in Canada as an exchange lecturer before joining her husband in South Africa in 1947. The couple farmed in Natal and Agnew taught at Harburg School 1947–52 before working as lecturer and then head of the geography department at Fort Hare University, where both she and her husband worked with the Church of Scotland Foundation (Steven 2000). At this time Fort Hare was the only university in Sub-Saharan Africa which awarded black students with university degrees and it is thought that Agnew taught the university's most famous graduate, Nelson Mandela (ibid.). The university was a target for the apartheid-driven government and was closed in 1960 when troops were brought onto campus and all staff dismissed for anti-apartheid activities. Whilst South African staff were put on trial, British citizens were exiled. The family was repatriated to the UK and Swanzie Agnew worked at Oxford University to complete *A Historical Geography of South Africa* (1960) with Fort Hare colleague Norman Pollock.

Agnew's career then took an interesting turn when she became academic head of the Royal Ballet School, Richmond 1960–5. After this five-year hiatus Agnew and family returned to Africa, where she took a post as professor of geography at Malawi University, the chancellor being a friend,[66] (Steven (2000) describes this as responding to the call of Africa).[67] She was to stay there 10 years 1965–75, but felt increasingly under intellectual and moral pressure by President Banda's regime, which required particular pedagogical approaches within the educational system and increasingly used repressive military force to control citizens. Agnew resigned in 1976 in protest at this repressive regime. She returned to Edinburgh for her retirement (her husband died in the 1970s), pursuing painting, pottery and poetry, donating many of her works to raise funds for cancer research. Lady Swanzie Agnew died in Edinburgh on 28 September 2000 and was buried in the Agnew family mausoleum at Locknaw Castle in Galloway. Swanzie Agnew and her educationalist husband had shared their liberal views, which became radical views in the context of the repressive regimes of apartheid South Africa and President Banda's Malawi. She is remembered as a person of style, elegance and energy, an original academic with strident principles, who 'never did compromise those fervently held views about humanity and a more tolerant society' (Steven 2000: 14). A poet herself, the outline of Agnew's biography in the order of service at her memorial concluded with a quote from Indian poet Tagore: 'Give me the strength never to disown the poor or to bend my knee to insolent might', indicative of her personal and political philosophy.

Publications

As is the case with numerous of the women studied here (and many of their male contemporaries), to count Swanzie Agnew's influence on geographical knowledge purely in terms of publications is at best limited. She played an important role through her principled life and teaching of university geography contra regimes which were repressive and racist. Nonetheless, she had several publications to her name. Three journal articles appeared during 1945–6 reflecting her Scottish and French place attachments: 'Montpellier: giver of names' (*SGM*), 'The Vine in Bas Languedoc' (*The Geographical Review*) and 'Brig o'Turk: a Highland aspect' (*SGM*). Her two related papers on Montpellier and the Languedoc looked at the historical geography of the town and the economic geography of the broader region, respectively, obviously based on her period of study at the College des Ecosses. In the Bas Languedoc problems with the stability of the economy and food production were identified as arising from dependence on a mono-culture which was vulnerable to international economic and political changes (including the Depression and the war) as well as the climate. Agnew concluded that with the innovations of cooperatives, refrigeration and some diversification, the vineyard economy could thrive during the post-war years. Her study of Brig o'Turk follows a similar regional description (starting with the geology and topography) and analysis of economic practices, limitations and opportunities for development.

A Historical Geography of South Africa (1963), jointly written with N.J. Pollock, was Agnew's most substantial publication; it appeared in a prestigious geographical series, but did not receive a good critical response. Whilst not doubting there was much of value in the book, Andrew Clark, writing in *The Geographical Review* (1965) criticised the book's 'elliptical style' and chaotic organisation and the lack of clear sense of place, he also criticised the maps, which had been a strong feature of Agnew's earlier papers. It is difficult to know whether the book fell foul of the history/geography divide (as Clark thinks) or whether the authors' summary dismissal from South Africa followed by relocation and new jobs took its toll. Holt's (1973) brief review of *Malawi in Maps* (1972) edited by Agnew and G.M. Stubbs, criticises its failure to include a map of water resources, but otherwise commends the clear maps in the atlas. The book is still found in the geography library at the University of Malawi and has never been updated.[68] In her own review (1978: 502) of A.J. Christopher's (1976) *Southern Africa*, Agnew's view of the inadequacy of a dehumanised predominantly archive-sourced white settler historical geography is made clear:

> In numerous figures of the various land holding and town plans extracted from archives, we are given the bones of the evolving cultural landscape with-out the substance of the men who responded to the structures placed upon them. This is a shortcoming of the classical approach in historical geography. But here, where the subject concerns the accommodation of conflicting racial and ethnic groups to the occupation of land, then the absence of all but a superficial consideration of the African claims becomes a limitation. The author recognises this and justifies it by appealing to the poor documentation

extant on other racial groups. He sees the Africans as a separate entity or buffer to settlement that the European diaspora encountered in southern Africa, and which diverted the 'classic wave' of the American model into new channels.

There is a clear resonance here with Gladys Hickman's contemporary work on South Africa (see Chapter Five), but a clash with the views of Monica Cole (see Chapter Nine).

G. Joan Fuller

Gwendoline Joan Fuller was born on 21 February 1909[69] at the Wesleyan Manse, County Road, Swindon, daughter of Ethel Gwendolen (née Hutchison) and the Reverend H.W. Fuller. The family moved five times with the minister's successive appointments while Joan grew up[70] and she attended the Central High School for Girls in Manchester 1921–5, before completing her schooling in Nottingham (1925–6) (Wheeler 1992). She studied at the University College Nottingham 1926–9 and attained a first class honours degree in geography (London External Board) – the first Nottingham student to do so. This was followed by a year at the Cambridge Training College for Women 1929–30 (her family were now in Barnstaple) and several teaching appointments (Abergavenny 1930, Wellington 1931–5 and Worthing Merchant Taylor's School 1935–40 (Wheeler 1992; Cole 2000)). During this time she obviously retained her interest in geography, the Le Play Society and a connection with the department as she completed an MA on Shrewsbury in 1940 under the supervision of the head of department K.C. Edwards.

Fuller was appointed temporary assistant lecturer in January 1942, one of four women who taught in the department through the war years: Gladys Hickman of Goldsmith's College, who had been evacuated with her students and small son, taught the much disliked regional course 'The remaining continents' 1940–1; a Miss M. Styan also taught in the same year; and Miss Phyllis A. Nicklin, a graduate of Birmingham University taught in the department 1944–5, and was described as 'having considerably strengthened' the teaching work of the department (Cole 2000; see Chapter Five for Hickman and Nicklin). Fuller's contribution was immediately recorded in the 1941–2 Annual Report as 'already proved of great value'. In the following year her innovative use of geographical films in teaching was noted as well as her industrial survey of East Midlands' towns, undertaken with Miss Illston for the Board of Trade, 'for which they were highly commended' (Annual Report 1942–3). Fuller assisted K.C. Edwards in 'investigations on behalf of the Ministry of Town and Country Planning' and took on some of his departmental administrative duties (1944–5) in order to free him up for his other part-time war work. At the same time Fuller was noted to be making good progress in her research, all of which gives a sense of the huge loads carried by academic geographers in the war years, given their teaching and extra departmental work, coupled with part-time specialist and generic war work and, in Fuller's case, research for a doctorate. The reward was promotion to the post of lecturer in 1946, with Edwards writing strongly in support of her permanent tenure (Wheeler 1992). In the same year that Fuller was promoted, **Miss W. Allen** and Mr H.A. Moisley were appointed as assistant lecturers[71] and **Miss S. Carter** was appointed as cartographic assistant 1946–7. Fuller continued her research in the immediate post-war years and was awarded her PhD (London) in 1949 for her thesis on 'A geographical study of the development of roads through the Surrey–Sussex Weald and to the South Coast during the period 1700–1900'.

After the war Fuller's teaching focused on historical geography, geographical exploration and North America and historical field courses for finalists (Cole 2000). Given her long attachment to and research on Sussex, it was very appropriate that Fuller should present a paper at the Brighton meeting of the BAAS in 1948 on her doctoral work on the historical geography of roads in Sussex (she was the only woman to present a paper in the Geographical Section that year). She went on to present a similar paper to the GA at Sheffield that year, the final paper being published in *TIBG* in 1953.[72] In total Fuller published about 20 papers and reports: her paper on 'The Peak lead industry' appeared in the BAAS volume on 'Nottingham and its region' (1966), but the core of her academic writing appeared in the *East Midland Geographer* (1954, 1955, 1957, 1965 and 1970), all on the historical geographies of the Fens, the Peak District and Sussex, with particular reference to roads and lead mining. She also wrote 'The Peak District National Park. Field Studies in the British Isles' for the 20th meeting of the IGC in 1964. *The Coming of the Europeans* (1962), co-authored with John Grenville from the Department of History at Nottingham, was Fuller's only book and its publication made her 'most happy' (Cole 2000). She was promoted to senior lecturer the same year the book appeared and was responsible for supervising postgraduates in historical geography.

In addition to this teaching and associated research fieldwork and publications, Fuller made three crucial contributions to the department and wider geographical community, namely through her work in geographical education and the GA, for the Geographical Field Group and the departmental newsletter. Having trained and worked as a teacher for 10 years before her appointment at Nottingham, Fuller used this experience within and beyond the university. In addition to her work in the geography department she was active in the university's Institute of Education and was chair of the Nottingham branch of the GA (Wheeler 1992). The Le Play Society and its student group (which Edwards headed) played a vital role in providing opportunities for training geographers in fieldwork at home and abroad in the pre-war years, given that so little fieldwork was being done at this time (Wheeler 1967). Administrative problems led to the separation of the Student Group from the Le Play Society and the Geographical Field Group (GFG) was formed to replace the Student Group in 1946 (still headed by Edwards) (Wheeler 1967). Fuller, who had been involved with the Student Group, continued and developed her involvement with the new GFG, and led at least six foreign field courses, and two British weekend field courses on the Fenlands (1960) and the Peak District National Park (1963). She went on to become chair of the GFG 1962–9 and was one of the longest serving members of the group (Wheeler 1992). The GFG and Fuller's involvement with it sheds interesting light on the question of women geographers and fieldwork. The GFG was not a masculine preserve: 12 of the foreign field courses run between 1949 and 1964 were led by women (see Wheeler 1967, Appendix 2); and the copies of group reports for courses headed by Fuller show group members to be overwhelmingly female (see Table 8.1). The purpose of the group was to investigate a small region in detail, using the specialist resources of the group. Whilst a party leader would have overall responsibility for the course, the group was based on very corporate principles, with a physical geographer leading a day's study on the locality's geomorphology and an economic geographer a day on local industry, for example. These parties were more than excursions, they involved serious study, what in today's terms might be described as 'continuous professional development' or 'serious leisure' (see Rojek 2000). All except one woman and one man in the 18-strong party of 1961 are listed as holding degrees, some group members can be identified as university teachers, but it is likely that the majority were school teachers or involved

Table 8.1   Foreign geography field group courses led by Joan Fuller

Year	Location	Gender of party members including leaders
1951	Les Marecottes, Valais, Switzerland	25 women; 10 men
1957	Saignelegier, Jura, Switzerland	15 women; 3 men
1958	Elba, Italy	16 women[*]; 1 man, 1 boy
1960	Vrsar, Istria Rosa, Yugoslavia	16 women; 2 men
1961	Salerno and Sele Plain, Italy	Not available
1964	Paris, France	Not available

[*] Records of group members show at least 12 of whom were single, including Fuller herself.
Source: data extracted from: Wheeler (1967, 1992) and Fuller papers at the University of Nottingham Library.

with geography teacher education, seeking to broaden their geographical knowledge and experience. The GFG provided a network for like-minded people to meet, travel and develop geographical understanding through corporate endeavour, within the safe environment of the group. Cuchlaine King, who attended a number of the GFG field parties in her early years at Nottingham, suggested that these were favoured by women because they were organised and offered opportunities for fieldwork and travel without the risks of doing these things independently in 'wild' areas (King's own fieldwork was groundbreaking for women geomorphologists – see above).[73] Once home, illustrated reports were complied based on research in the field and then circulated in turn to group members, who would hold the report for a period then pass it on. No doubt these provided the basis of many a local talk or materials for a university lecture or school class, but they were not formally published and were very limited in their circulation. Although enormously influential for the participants, this sort of geographical writing was informal in publishing terms and ephemeral in the historical records; they do not conform to the notion of the geographical canon. Yet the records retained (in this case by Nottingham University Library Archive) provide valuable insight to the production of this collective geographical knowledge. Fuller also travelled in North America when she was awarded a six-month sabbatical in 1955 courtesy of the American Embassy Smith-Mundt award. Some of her travel resulted in mainstream publications, such as her papers in *Geography* on the Trient Valley (1955) and the port of New York (1959) and a co-authored paper on the Upper Soča Valley in *Geographical Studies* (1955). Travel was a large feature of Joan Fuller's vacation activity and self-representation (in keeping with many other contemporary geographers) and her departmental newsletters usually reported her summer travels with the GFG or other friends, providing readers with tips about suitable places for 'a geographer on holiday'.

Fuller initiated the departmental newsletter in the 1950s, producing 11 editions, the last appearing in January 1969, shortly after her retirement. The newsletter included a brief report from each member of staff on their research and travel activities (usually from the summer vacation), provided information and reports on departmental events and listed student news including the occupation and whereabouts of the previous year's graduates. This newsletter provides a wonderfully detailed archive for today's histori-cal researcher, but it obviously played an important role in the social gel of the department, gave staff an opportunity to write informally about their interests and activities, and kept students and staff in touch with the careers of graduates. Fuller published the newsletters in January and it's easy to imagine former graduates sending her Christmas cards with their news. Fuller, who had been persuaded to give up her house in Sussex soon after her appointment, bought a house in Beeston near the campus[74] and certainly seemed to centre her life on the department and the GFG. Cole records that she felt she had 'missed out' by not marrying and noted that she always organised tutorial groups in mixed-sex groups, implying she was match-making. Matchmaking or not, it is clear she provided a core to the social network of the department, was very open to students, and in her newsletters captured something of the sense of 'family' that some staff/departments engendered in the 1950s and 1960s. Joan Fuller concluded in her last newsletter that 1968 had been a special year for her, notably in purchasing her new home in Stratford-upon-Avon but also the opportunities retirement brought. 'And having my house, I want to settle down and grow some roots there. I also want time without time-tables. Time to look at things, time to think and to write, and to take my geographical eyes on journeys in spring and autumn instead of cloudy August. Also – and this is something I feel strongly about – I want more time for friends. I have now a house in a beautiful place in the heart of England and if you want to call on me, its easy to find. But write in good time in case I am in the Highlands or the Alps. The address is: 11, Kipling Road, Stratford-upon-Avon'.[75] Cole noted that no one volunteered to continue producing the newsletter after Fuller retired in 1969 and it ceased to exist: it required a certain sort of relationship with students and represented 'unproductive' labour to more research-output focused colleagues.

After many years of retirement and travelling with friends, Joan Fuller spent her last years in nursing homes in Oswestry and Tittensor (near relatives), finally succumbing to long illness and dying in 1992 when Wheeler (1992) wrote: 'We, like her many students, shall remember her enthusiasm (and chuckle) with affection'. More formally the *Nottingham University Gazette* (1992: 19) recorded 'her major contribution to the geography department over nearly three decades'.

Jean Carter

Jean Carter was born in 1921 and graduated with first class honours in geography at the University of Manchester in 1942. During the remainder of the war she worked for the hydrographic department of the Admiralty. After the war she studied at the University of Oxford (1947–50), completing a BLitt entitled 'A functional classification of settlement in Breckland', followed by a year as research assistant at the Agricultural Economics Research Institute at Oxford. In 1951 Carter was appointed assistant lecturer at Manchester University and was promoted to lecturer in 1954 and senior lecturer in 1969. In her spare time she ran a troop of Girl Guides and would often spend Fridays in the department teaching in her full Guider's uniform.[76] At a time when geography students were split between the arts and science faculties, Carter, who taught biogeography, took responsibility for the BSc students. This division of students appears to have reproduced wider university politics within the department, with attendant loyalties and accusations of empire building. Anecdotally Carter is remembered as being too directive to students and compelling BSc students to take a subsidiary in geology, but departmental minutes in 1968 record her drawing attention to the dearth of subsidiary subjects available to second-year science students: 'this year the choice had, in effect, been between an unattractive geology course, archaeology and surveying. She felt that the Faculty of Science would have no objections to the field of choice being open to equivalent Arts subjects and possibly to subjects in the field of geography'.[77] One colleague, known to enjoy a robust discussion herself, found the Manchester department in the 1970s to be generally fractious and considered Carter to be a significant part of this atmosphere, but another colleague, Geoffrey North, represents a different perspective on Carter as 'a difficult but intensely human person, a very good tutor, very protective of her students, a product of Oxford rather than as a woman [per se]… [she was] a victim of an awful lot of chauvinism – she had to fight for her position, so she was a fighter'.[78]

Links with the University of Malaya, where the Manchester head of department Professor Crowe was external examiner, gave Manchester geographers the chance to spend a year in Singapore. Jean took this opportunity in 1956 and conducted studies of the tropical biogeography of mangrove swamps and visited Borneo. One colleague recollects that she may have had a love affair whilst in Singapore. Ill health (1969–72) caused Carter to opt out of tutorial supervision and take time off work for treatment. According to university records, Carter took early retirement in 1982, but continued as a part-time teaching assistant 1982–4.[79]

Alice Coleman

Alice Coleman was born on 8 June 1923, daughter of Elizabeth (née White) and Bertie Coleman. She attended the local council school in Broadstairs before going to Clarendon House, a grammar school in Ramsgate.[80] Leaving school during the war Alice Coleman felt compelled to seek some sort of practical work, so she started a two-year course in teacher training rather than go to university to pursue a degree. Her father worked on building sites and road-building projects, and as part of a large family on a limited income, Coleman opted for Furzedown Teacher Training College because it was the cheapest. Her family's modest income meant that Coleman was entitled to the full local authority grant of £25 per year, but she also had to take the official loan of £35 per year for the two years of the course. Financial constraints played a part in Coleman's initial choice to specialise in geography because a friend from school in the year ahead offered to share her geography books with Coleman and another friend: 'Since money was a real issue in those days we both decided to do geography'.[81] After completing her training, including an enjoyable change of scene when the college was evacuated to Cardiff, Coleman started teaching and looked to pursue further part-time studies at Birkbeck College. She was very interested in psychology, but it was difficult to get up to London in time for the Friday evening classes so had 'to think again' – the result was that she opted to take a degree in 'geography with a geology subsidiary – and have never regretted it'.[82]

Coleman worked during the week at a school at Northleigh, Gravesend and travelled up to London to study during the late war and early post-war years. She was taught by Eva Taylor in her first year, whom she considered to be 'eccentric … but she was very nice … we learnt a lot from her, she obviously had high standards'; she was also taught by Eila Campbell and S.W. Wooldridge, with the latter becoming Coleman's inspiration and guide through academia. Unusually, Wooldridge told her ahead of finals that he expected her to get a first class degree, which she did, despite the difficulties of her hearing impairment.[83] Not approving of PhDs, which he considered a 'foreign' innovation, Wooldridge encouraged her to undertake a Master's study looking at denudation chronology, following Ken Hare's study of the Thames.

Coleman chose to study the River Stour near her home area and was able to demonstrate that East Kent had sunk towards the North Sea Geosyncline. She was awarded her master's with distinction in the summer of 1951 and was only the sixth person in geography to have achieved this. Coleman noted, 'it doesn't mean a thing now, but at the time it was rather nice'.[84] On the basis of her achievements Coleman was encouraged by Wooldridge to apply for university posts and was offered a position by him at King's College. Coleman recounted that W.T. Gordon, professor of geology, hadn't wanted to appoint a woman, but he forgot to turn up for the interview, so Coleman was given the three-year post of assistant lecturer in 1948.[85] Gordon's attitude was not unusual; whilst the University of London had been open to women for over 70 years, different parts of the university were more receptive than others. At the historically male college of King's the ethos 'was very much like a nineteenth-century gentleman's club' (Balchin 1997: 19) and this reluctance to facilitate women was expressed spatially in a form of gendered apartheid of facilities. 'The whole of King's was anti-women. They didn't allow women staff into the senior common room and they didn't allow them into the dining room, but they did give them a little common room of their own and a little dining room of their own, which we used to invite men to come in – and a lot of them did! In fact for the first 10 years women were second-rate citizens, but when Noble came [as principal] he decided this wasn't good enough' and things began to change.[86]

In her third year Coleman was fortunate to be one of three from 16 assistant lecturers who were kept on and was subsequently promoted to lecturer in 1951 after the completion of her Master's course. Coleman was to spend her whole academic career at King's and whilst there taught 14 different courses, including the new subject of biogeography when first appointed, as well as regional courses on Europe and the Mediterranean and systematic courses on physical and economic geographies. Some years Coleman taught 30 hours of classes a week; she also ran the departmental library for 10 years. Coleman had always wanted to teach since she started school and enjoyed that part of her job as much as her research. Wooldridge recognised Coleman as an accomplished teacher, but she felt she had to persuade each subsequent head of department of this fact.[87] Coleman was also appreciated as a PhD supervisor. Philip Nicholls recorded that she was an excellent supervisor: 'I was extremely fortunate to have such an able, enthusiastic and helpful supervisor, who put me through my paces. I learnt more during this phase than any other in my Joint School experience' (Nicholls 1997: 65).

Reflecting the primacy given to fieldwork within the London geography departments and personified in Wooldridge's adage of 'learning geography through the soles of one's boots', Coleman undertook a rich variety of field courses often with William Balchin, and D.J. Sinclair from LSE (Coleman 1997a). Coleman herself had been deeply impressed by her first experience of a foreign field course in Paris with Harrison Church as a final year student and was keen to offer a similar opportunity to subsequent students who, like her, had little or no experience of the European places they studied. When she began running voluntary vacation field courses in continental Europe from 1953, they were so popular they had to be repeated two or three times over the summer (Balchin 1997). Students and staff came from within and beyond London University, including Audrey Lambert from LSE, Norman Pye from Leicester and David Large from Southampton, William Balchin and Gillian Groom from Swansea, and Miss Woodward, a teacher from Thanet (Coleman 1997a). Initially working from her research field study site at Boulonnais to the Vosges and Black Forest, subsequent trips used the train to reach Austria, Switzerland, Italy and Yugoslavia. All of this she found 'very rewarding for a teacher as the students were fascinated and appreciative' (ibid.: 25), a view echoed by Nicholls (1997: 65). Coleman was also co-leader with Stanley Beaver for a Le Play Society field course in Norway, for which she studied Norwegian. With the advent of charter flights, Coleman co-organised long-haul field schools with Balchin, Woodward and others to North America, the West Indies and East Africa until 'travel agents colonised our intercontinental niche' (ibid.). Changes in course subjects, especially the loss of regional courses, and in the structure of courses with half-yearly examinations also led to a decline in fieldwork in university geography departments in Coleman's view.

Wooldridge experienced a stroke in 1954, thought to have been at least in part as a result of overwork. He was compelled to take a term out from teaching and Balchin led the department temporarily before moving to a new post at Swansea. In Coleman's view Wooldridge was a changed man when he returned, she felt 'he turned against women'; there may have been other factors at play, but Coleman attributed this change to the fact that Wooldridge's wife was always trying to get him to do less as a result of his illness and he balked against her control. This resulted in the man who had appointed Coleman in the face of sexist opposition telling her and Eila Campbell that 'women should be banned from the university'. On a personal level, Coleman also attributed his change of stance to her own reorientation of research commitments as she moved from geomorphology to land use study: 'he regarded this I think as a stab in the back!'[88] and she felt that Wooldridge had withdrawn all support from her work.

In the first stage of her career Coleman's own research continued in geomorphological studies of rivers, applying her hypotheses to Boulonnais in France, where even more pronounced subsidence was identified, and Sunnfjord in Norway. Wooldridge envisaged her undertaking a long-term research project around the coasts of the North Sea, but Coleman opted for the mountainous setting of the River Salzach. She was conscious she could have developed a research career specialism in this work, but 'my generalist instincts were strong and I delved into mining subsidence in East Kent, and then into land use field mapping, which is broad enough to please the most avid generalist' (ibid.: 24). When the GA announced the intention to write accompanying 'memoirs' for each of the Land Utilisation Survey map sheets, Coleman was keen to write the text to accompany the East Kent map of her home area. Albeit principally a geomorphologist, in the London tradition, Coleman believed that 'Regional Geography was the peak of synthesis to which geography aspired' (Coleman 1997b: 37). Former student and human geographer, Kenneth Maggs, joined her in the endeavour. At Maggs' suggestion, they began a new survey of the area using the new 1:25000 series of maps which allowed more detail than the 1:63360 series used in the original survey. 'We replaced Stamp's two urban categories by manufacturing (red) with twelve types denoted by numbers, derelict land (black dots), tended open space (lime green), allotments (purple cross-hatch) and transport land (orange), but the scale still precluded separating commercial from residential uses. As far as possible we adhered to the international colour conventions ....We aimed to design maps whose information leapt to the eye spontaneously, instead of having to be laboriously dragged out' (ibid.). With the help of members from the Isle of Thanet GA, they were able to complete 99 field maps and unexpectedly gained not only moral but also financial support from Dudley Stamp. He was so impressed by the consistency of their maps when invited to view them that he donated £2000 to get the first maps printed and pump prime the project.

Within a week in early January 1960, volunteers had signed up, largely at the GA and IBG conferences, to survey one-third of England and Wales: teachers and sixth-formers made up 3000 of the volunteers, causing Coleman to comment, 'I cannot speak too highly of the enthusiastic way they made this huge enterprise possible' (ibid.). Coleman took groups of volunteers from the Joint School of Geography (King's and LSE) on youth hostel weekends to map the blank areas on the map. Whilst in the 1960s researchers on the project were usually made welcome and elicited interest in the project, Coleman recounted suspicion and racism in some urban areas in the 1970s: 'One day, after introducing a group to urban mapping, I was sought by the police who had arrested a black youth because residents thought his map annotations amounted to casing the joint for burglary' (ibid.: 38).

The whole of England and Wales were mapped in the 1960s, resulting in 6400 field maps. Serving on a sub-committee of the Labour government Ministry of Land and Natural Resources (at Stamp's suggestion), Coleman devised the best method for representing a national overview of the survey's findings, settling on regular point sampling, in order to produce a national map. The Ministry agreed to pay a then substantial £5000 for measuring all the maps within a year, but it was abolished just as the project was completed. A sabbatical term in Canada in 1965 led Coleman to study land *disuse* around Ottawa. The report was subsequently relevant to a case going through the Supreme Court and Coleman's findings were championed by the Mayor of Ottawa who had a copy sent to every member of the Ottawa Parliament. On return to London Professor Emrys Jones of LSE suggested applying the concept of disuse to the full land use typology, which resulted in complex system of mixed land uses. 'It took a long time to devise the precise method for delineating them, but eventually with the help of three dedicated geographers, **Mrs. Janet Shaw**, **Mrs. Yvonne Latimer** and **Miss Elizabeth England**, a dichotomous key was perfected on the flora principle and a national map was synthesised and printed' (Coleman 1997b: 39). The national map showed important post-war trends in land use/disuse including inner-city dereliction, peripheral sprawl, fragmented farmland and 'rurban' wasteland. The need for planning regulation had been the cornerstone of Dudley Stamp's recommendations based on the first Land Utilisation Survey, but Coleman argued that 30 years after the Town and Country Planning Act of 1947 the findings of the second survey suggested that the planning process had been both expensive and of limited value. Whilst her findings were not immediately popular with politicians, Coleman was 'deluged' with requests to speak and write articles after her paper at the RGS in May 1976 entitled 'Is planning really necessary?'. Friend and former colleague, William Balchin, chaired the meeting; he noted the Survey's innovative use of statistics and graphic representation and described the data gathered by the volunteers and collated and analysed by Coleman as of 'incalculable value to planners, local authorities and government agencies' (*GJ* 1976: 431). However, some discussants disagreed with Coleman's polemic that the free market might achieve the same result as planning agencies.

The RGS recognised Coleman's work on land use surveys with the Gill Memorial prize in 1963 and the Busk prize in 1987 which was given 'for innovations in land use analysis and practical contributions to inner city regeneration ... Professor Coleman is thus an academic geographer whose research is having a valuable applied effect on our communities' (*GJ* 1987: 446–7). In 1974 Coleman was awarded the *Times*/Veuve Clicquot prize as the 'most outstanding woman in a man's world'[89] for her work on the Second Land Utilisation Survey. This was followed by being the first holder of a visiting professorship for leading women in the social sciences at the University of Western Ontario. This recognition was not matched within the university, she had been promoted to Reader after Wooldridge was replaced by Professor Kenneth Hare, but had subsequently been turned down twice for promotion to a chair. Coleman believes that her promotions had been successively blocked by Eila Campbell, who played a significant role as the external on the panel: 'Unfortunately Eila Campbell stamped on it. She invited me for dinner and talked about how I had put in my application, then behind my back stamped on it. I cannot understand it at all because she was always talking about how unfair universities were to women and then she did that to me. I didn't in fact get it until she retired and then I got it straight away.'[90]

For a large part of her career Coleman was the only female lecturer in geography at King's and she attributes some of her slow progress to promotion to her socialised gender. 'Of course women aren't trained – or weren't trained then – to put themselves up for promotion. If you did good work, it was supposed to come as a reward for virtue, but it didn't. .... So I went to the Principal and I said this will affect my pension, I've put a lot of money into the land use survey and I'm not going to be very well off in retirement'.[91] Coleman was made professor of geography shortly before she retired and became emeritus professor on retirement. She maintains that gaining the professorial title was significant not only in terms of her pension but also to being commissioned by Prime Minister Margaret Thatcher to revive the worst of Britain's housing estates in the 1980s[92] when she used Newman's concept of 'defensible space' in a £50 million project for the Design Improvement Controlled Experiment (DICE). Coleman's emphasis on the privatisation of space and work for the Conservative local and national governments inevitably attracted political criticism and her inexperience in planning and related fields drew academic criticism (See Coleman

1997c for her own evaluation of the project). Coleman still houses the Second Land Utilisation Survey in her London home and continues to write on numerous subjects including literacy strategies.

Publications

Coleman estimates that she has some 400 publications spread across 95 different publication outlets.[93] Her early mainstream geographical publications were based on research in the field, including a paper co-written with Stanley Beaver on their Norwegian work: 'Changing geographical values in Dale-i-Sunnfjord' (1955), 'Land reclamation at a Kentish Colliery' (Coleman 1955) and 'The terraces and antecedence on part of the River Salzach' (Coleman 1958). Reflecting trends within the discipline, her writing became dominated by the Land Utilisation Survey. In addition to her reports and maps and a paper in *TIBG* (1965) on 'Cartographic methods for use in land-use memoirs', Coleman published updates on the Second Land Utilisation Survey in the *GJ* in 1961 and 1964, but this also developed into related themes such as 'The conservation of wildscape. A quest for facts' (*GJ* 1970). Later publications included *Utopia on Trial* (1985) and Coleman *et al.* (1988) *Altered Estates*, published by the Adam Smith Institute.

Writing for a wide range of audiences, Coleman's publications influenced the teaching of school geography through numerous school texts on map work and fieldwork for the GA, as well as her land use work. Coleman's early work on Thames-side cement manufacturing and coalfields, as well as her work on land disuse in Canada and land use and housing in the UK all attracted political interest and were used in public inquiries or state policy formation. In the case of mines she was called as an expert witness in the Vale of Belvoir inquiry. Yet despite widespread interest in her research Coleman (1997c) has argued that her work had not been widely cited by other geographers, suggesting the wide range of journals etc. in which she had published 'may account for the fact that I do not figure highly in the citation stakes whereby academics win prestige by citing each other's papers in the strictly geographical journals. One has to be in the fashionable mainstream to score there and I have always been off along original lines of my own' (1997c: 74).

Other women lecturers in the LSE/King's Joint School included **Audrey Lambert** who was appointed at LSE in 1952, where she taught until 1982, being promoted to senior lecturer in 1975 (a post which particularly acknowledged excellence in teaching). Her key publication was *The Making of the Dutch Landscape. An Historical Geography of the Netherlands* (1971), but it appears not to have been reviewed within geographical literature.

Lambert's involvement with field courses was noted above and in the 1970s she was personally responsible for reviving the lapsed Joint School Society which organised local field trips, career meetings and social events for geographers and an associated newsletter called *Horizon* (Balchin 1997), and in this seems to have played a similar departmental role to that of Fuller at Nottingham.

Eleanor Vollans was appointed to teach biogeography at Bedford College in 1948, but went on to specialise in historical geography, publishing on the Romney Marsh area, work that is still cited by Romney specialists today. She was remembered as a scholar who gave 'meticulous ordered lectures' that 'served the students very well' (Hilling 1994). After a period in which the triumvirate of Hosgood, Smee and Timberlake had covered a wide syllabus and provided fieldwork opportunities for undergraduates, Gordon Manley's appointment as professor in 1948 undoubtedly enhanced this fieldwork ethos (Hilling 1994), but the appointments of geomorphologist Clifford Embleton in 1954 and economic geographer Michael Chisholm in 1957 served to masculinise the department further. Manley initiated a new junior post of tutorial research fellow (first held by Michael Holland, then by **Jane Soons, Joan Kenworthy, Margaret Storrie** and **Brenda Turner**) and an externally funded research assistant (**Elizabeth Shaw**) (Hilling 1994), which suggests that there were only openings for women in these junior temporary posts at this time. Soons went on to an academic career at Christchurch University, New Zealand and Storrie to a lectureship at Queen Mary's College, University of London (see Chapter Nine); Kenworthy, an Oxford graduate, made her career at Liverpool and Durahm universities.

Gillian E. Groom

Gillian Groom was born on 9 May 1926, was educated at an independent school in Birmingham and studied at Birmingham University, where she completed a BSc. In October 1948, at the age of 22, she was appointed temporary assistant lecturer at the geography department at the University College, Swansea. She was promoted to lecturer in October 1951 and senior lecturer in 1964. Groom was a fellow of the Royal Geological Society and an active member of the IBG in the 1960s, serving on Council 1962–4. Groom retired in 1976 after a year of ill health which she spent in the Sancta Maria Hospital in Swansea run by Ursuline nuns. She never returned to academic work but went on to convert to Catholicism and take Holy Orders with the Ursulines, adopting the name Sister Gillian.[94] 'Recognising her academic ability the Order sent her to their mother house in northern France to be librarian and to catalogue the archives of the Order's founder, an appointment that she most enjoyed'. In the 1980s she returned to the convent in Swansea and, as Sister Gillian, became the RC chaplain to the university and joined the University's counselling service team. In 1986 she suffered a massive brain haemorrhage and died on 17 June.[95]

During her academic career Groom, a physical geographer who specialised in coastal and limestone geomorphology, published little; instead she committed her time primarily to teaching. She published two papers with Marjorie Sweeting (1956 and 1958) based on fieldwork in Spitzbergen; and a paper co-written with V.H. Williams appeared in a symposium on limestone in the *GJ* (1965). The findings were based on a six-year study and showed significant changes in calcium contents in the region's rivers reflecting seasonal and discharge variation, which were related to the region's limestone outcrop, karst development and rainfall regime. Marjorie Sweeting was also involved with the symposium and a letter from Sweeting to Eila Campbell is indicative of the friendship and research networks between these women: 'I hope you will have a good vacation. You don't want to go to Yugoslavia in September do you? I was going with Gill Groom but she's had a knee op. and thus can't go.'[96]

Whilst reviewing several books in the *GJ* and supervising research students, Groom seems to have published relatively little of her own work. Moving straight from her undergraduate degree into lecturing she did not undertake a master's or PhD, and although this was not unusual in the pre- and immediate post-war years, it may have influenced her status and/or confidence as a researcher. Gwyneth Davies, a colleague for a while in the 1960s, as well as a neighbour and friend, records that although Groom was supported in the department by fellow physical geographer Professor Oliver until he left in the late 1960s, 'I was aware that there were some unkind remarks about her contribution by some of the male members of staff'.[97] Colleague Jack Davies noted: 'After her death I was asked to go through her various papers and documents. It was plain from this that she had amassed much very useful academic material relating particularly to limestone geomorphology which could easily have been published with acclaim. The one publication which demonstrated her ability was the chapter she contributed for *Swansea and its Region*, a book published in 1971 in association with the BAAS meeting in Swansea. Sadly, her experience as a young lecturer giving a paper at an international conference when she was subjected to unfair criticism eroded her confidence. This, plus in modern terms the lack of a mentor, leads me to conclude that her time as a lecturer was not as happy as it should have been'.[98]

Colleagues and students remember Groom as an accomplished and appreciated lecturer. One postgraduate at Swansea in the 1970s described Groom: '[She was] an excellent teacher and very attentive research supervisor. She also did a lot of administration for the department. I remember her saying once that she had eyes in the back of her head. She knew all that was going on in the department'.[99] The combination of devotion to teaching, rigorous supervision of research students whom she saw weekly and a heavy departmental administrative load would have made for more than full occupation. Groom also had health problems: the knee operation cited by Sweeting and back problems in later years, which may also have affected the possibilities for field research.

Jean Sidebotham

Jean Margaret Sidebotham BSc (University of Manchester) was appointed as an assistant in the department of geography at Queen's University Belfast for the academic year 1947–8, one of two temporary assistants appointed after the university failed to fill a new lectureship and another member of staff resigned (Campbell 1978). By 1951 the two lectureships were filled by E. Jones and N. Stephens after the university had gained a sounder financial footing, giving rise to the prevailing view at the time that these appointments 'served to endorse the general contention that implementation of the University Grants Committee's recommendations had made it easier to attract *able men* to Belfast' (Campbell (1978: 31); my emphasis). Sidebotham was reappointed as a temporary assistant lecturer on a year-to-year basis until December 1954. The Senate minutes note that her name changed in 1950 to Mrs J.M. Graham and there appears to be a hiatus in Jean Graham's employment. A Senate minute in May 1960 recorded that she was to give assistance in the department of geography two afternoons a week during the lecture term commencing 21 April 1960, but this appears to have been an isolated contract.[100] **Rosemary Harris**, a UCL geography graduate (MA with distinction 1955, PhD 1959), held the Sir Robert Hart scholarship at Queen's 1952–3, was the research secretary to the Committee on Ulster Folk Life and Tradition, chaired by Estyn Evans (1954–5), and the first woman to hold a full lectureship at the Queen's department when she was appointed lecturer in geography (social anthropology) in 1960. However, this was to be short lived as in 1962 she left geography to found the university's new department of social anthropology (Campbell 1978).[101] With the exception of these brief appointments, the Queen's geography department in the period 1940–70 was very much a masculine domain.

Conclusion

Women geographers' war work was varied and widespread (see Maddrell (2008) for a fuller discussion). The biographies discussed here show how women were called upon, alongside their male colleagues, to teach geographical skills such as map reading and write accounts of places for intelligence purposes. Alongside this and their mainstream university teaching they were often maintaining geographical societies and other civil society groups, as well undertaking home defence roles such as fire prevention. Other women prepared maps for the Ordnance Survey or the Admiralty (e.g. Carter) and joined the women's auxiliary forces (e.g. King) (ibid.). Women, by dint of their university qualifications, were recognised as a source of authoritative knowledge, providing training and information or the maps on which their trainees would depend upon in active service. Whether by providing training, maps, intelligence or auxiliary services, women geographers' war work represents a blurring of the gendered boundaries between home-based civilian women and frontline combatant men (ibid.). For some women the war offered experience and job opportunities, others lost opportunities and health. Those who were air raid wardens etc. (like Ormsby) were at direct risk, but none lost their lives (as happened to some of their male counterparts on active service) because of their gendered exclusion from armed service. Women geographers were also involved with non-wartime public work. The value of Dudley Stamp's contribution to national life through the interwar Land Utilisation Survey has been recognised within and beyond geography, but Alice Coleman's post-war Second Land Utilisation Survey has received little critical attention and Margaret Davies's work as a public geographer, especially in relation to national parks, is little known and merits further engagement.

Chapter Nine

University Expansion, Specialisation and Quantification: 1950–70

Introduction

The period 1950–70 includes important phases of post-war austerity followed by an economic upturn in the mid-1950s, the social changes of the 1960s and the emergence of second-wave feminism, which was to come to the fore in the 1970s. While female employment grew throughout the twentieth century, the Second World War represented a significant shift with 13 percent more women remaining in employment in 1946 than before the war started (Walsh & Wrigley 2001). Rates of female employment accelerated after 1950, with 51 percent of women of working age being in employment by 1965 (increasing slightly by 1970), a pattern echoed in the USA (ibid.).

The expansion of the university sector at this time was a reflection of that prosperity and political commitment to education, and included new universities which had gained their charter, e.g. Nottingham (1948), Southampton (1952) and Keele (1962) (Curtis & Boultwood 1970) with a resulting increase in student numbers. As social attitudes and employment practices changed, it became more common for academic women to combine work with marriage and family responsibilities (caring for children and/or elderly relatives); however, acceptance of these changes was far from universal and the varied experiences and strategies adopted by different women are examined below. Women geographers' biographies and careers are placed in the context of these social trends, as well as changes within the discipline, which included increased specialisation within sub-disciplines and the so-called 'Quantitative Revolution' in the 1960s. Opportunities for and barriers to promotion are also discussed in relation to publications, constructions of gender and institutional practices.

Geographers beginning their careers during these years include Monica Cole, Marjorie Sweeting, Jean Grove, Helen Wallis, Joy Tivy, Catherine Delano-Smith, Barbara Kennedy, Nicola Crosbie, Gwyneth Davies, Margaret Wilkes, Margaret Storrie, Christine McCulloch, Janet Momsen, Doreen Massey and Anne Buttimer. These women have been selected to represent a wide range of geographical interests and locations around the country. Most discussed here continued to work beyond 1970, the 'cut-off' for this volume. The biographies are continued for those women whose careers are bisected by 1970; for those who were appointed to posts in the mid- to late 1960s their personal and career biographies are outlined only briefly, with an emphasis on their experience of the geographical community up to and around 1970.

Monica Cole

Monica Mary Cole was born on 5 May 1922, in Clapham Common, Battersea, daughter to Dorothy (née Thomas) and William Cole, a bank clerk.[1] She went to school in Wimbledon and was a high achiever, good at academic work and sports (Potter & Catt 1994). She stayed relatively close to home and opted to study at Bedford College. She was taught by Hosgood, Smee and Timberlake and spent all of her undergraduate years in the department evacuated to Cambridge. True to academic form, Cole gained a first class honours BSc in geography, with a subsidiary in geology in 1943, went on to be awarded the Busk-Howell Postgraduate Scholarship from Bedford College and in the following year a University of London Postgraduate Studentship. This she held while simultaneously completing war work as a research assistant at the Ministry of Town and Country Planning (1944–5). Clearly she did not want her career trajectory to be stalled by the war as she took the opportunity afforded by her base at the Ministry to undertake research for her doctorate in the economic geography of building materials. She was awarded her PhD in 1947 and the practice of combining research with applied work would characterise much of her subsequent research career.

In what was a common career pattern for British academics in post-war years, Cole spent some time lecturing overseas within the Commonwealth network.[2] She took her first lecturing post at the University of Cape Town in 1947 (Potter 1994a); the following year she moved to the University of the Witwatersrand where she stayed for three years (1948–51) working in the department headed by Professor John Wellington. It was here that Cole became interested in savannas and undertook extensive fieldwork, reflecting her belief that 'geographical research should begin in the field' (Jackson 1994: 244). Her study of the relationship between soils, climate and crop

Figure 9.1   Monica Cole.
Source: Courtesy of Department of Geography, Royal Holloway University of London

yields in the geologically interesting and agriculturally important Elgin District is described by Jackson (1994) as 'one of the most thorough and useful land utilisation surveys carried out anywhere in South Africa' (see Cole 1949). This work is notable because it is clearly based on the fieldwork experience she had gained at Bedford College and mimics both the Land Utilisation Survey work undertaken by Bedford students and the agricultural soil surveying undertaken at Haselbech, Northamptonshire under the direction of Smee and Timberlake. However, Cole was also to develop new interests at the Witwatersrand, and Jackson records her particular interest in vegetation anomalies in the savanna and how plants might be used as indicators of mineralisation. Needless to say, South African mining companies were very interested in this research, as no doubt farmers were in the earlier study, and they gave her 'support and encouragement' (Jackson 1994).

In the four-year period that Cole spent in South Africa, she developed her expertise in geobotany and mineral exploration and began to publish on the region, thereby establishing her reputation as a South African geographer (*The Times*, 7 February 1994: 19). At the age of 29 she took up a post at the new University of Keele in 1951, where she stayed for 13 years (1951–64). During this time she joined the Geographical Field Group and attended a field trip to Yugoslavia in August 1952, led by Joan Fuller; she was also in friendly correspondence with Eila Campbell, which suggests a degree of interaction and networking with other women geographers.[3]

Cole gained a reputation early in her career in South Africa for transgressing boundaries of gender and age in her pursuit of primary data in the field. A testimonial written by a Witwatersrand colleague for Cole when she taught a summer school at the University of Idaho in 1952 was reproduced by the campus journal and again in Cole's obituary: 'Miss Cole [sic] carried out more field work than any geographer in the country, though this involved many weeks of travelling alone. It is usually considered unsafe for young women to do that sort of thing out here! In these days when geographers are too inclined to write as much as they can and travel as little as they can, Miss Cole [sic] was eager to undertake arduous and extensive field work because she felt that geography should be advanced in that way' (cited by Potter & Catt 1994: 374). The racial context is implicit in this statement, with Cole considered to be taking risks not simply as a woman but as a white woman 'out here': as with earlier women travellers, her race compounded her perceived vulnerabilities of gender and age, but was also the means to transgress these constraints. As a woman (a bodily female presence) she had not received an immediate warm welcome from senior field geologists in South Africa, but 'she endured the discomforts of fieldwork as well as they did and soon gained their respect' (Jackson 1994: 244). These sentiments, intended as complementary, are very telling of the gendered experience of the field in a masculinised and racialised environment. She was accepted because she proved her body and intellect could undergo the rigours and work defined as 'man's work'. Her work on plants as indicators of mineral sources brought her to work collaboratively with the mining industry, principally in South Africa and Australia, but also closer to home in Cornwall for the English China Clay Company, indeed the majority of her research was commercially sponsored (Potter & Catt 1994). An early adopter of remote sensing, she was keen to apply technological developments to research problems (ibid.). In later work she also looked at plants as indicators of environmental pollution and in 1972 was an enthusiastic member of the Department of Transport Advisory Committee on Landscaping Trunk Roads (ibid.).

Cole moved to Bedford College in 1964 to take up the post of professorial chair and head of department, replacing Professor Gordon Manley, but she was forced to take a sideways move in 1975–87 when she became director of research in geobotany. All accounts of geography at Bedford College describe the 1960s as a period of great change. This was partly related to the expansion of student numbers experienced across the university sector at this time and the political radicalisation of some of those students; more locally it was attributed to the Monica Cole's leadership. Cole, at the age of 42, had been chosen in preference to the senior candidate Eila Campbell and 'From 1965 to 1975 she served as Head of Department, one of only two female Professorial Heads nationally. During this period she presided over the early expansion of what today is one of the leading departments in the country' (Potter 1994a). During this time Cole made a distinctive contribution to geography which was characterised by three commitments: first, that geography should operate as part of the wider scientific community;

secondly, that there should be no geography without fieldwork; and thirdly, the value of overseas fieldwork (ibid.). Cole is credited with being an excellent teacher, borne out by the number of her students who went on to further research (Potter 1994a); but is also remembered as overdirective if not dictatorial regarding student's options.[4]

Potter (1994a: 15) records:

> By the early 1970s, Monica Cole had engineered the development of one of the most extensive suites of analytical labs to be found in a British geography department. This really highlights the fact that Monica was one of the first British geographers to bring work in the discipline to the attention of the wider scientific community. This is also exemplified by the fact that following her early studies of plant indicators of mineralisation, she was swift to recognise the utility of the techniques of remote sensing, and from 1971 onwards, she carried out investigations which were sponsored by the UK Department of Industry and the Ministry of Technology, in collaboration with the Australian Bureau of Mineral Resources, and occasionally under contract to NASA.

These accounts of Cole's departmental leadership and innovative research techniques belie the tension in the geography department generated by friction between her and both colleagues and students. Part of this was a clash of cultures between Cole who was assertive and ambitious for herself and her students, while at the same time being very feminine in appearance and socially conservative. Jacky Tivers, who began studying geography at Bedford in the same year as Cole's appointment, remembers her banning the then all-female students from wearing trousers to her lectures. Cole also reduced many students to tears (including Tivers) during pre-finals interviews: 'We all queued up outside her room, nervously waiting. Other girls emerged in tears, so I knew what to expect. When it was my turn, she simply asked me what I intended to do next after graduation. I told her I was going to be a school teacher (I had a place to do a PGCE at Leicester). She got really angry about this and told me I was wasting my time and my future life. I also emerged from the room in tears'.[5] However, Cole's friction with students also related to her support for apartheid and conservative views. Tivers recalls the outrage of first-year students being forced to sit through a farmer's defence of fox hunting while on a field course in Appleby, as well as Cole's justification for apartheid and associated policies of separate development and townships during lectures on South Africa. Tivers and other Bedford students were attending anti-apartheid demonstrations in London at this time and it is hardly surprising that student attention would turn to scrutinise Cole's work on campus given her public justification of apartheid in lectures.

In the late 1960s Cole became a focus for student protest, principally for undermining the principles of liberal education by taking commercial sponsorship and for her work, past and present, in South Africa. These views were articulated in detail in the Bedford College Students' Union magazine *Inner Circular*. In November 1971 the magazine's 'political slot' was an unsigned piece entitled 'A question of competence', which called for Cole's removal. The article accused Cole of 'deliberately perverting the educational purpose of the college' through inputs of commercial capital into the department, as well as supporting apartheid and having 'deplorable relations' with both students and staff. The student initiative to remove Cole was not successful, but was followed by formal complaints by members of the geography department staff (1973–4), centring on the forced retirement of G.A. Worrall and included a letter signed by six members of staff expressing a 'deep seated lack of confidence in her capacity to act as Head of Department'. A formal inquiry was initiated at Bedford College to ascertain whether the interests of all members of the department, staff and students, were being 'safe-guarded' and the Council minutes record the outcome that Cole was relieved of her post as head of department but kept her own research chair. Another chair was funded, and Ron Cooke appointed as head of department. Worrall's appointment was renewed until the end of the academic year (1976–77), but was to be held in the geology department.[6] Cole became director of research in geobotany, terrain analysis and related resource use, a post she held until her formal retirement in 1987. There has not been another female head of department at Bedford or the merged Royal Holloway and Bedford New College since Monica Cole. Known for her forthright views, Cole 'had drive and ability in abundance' and was a 'formidable adversary' – characteristics which no doubt brought her the successes she achieved – but she clearly had limited ability to persuade others and carry them with her rather than simply direct them. Dominance and inflexibility were characteristics other male heads of departments shared, but they were not removed from their posts. Her politics and personality to one side, one has to ask whether her removal was at least in part a gendered response to strong leadership and/or that she was judged according to gendered criteria – unpopular policies could be pushed through by a man but not a woman?

In retirement she became a Leverhulme Emeritus Professorial Research Fellow (1988–94); she also took up watercolour painting, her pictures of landscapes and flowers reflecting her professional interests (Potter 1994b). At 69 years of age, Cole was awarded the RGS Murchison medal in 1991 for 'major contributions to the geography of South Africa and to the understanding of savannas, during 30 years of meticulous research .... [as] a pioneer in the application of remote sensing to the understanding of global character and distribution of plant cover'. She was also honoured by the South African Geographical Society as an honorary life member in 1993. Monica Cole died on 8 January 1994, after being ill for some time: 'Monica faced up to her illness, the operations and arduous treatment with her customary resilience and fortitude' and continued to work on academic projects to the end (Potter 1994b).

Rob Potter, writing as a former student and then head of geography at Royal Holloway and Bedford New College, clearly sought to rehabilitate Cole's reputation: 'Despite all that had happened *I* certainly always thought of the Department at Regent's Park as *her* Department. ... As all who knew Monica were aware, she was determined, forthright and tenacious. She was undoubtedly a larger than life character ... Monica had both drive and ability in abundance. This is most clearly attested by the fact that by the mid-1960s, she had scaled the heights of a profession, that even today, finds all too few women as the incumbents of chairs. Quite simply, her contribution has been outstanding and truly memorable' (Potter 1994b). An RGS grant for women's fieldwork in physical geography was established in Cole's memory and has been awarded on a triennial basis since 1995. Jacky Tivers recalls meeting Cole at the RGS-IBG conference shortly before she died: 'I introduced myself to her and reminded her that I had been her student at Bedford more than 30 years before. She said that she knew my name – whether from student days or from subsequent work, I don't know – and she was very friendly to me, asking me about what I had done since the 1960s. I am glad that the last time I met her she was such a different person – much less abrasive – to the one I had remembered all those years ago'.[7]

Research and publications

During her time at Keele, Cole internationalised her reputation for savanna research as a government-sponsored scholar in Brazil (1956) and as a part commercially and part Commonwealth-sponsored researcher in Australia in 1960, as well as continuing work in Southern Africa (including the then Rhodesia and Northern Rhodesia), where she was president of the South Africa Geographical Society (SAGS) (1963). Her extensive research in Brazil showed the origin and distribution of different forms of savanna vegetation (cerrado, caatinga and pantanal), stressing the need to appreciate the roles of geomorphological and climate change in understanding the evolving landscape. Her experience of intercontinental research prompted Cole to call for an internationally uniform and recognised nomenclature for savanna vegetation (Jackson 1994; see Cole's SAGS presidential address 1963).

Cole produced over 50 publications from 1949 to 1991 (see Potter & Catt 1994 for a comprehensive list) and 'There can be no doubting the fact that she was a leader in her chosen field of academic research' (ibid.: 373; also see Jackson 1994). In pursuing biogeography she continued a field populated by distinguished predecessors, such as Marion Newbigin, which

Cole acknowledged in her preface to the seventh edition of Newbigin's *Plant and Animal Geography* (1968). One of her most influential books was her regional text *South Africa* (1961), which quickly went to a second edition and was 'well received academically and widely read in South Africa' (Jackson 1994: 244). While congratulated for the scope of her work and perseverance with such a large topic, one reviewer was surprised to see few references to apartheid, especially as the book aimed to provide a background to the racial problems in South Africa. Concluding the review, Board (1962: 222) implies a lack of emotional engagement with the country: '... one is left with a slight impression that South Africa is not so much a real place, nor the home of real people, for years in a political backwater, but now in the forefront of international concern, but merely a far-off land with a geography of its own'. Her 1986 book, *The Savannas. Biogeography and Geobotany*, published the year before she retired, was based on 30 years of extensive research, causing one reviewer to comment: 'there are few people more competent to give a comparative view of the world's savannas than Professor Cole' (Furley 1989: 122).

Marjorie M. Sweeting

Marjorie Mary Sweeting was born on 24 July 1923 in London, daughter of Ellen Louisa (née Liddiard) and George Scotland Sweeting, university lecturer in geology at University College London. Educated at Mayfield School in Putney, Sweeting went on to study at Newnham College, Cambridge (1938–41), where she was College prizewinner in 1939. She obtained a first class honours degree, as well as being actively involved with several university societies: she was president of the Women's Boat Club, president of the Sedgwick Club, secretary of the Geography Club and founder member and president of the Caving Club. Sweeting received a string of university awards after graduation: a Caroline Turle Scholar 1941–2; she was awarded the Marion Kennedy Scholarship in 1942, but deferred taking this up until 1945–7 after her war-service teaching at Howell's School, Denbigh, in North Wales; this was followed by three years as Old Students' Research Fellow (1948–51).[8] She gained her MA in 1945 and her PhD in 1949; her thesis was entitled 'The landforms of the carboniferous limestone of the Ingleborough District, NW Yorkshire'. In a period of extended post-war rationing, Sweeting did much of her travelling by bicycle and this left her with an abiding love of the area (Viles 1996). Having originally been encouraged to research limestone relief by Alfred Steers (Goudie 1995), limestone landscapes were to be the focus of Sweeting's lifelong research. She became an internationally renowned expert in karst geomorphology and 'Reading through her list of publications is rather like doing an armchair tour of the

Figure 9.2   Marjorie Sweeting.
Source: TIBG 1995: 429

best karst sites in the world': Jamaica, Mexico, Canada, New Mexico, Borneo, Czechoslovakia, Australia and China (Viles 1996: 429).

Whilst working in Cambridge as a research fellow, she was fellow of Newnham College, and she became an associate during 1956–69. In 1951 Sweeting began her long career at Oxford University when she was appointed lecturer at St Hugh's College (1951–4) and then a lecturer of the university from 1953. This was followed by becoming a fellow and tutor in geography at St Hugh's in 1954, a post she held until her retirement. Her move from Cambridge to Oxford has been described as 'missionary', with Sweeting working in the cause of updating and developing physical geography in her new academic home, something which was not always an easy task (Goudie 1995; Viles 1996). Prior to the 1970s, most of the appointments at Oxford went to Oxford graduates, 'So, in a sense, Marjorie Sweeting was the only real outsider' (Scargill 1999: 27) and '… for many years she ploughed a pretty lonely furrow at Oxford, not only as one of the few women to hold both university and college posts in Geography, but also as an advocate of "scientific" studies in a school which was, to be charitable, not exactly world-renowned for its dynamic and research-oriented views' (Kennedy 1995). Despite the difficulties, Sweeting, in part through her network of graduates, 'did much to keep Oxford Geography in touch with international trends' (ibid.).

During her career Sweeting took on the major college roles of dean and senior tutor, and served on the University General Board. Despite her administrative responsibilities and commitment to international research, Sweeting was very much a committed teacher of geography at undergraduate and graduate level. 'She acquired a high reputation for her work on

Karst Geomorphology and contributed importantly to the attraction of good graduate students as well as building up undergraduate studies at her college, St Hugh's, all of which contributed to the growing strength of the School [of Geography at Oxford] at this time (Scargill 1999: 25). Over her career she supervised some 30 graduate students '... who benefited from her international connections and her *laissez-faire* approach to supervision. As a graduate, one always knew that Marjorie was a true supporter, ready to visit field sites, proffer advice and provide solid meals in college but she was no hand holder. She expected determination and independence' (Viles 1996: 429). Whilst Viles and Goudie describe Sweeting's 'light touch' in graduate supervision, Kennedy (1995) gives a rather different picture of her competitively 'grooming her charges for Final Honour School in much the same protective bullying vein as an athletic coach'. Undoubtedly, Sweeting's influence on generations of students was a major feature of her career (Goudie 1995). Within the university department Sweeting was promoted to Reader in geography *ad hominem* in 1978[9] and, after the retirement of Jean Gottman, Sweeting became acting head of department, as Kennedy (1995) wrote pointedly, 'in the interregnum between two (male) professors'. Kennedy considered it a disgrace that Sweeting was not promoted to professor given her international reputation and publication record.[10] Sweeting retired in October 1987 as emeritus fellow of St Hugh's and although slowed by arthritis, continued her travels and research, notably in China, until a year before her death (Goudie 1995).

Sweeting received numerous awards and honours throughout her career. These included the RGS Gill memorial prize (1955) for her work on limestone, a Certificate of Merit from the National Speleological Society of the USA (1959), a medal from the University of Olomonč in Czechoslovakia (1973) and the RGS Busk medal (1980) for fieldwork abroad, particularly her leadership of the Society's geomorphology expedition to Sarawak. She received a Leverhulme grant (1955) and was visiting fellow at the National University of Canberra, Australia (1958–9).[11] She served on the Karst Commission of the International Geographical Union (IGU), led an 'outstandingly successful' field symposium for the group in 1964 northern England and subsequently worked on the new commission on human impact on karst (Goudie 1995; also see Sweeting 1965). She also played her part in national organisations, serving on the IBG and RGS Councils, the Executive Committee of the Field Studies Council and the British Geomorphological Research Group (BGRG). She was the first female chair of the latter and the Group's annual undergraduate dissertation prize was named in memory of Sweeting (Viles 1996). Marjorie Sweeting died on 31 December 1994, after being treated for cancer for a year and was buried in Church Enstone. Her final book on karst in China was at proof stage and she was still learning Chinese in hospital (Goudie 1995). A long supporter of the RGS Henrietta Hutton travel grants, Sweeting left a bequest to support undergraduate geomorphology expeditions (Goudie 1995). Sweeting was remembered as a great enthusiast for the things she enjoyed: offering hospitality, opera, friends, caving, karst, travel, sports and breakfast television (Kennedy 1995; Vincent 1995; Viles 1996). The special session on tropical karst at the 1995 International Association of Geomorphology Southeast Asia Conference in Singapore was dedicated to her memory.[12] Unusually for a woman geographer, a Festschrift was published in honour of Sweeting's work, entitled *Tropical and Subtropical Karst – Essays Dedicated to the Memory of Dr Marjorie Sweeting* (Williams 1997), indicative of the school of graduates and the international network of scholars she had galvanised around her work.

Publications

Throughout her career Sweeting wrote some 70 single and co-authored scholarly articles and book chapters as well as two key texts *Karst Landforms* (1972) and *Karst in China: its Geomorphology and Environment* (1995). She also edited a volume on *Karst Geomorphology* (1980) for the Benchmark Series in Geomorphology, assisted A.C. Waltham in editing *The Limestones and Caves of NW England* (1974) and co-edited *New Direction in Karst* (1986) with K. Paterson. The comprehensive list provided by Viles (1995) demonstrates the international range of Sweeting's fieldwork, collaborations and publications. Sweeting's first paper was based on her master's work on wave trough experiments on beach profiles and was presented to the RGS in April 1943, appearing in the *GJ* later the same year. Hers was the first paper to be given on research work in the new Cambridge University physical geography laboratory and she expressed her appreciation to Vaughan Lewis in the preparation of the paper. Sweeting's first co-author was Gillian Groom from Swansea (Sweeting & Groom 1956, 1958) writing on Spitzbergen,[13] but she was to publish with a wide range of collaborators, including her father (1969), graduate students, colleagues and scholars from around the world e.g. Sweeting *et al.* (1985).

Sweeting's work was not immune to criticism and her paper on Jamaican karst (1958) was criticised by Vaughn Lewis (1959) for its inappropriate description of landforms as cockpits. Sweeting (1959) gave robust reply in which she acknowledged her pictures would have been clearer for use of stereographic pairs, and highlighted the problems of nomenclature (suggesting the area traditionally described as 'cockpit country' might be better represented as 'cone and cockpit'), but she stuck to her analysis and search for generalisations. In the early 1970s Sweeting's work was cited as an example of process-oriented geomorphology (Cooke & Robson 1973). *Karst Landforms* (1972) received high praise from Alfred Steers (1973: 143), who considered it a 'work of distinction and scholarship'. He appreciated the extent of Sweeting's field experience across the world and the way in which she discreetly combined her own findings and conclusions with wider literature, as well as the range of types of karst and associated processes covered (e.g. tropical and glacio-karst and karst hydrology). Sweeting went to study the karstlands of China in 1977, as soon as Chinese state policy allowed visitors and she made 11 visits over 15 years, combining her own findings and suggestions for updating methods with a respectful appreciation of Chinese work. This was exemplified in her important and much valued contribution to the month-long meeting formalising the *Glossary of Karstology in China*. Her work was welcomed for its integration of western and Chinese work and 'highly appraised by former Chinese collaborators'.[14] Vincent (1995: 273), writing after her death, dedicated his review of 'Limestone Pavements in the British Isles' to the memory of Sweeting 'whose enthusiasm for all things karst was infectious'. As Vincent stressed, her early work was hugely important to the field of limestone studies, even if some findings had been overtaken by developments in related fields.

Hilary C. Chew

Hilary Christine Chew was born on 9 October 1924; she haled from Wilmslow and attended the Manchester High School for Girls (1936–43). She attained a first class honours degree at Bedford College, University of London in 1946 (studying under Dora Smee and Eunice Timberlake), with special options in economic geography and geomorphology. She then completed a professional teaching course at the London Institute of Education (1946–7) and taught for three years at Howell's School, Denbigh (a school also significant in the lives and geographical careers of Marjorie Sweeting and Jean Grove). She was awarded a two-year research scholarship at University College, Leicester where she completed an MA on 'Recent trends in the agricultural geography of Leicestershire' in 1952, having been supervised by Charles Fisher. This was followed by two years in the non-progressive fixed-term post of tutor in the department of geography at the University of Liverpool (1952–4), where she was also resident tutor at the Rankin hall of residence for women students. This was followed by a one-year temporary assistant lectureship at the University of Hull (1954–5) while Herbert King spent a year in the USA.[15] In 1957 Chew was appointed as lecturer at the County of Stafford Training College for Teachers, making a sideways career move into teacher education, a course paralleled by a number of other women geographers and reflecting both Chew's training and the opportunities open to her.

Referees' reports sent to Hull on her application also give a good sense of the criteria by which academic staff were selected and what was considered acceptable for temporary assistant lecturers as opposed to permanent staff. In his reference W. Smith from Liverpool emphasised her qualifications and described her as a very good judge of the intellectual qualities of students but she herself was rather 'rigid and tense' with students, which he attributed to her experience of school teaching: 'I put much of this down to her two years as a schoolmistress for I find that those who come to University work from school teaching do not readily unbend and do not easily treat students naturally'. If his views were commonly held then it is a wonder any of the women (and a significant number of men) who moved from teaching into lecturing were able to acquire posts. He attributed this in part to Chew being shy and reserved and described her as intellectually 'good'. Gordon Manley at Bedford, who had not taught her but had examined her MA with Dudley Stamp wrote at length, stressing her first class honours degree had been achieved through very hard work and her MA was 'a sound painstaking job'. He acknowledged that she had done well in the MA examinations but added 'I must say, however, that in the viva, and from subsequent correspondence, I could see no evidence of brilliance or of any outstanding personality'. He went on to make gross regional stereotypes about both Chew and the students she might teach: 'She seemed to me a fairly typical Northern-urban product, wiry and tenacious, rather pallid, very determined and conscientious (Manchester). I think there is no doubt that she would tackle any work she was given in a most conscientious manner; and there are quite a number of students, more particularly Northerners, who would undoubtedly respect her painstaking efforts and gain, from them, a good example'. He later adds that he knows nothing of her teaching capacity and reports that he had told her that at 30 years of age she would have trouble gaining the junior university posts in her sight. Whilst acknowledging her tenacity and ability to work hard, 'I do feel however, that she lacks the personality and breadth of interest which seems to *me* (perhaps wrongly?) a desirable attribute in a university lecturer, although it may be that at interview she shows up better; after all, it is upwards of two years since I saw her, and her recent experience may commend itself'. He seems to have allowed a degree of caveat in this statement, but then concludes with a comment that many students with second class degrees would give a better personal impression than Chew. He adds a personal postscript to King further undermining Chew's acquisition of her first, describing her as 'unquestionably a dull young woman' and expressing regret that she had been encouraged to leave a good teaching job for the fringes of university teaching, but that she could return to school teaching.

His comments both public and private illustrate the importance of verbal skills and social confidence in the acquisition of university posts (even in preference to examination results) and the importance of being perceived as 'brilliant' or 'outstanding' as opposed to hard working and competent. Ironically (in relation to Chew's qualifications) the same characteristics have been associated with the criteria for first class degrees, which have been associated with masculine forms of confidence, hence the preponderance of firsts awarded to male students. In contrast to Manley, P.W. Bryan at University College, Leicester described her research as 'fine', and herself as a 'pleasant colleague' whom he recommended not only for the assistant's job but also for any permanent lectureship which might arise. Bryan, evidenced his views from Chew's work as a tutor on two-weeks' fieldwork in Cornwall, where she was responsible for the work on agriculture including soil surveys; 'she did a very good job; gave a good general lecture [on agriculture] and was quite good with talks in the field. Her lectures were models of clarity and conciseness'.[16] Charles Fisher, who had supervised her MA research for a year, wrote an open testimonial for Chew in which he reiterated her pleasant character and stated she would 'fully justify her place on a university staff'. A.T.A. Learmouth, who Chew worked with most closely at Liverpool, also described her abilities in leading fieldwork, her perceptive assessment of students' work and her skill at taking students from demonstrated techniques to independent work; he noted her ability to put students *at their ease* and that 'her basic earnestness and firmness of character and opinion are relieved, on all appropriate occasions, by *a most lively sense of humour*' (my emphasis). By the time of her application at Hull, Chew had experience of varied teaching and fieldwork, including co-leading a fortnight's field course in Brittany with another Liverpool colleague in 1953, when she taught all the human geography components. Chew had previously visited Brittany in 1948 on a field study party (as well as being a member of similar groups in the Bernese Oberland, the Auvergne and the Austrian Tyrol)[17] demonstrating the importance of these field study groups in getting to know suitable areas for study outside the UK and gaining the confidence to lead field courses abroad. Learmouth's testimonial also stressed the importance of Chew's research work.

Chew presented a paper on 'The agricultural geography of eastern Leicestershire – the revolutionary changes of our times' to Section E of the BAAS in 1953 and developed her research theme to examine wider post-war trends in England and Wales, particularly the extent to which practices introduced as wartime measures were continued. She published papers on this work in 1953 and 1956 in which she mapped and articulated important shifts in agricultural practices, notably permanent grasslands being given over to fodder and cash crops; Coppock (1964) credited Chew as one of the two key writers on overviews of post-war agriculture.

Jean Grove

Jean Mary Clark was born on 10 March 1927, at Teddington, Middlesex, daughter of Mary (née Johnson) and Leslie Clark. Her mother had been a pioneer in research chemistry at Cambridge but had given up her research to bring up her family. At the time of Jean's birth her father was a research chemist at the National Physical Laboratory, and went on to become technical services manager at ICI and co-designer of the Normandy Landings Pluto Pipeline (Allen 2003; Whalley 2004). Jean Clark was educated at the Queen's School, Chester and then at Howell's School, Denbigh (when the family moved to Wales during the war), before going up to Cambridge (1945–51). Her parents were both mountaineers and her father favoured fishing holidays in Scotland and Norway; this family experience obviously influenced Jean's love of the outdoors and things Alpine. Marjorie Sweeting, another Cambridge geography graduate, played an important part in Grove's career during Sweeting's war-service posting at Howell's School (1942–5), a time when 'Jean was inculcated with a lifelong love of mountain landforms and physical geography' (Goudie 2001).

At Cambridge she followed in her mother's footsteps becoming a member of Newnham College and studied geography tripos, in which she attained a II:I in Part 1 (1947) and a II:II in Part 2 (1948); she also served as secretary and then president of the University Geographical Society. After graduating, Clark combined work as an assistant lecturer in geography at Bedford College (1951–3) with research for her master's degree. She was a Bathurst

Figure 9.3   Jean and Dick Grove, with son Richard on research fieldwork in Norway 1957.
Source: Courtesy of Dick Grove

student (1951–2) and was awarded a further grant from the University of Cambridge Philip Lake, Worts and Scand Studies Fund; she completed her MA in 1953.

Goudie records: 'Debenham persuaded her that women should lead scientific exploring expeditions, and in 1947, as Miss J.M. Clark, she went on a third-year field trip to the Jotunheimen mountain range in Norway with the glaciologist Vaughan Lewis, [Gordon Manley and Ronald Peel]'. In the three years following this experience, she led small groups of students in Norway working on glacier mapping and velocity. She joined several of the Cambridge University glaciology research expeditions from 1951, when a team organised by Vaughan Lewis and John McCall, excavated a tunnel into the Vesl-Skautbreen cirque, in order to investigate the structure and flow of the glacier. This achievement 'provided graduates in geography, geology and mineralogy with the opportunity to make observations that laid the basis for post-war British glaciological research'.[18] She also undertook fieldwork in Norway in 1952 and 1953, researching the banding on Veslskautbreen and Veslgjuvbreen for her doctorate on 'A study of the physiography of certain glaciers in Norway'.

After converting to Catholicism, Clark married Alfred (Dick) Grove, lecturer in geography at Cambridge, on 17 July 1954 when she took his family name; Jean Grove took her unfinished thesis with her on their six-week honeymoon in Achiltibuie. Their first child, Richard, was born in 1955 and Jean was awarded her PhD in 1956. Jean, accompanied by Dick and Richard (and Rosemary Moore – later Chorley), returned to Norway in 1957 (see Figure 9.3); this was followed by a month in the summer of 1958 with their now two children and a small group of students at Mont Blanc, returning with their three children (Richard, Jane and Lucy) in 1959 and 1960, and with 20 undergraduates at Les Haudères in 1961. It was during this time that Jean turned to the historical studies of glaciers as something more practicable with a growing family (ultimately the couple had three sons and three daughters between 1955 and 1971).[19]

Appointed lecturer and director of studies at Girton College, Cambridge in 1953, Jean Clark became a full fellow of the College in 1960 (Newnham College Register). She remained director of studies until 1994 and took her teaching role very seriously, influencing generations of women geographers at Girton (Allen 2003). Grove was chair of admissions at Girton (1971–89), was part of the committee recommending that Girton admitted men students in 1979, and worked to improve links with and applications from state-sector schools. Having a love of travel and an 'adventurous, experimental approach' herself, Grove encouraged Girton students to 'broaden their horizons' and undertake vacation fieldwork (Goudie 2001; Allen 2003).

Goudie (2001) described Jean Grove as 'a remarkable teacher, parent and researcher' and added 'She shared these roles with her husband and fellow Cambridge geographer, A.T. Grove'. Goudie goes on to summarise Grove's achievements in juggling family and career: 'Jean Grove was a mother of six. The children were carried on her back up Alpine valleys, and when nappies needed to be changed bottoms were dunked into a meltwater stream. The Grove household in Cambridge was homely, with streams of visitors. It required a person of great energy, organisation and dedication to run a large family, to do research, to fulfil the role of teacher and college administrator' (ibid.). Dick Grove's account of their Easter vacation in 1962 gives a sense of the juggling of two career interests and family commitments and the support networks called on to manage this: 'Jean and I met in Delft, where I had been on a short air photo course, and travelled together to Sion by train. After a visit to the Val de Bagnes, Jean went on to attend a glaciological conference at Obergurgl while I returned to Cambridge to help an au pair, plus Jean's sister and her husband (now history Professors Margaret and Peter Spufford), look after the three children who had contracted measles'.[20] The following year they were to spend six months in Ghana, while Dick studied the Volta Delta and Jean took the opportunity to research irrigation in the area. They both taught at Legon University and individually went on field trips with students to northern Ghana, as well as meeting with Cambridge students at their respective research sites en route home. After a return visit to the Volta Delta in 1966, with Shirley Hewitt but sans famille, Jean wrote two long papers (1966, 1968) on the indigenous irrigation practice of the Ewe in the delta; she also contracted dengue fever.

It was during the late 1960s that Jean's interest in historical climatology was consolidated, and this theme was to dominate her research for the rest of her career. Influenced by Gordon Manley and Hubert Lamb's work on glacial advance and retreat in the Alps 1600–1900, she looked for further evidence in the Alps and Scandinavia. Arthur Battagel (a relation by marriage) and Grove studied Norwegian land tax records as sources of data on glacier damage to farmland and climate change and the family vacation in 1969 was spent touring the areas affected, resulting in 'The incidence of landslides, avalanches and floods in western Norway during the Little Ice Age' (1972). Jean and their five children accompanied Dick for a semester at UCLA where he was visiting professor, during which time she gave lectures on glacial landforms in his geomorphology course (regardless of the near university-wide strike in opposition to the Vietnam War). This was followed by a typical family holiday, travelling by car to visit glaciers at Yosemite, British Columbia, Glacier National Park and Colorado.

Grove supervised Brian Whalley for a year while Cuchlaine King (of Nottingham) was on sabbatical, an arrangement which was beneficial to both. Brian drew Jean's attention to one of the current Swiss papers in *die Alpen* on Holocene glacier fluctuations in Switzerland which stimulated her to write in 1979 'The glacial history of the Holocene'.[21] Jean's appointment at Girton College did not include lecturing at the University, but she was made a member of faculty after Clifford Darby's retirement as head of department[22] and was invited to give a lecture course on glaciers and climate history for a few years in the early 1980s. Jean Grove died on 17 January 2001 at Addenbroke's hospital after an aortic aneurism.[23]

In his obituary Andrew Goudie wrote: 'Grove, never an ardent feminist, demonstrated what a married woman can achieve' (Goudie 2001). In a slightly different take, Brian Whalley (2004) wrote 'Though never an ardent feminist Jean Grove was true to her mother's upbringing. She showed in her teaching and research, her commitment to college, and support of friends and students, that a mother of three sons and three daughters (with the valued assistance, it might be added, of au pairs) could continue to pursue a productive academic career'. Jean Grove had always been conscious of her mother's disappointment that she was not able to continue her academic career after marriage[24] and clearly did not want to suffer the same fate, and employment law and practice had changed sufficiently by the late 1950s and early 1960s for her to find a way to combine not only marriage and work but also an immediate and extended period of childbearing. This was less a case of demonstrating what a 'married woman' could do, but rather what a middle-class mother of six children beginning a career in 1950s' Cambridge could achieve. Whilst her lack of a university lectureship and her husband's tenure of the full-time post can be seen as illustrating a patriarchal system combined with the exploitation of female reserve labour, it can also be read as a workable compromise between two supportive partners, coupled with the Cambridge collegiate system which accommodated a meshing of family with part-time teaching and research: a degree of agency found within a framework of structural constraints and opportunities.

Publications

Grove contributed two chapters to the RGS Research Series volume on *Investigation on Norwegian Cirque Glaciers* (1960) edited by Vaughan Lewis and went on to publish other papers on glaciers including 'The Little Ice Age in the Massif of Mt. Blanc' (1966) and a paper on 'The incidence of landslides, avalanches and floods in western Norway during the Little Ice Age' (1972). Her research on glaciers led her to the study of long-term climate change, long before such topics were fashionable and she became 'one of the most perceptive of environmental historians' (Goudie 2001). Later work included collaborations with Anna-Lisa Conterio, Anne Gellatly and with her husband. Jean Grove's research culminated in her book *The Little Ice Age* (1988), which demonstrated the extent to which the climate could vary in response to natural rather than human causes, exemplified in the medieval warm period. A revised version was edited in her retirement and completed posthumously by her husband (Allen 2003), retitled *Little Ice Ages: Ancient and Modern* (2004). Goudie describes this as the 'triumphant' result of Grove's long work, in which she challenged current assumptions that major climate change must result from human causes. As a result of her expertise, Grove served on the Intergovernmental Panel on Climate Change in the 1990s (Goudie 2001; Allen 2003). Although Grove's publications were not great in number compared with today's expectations, 'her influence was substantial' (Goudie 2001; Allen 2003), in part through those publications, and through her many Girton students who were so well versed in fieldwork.

Helen Wallis

Helen Margaret Wallis was born on 17 August 1924 at her parents' home in Park Road, Barnet in Hertfordshire, one of twins, daughter of Mary (née Jones) and Leonard Wallis, both school teachers.[25] Her parents were described as 'of high principles and integrity, but wise and tolerant' (Ravenhill 1996). Initially educated at Colet's Girls' School, she went on to St Paul's School (1934–43), where she won junior and senior foundation scholarships, and prizes for classics and piano (1942). She was exhibitioner at St Hugh's College, Oxford where she studied from 1945 to 1948. During her time at Oxford she was baptised into the Christian church and maintained a strong faith from that time (Campbell 2004).

Whilst at Oxford Wallis became deeply influenced by historical geographer J.N.L. Baker, who encouraged her involvement with the IBG and her orientation to cartography (Ravenhill 1996). In her final year Wallis won the 1948 RGS University essay prize and turned to postgraduate work. As a graduate Wallis held the Mary Grey Allen Senior Scholarship at St Hugh's and was jointly supervised by Baker and Peter Skelton, a leading scholar of the history cartography and superintendent of the Map Room at the British Museum. Wallis successfully completed her thesis on 'The exploration of the South Sea, 1519 to 1644' and gained her DPhil in 1954, three years after she was appointed assistant keeper of the Map Room in 1951. However, working for Skelton proved difficult and her relationship with him became strained. The Map Room had a very masculine environment where staff were only ever referred to by their family names, public-school style.[26] Wallis tried very hard to fit in, working to 'overcome her natural diffidence, even to the extent of radically changing her appearance' (Campbell 2004). Wallis's 'feminine charm' of later years was only espoused after years of suppressing her gender in the highly masculinised world of the British Museum/ Library Map Room under Skelton's leadership. Indeed Wallis was advised by her analyst to give vent to her feminine characteristics in order to avoid a breakdown, after this she could be 'outrageously' feminine whilst being highly organised and professional.[27] Although the obvious candidate to succeed Skelton, his poor opinion of her initially blocked her promotion in a 'morale-sagging delay' (Campbell 2004). Eventually she attained the post of superintendent of the Map Room in 1967 becoming the first woman map librarian of the British Museum and later British Library.[28] 'Wallis's pent-up energies were then released and by the time she retired in 1986 she had established herself as the leading figure in the world of map librarianship in the country and, probably, internationally' (Campbell 2004). The following excerpts from obituaries are rather hagiographic but are evidence of the wider esteem and affection in which Wallis was held: Ravenhill (1996: 300) states that throughout her tenure 'She wove the three threads geography, cartography and librarianship to perfection during the rest of her life'; and Campbell (2004) adds that she and the Map Library have been described as 'indivisible'. She was particularly effective in making the research material of the Map Room accessible to national and international scholars (Ravenhill 1996). During her career she made two particularly significant acquisitions for the collection, namely the only known copy of Giacomo Gastaldi's (1561) wood-cut wall map of the world and the map collection of the Royal United Services Institution, which in 1968 was the museum's largest map purchase (Campbell 2004). She also did much to publicise the Map Room collections and related work through exhibitions and she used her international network to facilitate three being taken to North America (ibid.). After 35 years of service Wallis retired in 1986. She continued to be active on the international conference circuit, often joining forces with Eila Campbell to travel to IGU and cartographic meetings around the world.

Wallis was an active member of both the RGS and IBG. She was elected a fellow of the RGS in 1946, whilst an undergraduate. She served on the RGS Library and Maps Committee from 1967, which she chaired from 1986 to 1994; she also sat on the Society's Council twice and was elected vice-president (1982–5) and honorary vice-president from 1988. She served on the IBG Council for three years (1970–72) and for many years on the cartographic sub-committee of the Royal Society's National Committee for Geography. An inveterate committee worker and leader, she also took a lead in the development of the International Cartographic Association and was first chair of its working group on the history of cartography (1972–87). A long-standing member of the Society for Nautical Research, she was asked to take over the chair temporarily in 1971, a post she held until 1988, but her longest service was on the Council of the Hakluyt Society which amounted to almost 50 years. Described as a 'nucleus around which a world community of map-lovers was able to coalesce' (Ravenhill 1996), Wallis was also someone who had the vision to see the necessity for institutional structures for the preservation and development of cartographic studies and 'was the dynamo that powered' the creation of the British Cartographic Society's Map Curator's Group, the Geography and Map Library Section of the Library Associations and the British Committee for Map Information and Catalogue Systems (Ravenhill 1996). Her various committee roles demonstrate 'the many instances of her thriving in a man's world. She was the first woman to hold several of the posts she occupied' however, 'these distinctions rewarded merit, not assertiveness' (Campbell 2004). She died of cancer on 7 February 1995, aged 70, six months after her great friend Eila Campbell. Her funeral was held at St John's Wood Church (Campbell 2004). A volume of tributes to her was produced under the title *The Globe my World* (1995) (ibid.). She was remembered for her wide knowledge and professional achievements, her support of junior colleagues and researchers, her enthusiasm, sociability and humour, her faith and cheerfulness. She loved choral singing and the British Library and Museum Singers performed at the thanksgiving service for her life.[29] During her lifetime Wallis was honoured with numerous awards (see Campbell 2004), culminating in an OBE in 1986 when she retired. She was also awarded the British Cartographic Society's medal for services to cartography (1988) – the first woman to receive this award and the first woman president of the Society (1972–4). Sadly she died before receiving the 'much-deserved' RGS Victoria medal in 1995 (Ravenhill 1996).

Fellow historical geographer and historian of cartography, Professor Bill Ravenhill (who was himself to die within a few months of Wallis) wrote: 'Painstakingly and unfalteringly over the decades, Helen has done as much as any recent scholar to redeem the scholarly stature and academic acceptance of cartography and particularly its history in spite, it must be said sadly, of the neglect of the calling in some sectors of geography where, at best, the subject has been marginalized and, at worst, ostracized' (1996: 300–1). On her retirement, colleagues produced a bibliography of Wallis' writing which included 250 entries. It has been suggested that 'it may not be fanciful to see, in her relentless drive represented a determination to overturn her predecessor's judgment on her' (Campbell 2004). Wallis also followed in the wake of the prolific Eva Taylor and may have emulated her role model. Wallis's first publication, 'The first English globe. A recent discovery' (1951), discussed her discovery of the earliest version of England's first globe by Emery Molyneux at Petworth House. Wallis published numerous other papers and reviews in the *GJ* and *Imago Mundi*. She edited *Carteret's Voyage Round the World, 1766–1769* (1965) for the Hakluyt Society, the International Cartographic Association's *Map-Making to 1900: An Historical Glossary* (1976) and co-edited its successor volume *Cartographic Innovations* (1987) with A. H. Robinson. The facsimile of Jean Rotz's *Boke of Idrography* (1542) is considered her 'most splendid publication … on which her long-term reputation will rest' (Campbell 2004); a full list of Wallis's publications can be found in *Imago Mundi* (1995: 188–92).

Other women who moved from a university training in geography to map curation include those who worked as university department cartographers and librarians. For example at the Sheffield University geography department **M. Lesley Babington** (BA London) was map curator and librarian for the department (1959–61), followed by **Margaret Wilkes** (1961–3). Wilkes studied at Leicester and graduated with a BA in geography from London University in 1959, and an MA from Leicester in 1961. At Garnett's suggestion Wilkes was appointed research assistant for the remainder of the 1960s until April 1973 and was succeeded as map curator and librarian by **Joan Chibnall** (BSc Bristol, MSc London) in 1964, a post Chibnall held until retirement in 1989. Wilkes moved to the National Library of Scotland in Edinburgh, where ultimately she became head of the map division at the Scottish National Library. She attributes her interest in maps to her childhood: 'I was attracted into geography mainly by my father's influence as a child, including immersion in maps from an early age in the home environment. I also possessed a strong visual and graphic sense – and was far more attracted to maps and pictures than text, had an early natural curiosity about places, their locations relative to one another on the map and a compulsion to know what places looked like and how they were structured … Map curation was the natural concomitant of this as I felt information portrayed graphically in map form was so much more exciting, focused and concise than textual attempts to describe the same thing'.[30] She went on to receive a number of awards for her work: the Tooley award from the International Map Collectors' Society in 1991, an honorary fellowship from the RSGS in 1992, followed by the Society's Bartholomew globe award in 2001 for her 'exceptional contribution to cartography in Scotland'; and the British Cartographic Society's Alan Godfrey award in 1998.[31]

Joy Tivy

Joy Tivy was born in 1924 in Carlow, Eire and studied at Trinity College Dublin, when she read natural sciences, majoring in geography with botany and geology, which she attained with first class honours. Tivy then spent a year as a scholar at Trinity College Dublin, before moving into school teaching where she taught geography and biology at Leeds Girls' High School and whilst working completed the Cambridge Teacher's Certificate as an external candidate. This was followed by a short time working as a research assistant in West Riding's Town and Country Planning Department before being appointed as assistant in geography at Edinburgh in 1950. There she researched her doctorate on 'A study of the effect of physical factors on the vegetation of hill grazings in selected areas of southern Scotland', famously conducting some of her research on horseback! She was appointed lecturer in geography at the department of geography, University of Glasgow in 1956 and, after the peripatetic character of her early career, was to stay there for the rest of her working life. In her early years at Glasgow, Tivy built on her doctoral work, undertaking further research (with the help of students and colleagues) in the Southern Uplands, publishing a detailed paper on slope deposits and the tendency for gully formation in the Lowther Hills in *TIBG* in 1962. In her conclusion she articulates the diagnostic purpose of such work: 'Fuller knowledge of the nature of these deposits is essential to an elucidation of the processes which produced them and their many associated surface forms: it is essential also to a clearer understanding of the ecology and of the effects of man's utilization of such hill land' (1962a: 71). She was promoted to senior lecturer in 1965, Reader in 1973 and professor in 1976. She served as head of the department of geography and topographic science from 1985 to 1989 (during what was a difficult time in British higher education) and sat on all the main committees of the university during her career. Tivy became a senior research fellow after retiring in 1989 (Diamond 1995; Caird 1996), continuing to research and write alongside her interest in music and water colour painting, but died suddenly on 10 July 1995.

Tivy's early combination of geography with botany and geology and her (albeit relatively brief) experience of teaching and planning were to influence her work throughout her career. She became a great proponent of the Field Studies Association (FSA) and undertook a lot of work with schools and adult education classes, including those at the FSA centre at Kindrogan and Glasgow University's extension classes along the length of the west of Scotland (Diamond 1995; Caird 1996). In the early 1960s, she established honours courses at Glasgow in biogeography and the USA, making strong links between her teaching and research and is remembered as a gifted teacher who enthused her students and acknowledged their input in her own work. This enthusiasm for her subject and sharing her views can be seen in her numerous contributions to the Glasgow department of geography's student magazine *The Drumlin*. As noted in Chapter One, different forms of writing allow a different authorial voice as well as range of subjects, and in the pages of *The Drumlin*, addressing current and former students and immediate colleagues in an informal publication, Tivy offered forthright views on the nature of geography and biogeography, the relationship between teaching and research, and the importance of fieldwork and sound methodology. In a 1962 article on 'Some thoughts on field work' Tivy stressed observation as the cornerstone of work in the field: 'acute, objective, systematic observation' which could only be acquired with practise in the field, and which led to an understanding of interrelationships: 'While the geographer is, and must be, concerned with the nature of the phenomena he is studying, this concern must be for him a means to an end – that of understanding the way in which the phenomena interact, one upon the other, through time, to produce given patterns, distributions or "regions" on the earth's surface at the present moment ....Within the blinkers of the classroom, these inter-relationships can only be studied at "second hand"' (1962b: 24). She went on to stress the importance of analysis, appropriate units of measurement, use of both quantitative and qualitative methods, and candidly expressed distrust for 'deep-rooted reverence for the documented fact, the glitter of statistics and the authority of the specialist opinion' (ibid.: 25).

Joy Tivy published some 40 articles, chapters in books, project reports and scholarly works (Caird 1996), in addition to several books. Her key publication was the text *Biogeography: a Study of Plants in the Ecosystem*, first published in 1971, and revised in 1982 and 1993. She wrote on the carrying capacity of recreational land (1972), edited *The Organic Resources of Scotland* (1973) and co-wrote *The Human Impact on the Ecosystem* (1981) with Greg O'Hare (a former student of Tivy's). Critical engagement with Tivy's work was mixed. Books she wrote or edited were criticised for numerous minor errors and the more significant relative lack of attention to animals within the ecosystem (Grace 1974; Taylor 1994); her *Biogeography* text was criticised for lack of structure (Kuchler 1973) and its casual approach to the use of technical terms (Carlquist 1978: 187–8). Whilst recognising the difficulties of covering such a large interdisciplinary field, Carlquist criticised her inclusion of conservation-related material at the expense of more foundational topics and went on to argue that 'the book is too elementary to be useful'. In fact *Biogeography* became the 'world's best-selling biogeography textbook' and was in print for over 20 years (Diamond 1995: 356). Such mixed messages raise questions regarding the evaluation of the reception of work, but as noted in Chapter One, reception can vary enormously between individuals, groups, places and times (Livingstone 2005; Keighren 2007). It is not surprising therefore that students might favour a readable book (and some tutors must have been recommending it), whilst, at the same time, specialists from a variety of disciplines might criticise perceived omissions. Still an active researcher in retirement, Tivy published *Agricultural Ecology* in 1990. Tivy considered biogeography was the '"Cinderella" of geography', and that biogeography specialists were considered 'odd birds', and students' lack of familiarity with the subject was a psychological barrier to recruitment to courses (1965: 32). Despite this she was credited with putting 'biogeography on the academic map for the first time, by dint of scholarly research and writing ...' (Diamond,

1995). Whilst this view may disregard the earlier work of Newbigin and the parallel contemporary work of Monica Cole, it gives a clear sense of the esteem held for Tivy's role in establishing biogeography as a modern sub-discipline, even if this was not always recognised by reviewers such as Taylor. Tivy (1965) wrote that the main foci for biogeography were two-fold, first the relationship between organic and non-organic in the environment and secondly the reciprocal relationships between people and the biosphere, the biosphere being the vital link between people and the physical environment. It was this view which coloured her teaching, publications and consultancies. Ultimately Tivy, in her grounded-research approach to biogeography, is credited with acting locally and thinking globally ahead of her time (Diamond 1995); reviewers may have wondered why she devoted valuable space to conservation issues in her books in the 1960s and 1970s, but this now seems emblematic of insight of what is at the heart of contemporary biogeography (ibid.).

Tivy's research attracted numerous fellowships and awards including time at the universities of Syracuse in 1963 and Cornell in 1973. She had an international reputation and was consulted by the European Union on set-aside agricultural land, and at home 'Every Scottish agency concerned with the environment sought her advice at some time' (Diamond 1995: 356), e.g. the North of Scotland Hydro Electric Board and The Countryside Commission for Scotland (Caird 1996). 'She served her University with great distinction' (Caird 1996: 355), although anecdotally she was not always popular as head of department. She also served the wider community as editor of the *Scottish Geographical Magazine* from 1955 to 1965 and was honoured by the RSGS with an honorary fellowship and later a life membership. She was honorary secretary and editor of the FSA, and chair of its Executive Committee from 1984 to 1987. She was also awarded fellowships of the Institute of Biology and Royal Society of Edinburgh.

Catherine Delano-Smith

Catherine Delano-Smith was born in 1940 in Wallingford, Oxfordshire, she grew up in London and attended Wycombe Abbey School in Higher Wycombe. Until 1940 her parents had lived in Paris. Her Russian emigré mother (naturalised French) encouraged her to speak French, and her father, a graduate of Arabic at the School of Oriental and African Studies (SOAS), taught her to be observant. Sadly he died shortly before she went to university, leaving her, as an only child, with responsibility for her mother. She gained a place at Lady Margaret Hall (LMH), University of Oxford in 1959, owing much, she feels, to Mary Marshall's faith in her potential, but Marshall was to be a short-lived tutor and Delano-Smith found herself without a college-based geography tutor and the only geography student in the college. As an undergraduate in the geography department, she was one of about 10 women out of a total of 60–70 students. With her LMH friends she had a strong sense of the privilege of not just being at university, but being a woman at university: 'only 3% of our generation went to university ... we did feel as women that we were an exceedingly privileged minority, we were a minority and we knew for every one who got in there were two or three who might equally well have been given a place'.[32] In her second year Delano-Smith organised an interdisciplinary expedition to Madeira (having been barred from their intended destination, the Cape Verde Islands), where the three geographers in the team produced a land use map, which Delano-Smith wrote up for publication by Dudley Stamp. The expedition group had to be single sex in order to gain university recognition and therefore be eligible to apply for RGS funding and theirs was only the second women's expedition to take place. For her final year regional description Delano-Smith opted to study the cantons of Beaucaire and Tarascon in France, in part through her love of the Mediterranean, already formed through family connections in Nice, and in part for convenience (her mother was at that time back in Paris). Newly arrived in Tarascon, she was quizzed at length by a gendarme who – confronted with an unaccompanied young woman – was doubtful about her story that she was undertaking fieldwork for her university degree. She was awarded an upper second class honours degree, having been warned that organising the expedition could cost her a grade in her final examinations. After finals, waiting for news of possible funding for postgraduate work, Delano-Smith joined an 'informal' self-funded (and therefore mixed-gender) field group to study agriculture in southern Greece. It was on this trip that she gained a sense of direction for her future research. 'We went to Tiryns' tomato canning factory and we came out of that and I thought "So what?, cans whizzing around" and then we went to visit the ancient site of Tiryns itself, which was right outside the factory. I remember distinctly standing on those ruins and saying that I'm much more interested in Tiryns, and the history of this mound than I am in tomatoes ... and that's when I consciously started being a historical geographer.'[33]

Delano-Smith gained government funding and a place to continue at Oxford and undertook fieldwork on the Garrigues in the Languedoc, completing her BLitt in 1965. In retrospect, she recognises that she was groping her way into environmental archaeology, a subject not yet invented. From southern France, Delano-Smith worked in southern Italy with the Society of Antiquaries, working as an environmental geographer with a group of archaeologists, who had taken over John Bradford's pioneering project. The long-lasting, but never fully completed project, an integrated account of the Tavoliere of Apulia in

Neolithic, Roman and Medieval times, was based on the remarkable evidence on RAF wartime photographs. In Foggia, Delano-Smith stood out, it seemed to her, as the only woman driver in city or countryside.[34]

From 1964–7 Delano-Smith held the post of assistant lecturer at the University of Durham, where she overlapped with **Ann Graham** (who had been an RSGS University prize winner in 1959 and had a MA from St Andrews and a PhD from London University), but the only other woman in the department was the librarian. As a young lecturer-elect Delano-Smith attended the IGU conference in London (1964) and her account illustrates the intimidation many can experience at large academic gatherings for the first time: 'I do remember I went to that not knowing much about it and was very shy, not knowing how to mix, or what to do, although I did meet a couple of my future colleagues who were very welcoming … I went only because I had been advised by my future head of department to go'. However, once established at Durham, within a network of colleagues, the experience a few months later at the IBG was radically different: 'a lot of us went from Durham as cheerful young tykes, and I met people like Brian Harley and Alan Baker at what was then called the Agrarian Landscapes Research Group from which the Historical Geography Research Group was born'.[35]

Declining the opportunity to stay on at the end of her three-year tenure at Durham, Delano-Smith then applied for a post at Nottingham University in 1967. She wanted to be closer to her researches in southern England and her widowed mother in London, and the London to Nottingham motorway had just opened making regular journeys realistic. Soon after arriving at Nottingham, Delano-Smith was called in by the two professors K.C. Edwards and Dick Osborne, and told 'we'd like everyone on a level playing field, everyone has to have a doctorate'.[36] Consequently she registered as an external student at Oxford and undertook to channel her Italian research into a DPhil (gained in 1974). Her thesis was translated into Italian and published by the Foggia municipality where her research was conducted (a mistake she soon discovered in terms of academic strategy as this, and her other non-English language work, remained little known in the UK among geographers). At Nottingham Delano-Smith took responsibility for all historical geography teaching when Joan Fuller retired in 1969 and helped to run field trips to Tuscany and Montpellier. For Delano-Smith the value of fieldwork was about training students to 'ask the right questions' in response to what they saw. Eventually she was promoted to Reader and took early retirement in 1990. Initially she taught in the Interdisciplinary Mediterranean Department at QMWC, London, then focused full time on the history of cartography, to which her interests had moved c.1980. In 1994, she took over the voluntary job of editing *Imago Mundi* at Eila Campbell's invitation, which she continues at the time of writing, and is currently a Senior Research Fellow at the Institute of Historical Research, University of London.

Although Delano-Smith didn't consider herself to have suffered from sexual discrimination, in the course of our interview she described a number of incidents in her career that suggest the difficulties of navigating the cross-currents of gendered social norms and academic equality, as well as possible and explicit institutional and individual sexism. 'The embarrassment at Durham when I first went there was social rather than gender, in the sense that a lady was supposed to be paid for; I was not married and when you're among young married colleagues all earning around the same amount of money and it was quite hard – for everybody – to make that [transition]'.

One experience she recalls as odd, although she herself has never ascribed it to a gender issue, was when she applied for a newly advertised academic post and went for the interview only to be told 'I'm terribly sorry the job doesn't exist'.[37] She also recorded the seeming trivialisation of her work in historical geography by colleagues as 'hysterical geography' in both formal and informal meetings ('once is amusing; repeated over the years, it started to come over as unsupportive'). The notion of hysteria has long been shown to have very strong gendered overtones (Bankey 2001) and, although Delano-Smith may not agree, this treatment represents informal individual sexism: it is hard to imagine the same comments being addressed to senior male historical geographers. Delano-Smith also described how it could be difficult to make yourself heard in informal meetings: 'men will barge in at a round table discussion and women either won't be listened to unless they bark or you'll be listened to in an exaggerated silence which is embarrassing'.[38] In the final analysis, Delano-Smith felt she had always worked for individual rather than gender recognition. Only once (and very new at Nottingham and in the heat of the 1968 political protests) did she argue for women's representation on a university committee – only to be acutely embarrassed to be appointed as that representative. She stresses, however, that 'I've never been a feminist as such ... I've never pushed the "women" line. That was probably the only remark [of that type] I've ever made in my life, I've fought my battles on personal grounds, not on any [gendered basis]'.[39]

Publications

Delano-Smith began publishing when she was an undergraduate, but the fact that she often published outside geographical journals and in non-English language media means that much of that work was unknown to those reading only mainstream English language geography journals. Most of her publications in the field of historical geography appear after the completion of her doctorate in 1974 and continue after her early retirement. At a time when historical geography in England was centred on Clifford Darby's work on Domesday, Delano-Smith felt there was a minimal interest in foreign prehistoric geographies within the sub-discipline. In her 1967 paper in *TIBG* she argued that tracing links between prehistoric and recent history or contemporary landscapes could be difficult, 'But if a study of geographical change throughout time, from prehistoric to the nineteenth century, could illuminate some of the problems of man–land relationships, those of today as of the past, such studies may prove more than merely absorbing academic exercises' (1967: 208). Delano-Smith went on to publish *Western Mediterranean Europe. A Historical Geography of Italy, Spain and Southern France since the Neolithic* (1979), which still appears on university reading lists (although not necessarily in geography) and is cited in recent work on historic evidence for climate change. She continues to research and publish widely on the history of cartography including *English Maps: a History* (1999), co-written with J.P. Kain.

Anne Buttimer

Anne Buttimer came to a career in geography through her work as a teacher in Holy Orders in the Catholic Church. Born on 31 October 1938, daughter of Eileen (née Kelleher) and Jeremiah Buttimer, she grew up in a farming family in County Cork, Ireland. Having studied geography alongside mathematics and Latin and graduating with a BA from the University College, Cork in 1957, she went on to take an MA in geography there in 1959, followed by teaching in a secondary school for a year. It was then she decided to take religious orders, becoming a Dominican nun in the USA in the same order as her sister Mary. There she became part of a group of nuns reviewing new developments in the social sciences for the wider school curriculum. Tasked to examine the potential place of social geography within the school curriculum, Buttimer undertook doctoral research 'so I could represent my community on a Seattle University Faculty team of (nun) professors who were to implement an integrated curriculum for the training of (nun) teachers' (1987: 308). Buttimer completed her PhD at the University of Washington (1965) under the supervision of Morgan Thomas, exploring the French geographical tradition as an entrée to 'social geography', at a time when many geographers – especially at Washington – were passionately promoting the merits of spatial science and quantitative techniques. Indeed Buttimer records Professor Ed Ullman telling her in relation to her own work: 'We got all that stuff out of our systems long ago – with Ellen Churchill Semple'. However, 'Much as I admired Ullman, I admired

Figure 9.4   Anne Buttimer c. 1968.
Source: Anne Buttimer

Semple even more, her Mediterranean work [Semple 1931] having built bridges for me between classical history and physical geography' (Buttimer 2001: 24). By the late 1970s social geography was established within university curricula internationally.

From 1965 to 1966 Buttimer was a postdoctoral fellow at the University of Louvain, Belgium, which was followed by two years as assistant professor of geography at the University of Seattle, Washington, USA. She was then recruited to a post as lecturer in urban studies at the University of Glasgow (1968–70), where she was part of an interdisciplinary team evaluating urban planning processes, funded by the Social Science Research Council. She drew on her US training in quantitative techniques and her interdisciplinary perspective, including psychology and philosophy, and the ideas of social space from French geographer Maximilien Sorre and sociologist/ethnologist Paul-Henry Chombart de Lauwe (see Buttimer 1969). Indeed much of her life's work has been framed by what she describes as Chombart de Lauwe's two cardinal concepts: social space and social milieu. 'De Lauwe identified a hierarchy of urban social spaces: familial/domestic networks and social interaction; the neighbourhood space of daily interaction and movement; the economic space (largely relating to employment foci); and the urban regional'. This framework allowed the identification of areas where people felt comfortable to move in and those where they felt tension or anomie (Buttimer 1969: 421).

The interdisciplinary project evaluated the relocation of working-class families from the slums to the periphery in Glasgow and Buttimer looked at planning standards and health and welfare provision. 'I was assigned the residential area design part of it, the others really reviewed planning litera-ture but I felt I had to go and interview people about their everyday life experiences before the move and after the move and then to see what could be learned about planning standards. I took little sections of the city where there were planning standards applied and those where they were not, to see if they made a difference'.[40] She looked at three specific aspects of everyday life: how people imaged/perceived space; activity networks of interaction; and territorial spaces. She asked: 'Where do you go, how far away is it, how far away is too far away? So I took a great number of destinations of volun-tary associations, family, friends, shopping, school, business, the whole lot, then I looked at the contrast in the shapes of these activity spaces, the size of them, depending on whether people were close to their traditional home area or whether they were on the periphery, and it was very clear that those who had been displaced, they would go back to the downtown for the doctor, for church, for various things, they never became integrated into the new environment ... I asked people to circumscribe the area that they felt at home in and some had very very small territories in the new high rise and some in the old Gorbals, it was very clear that there was consensus as to what constituted home as well ... Each of the profiles had a different set of lessons, but it was clear that afterwards that planners did think differently about rehousing people closer to home rather than scattering them out ... my concern about the Glasgow housewives was related to the fact that I could see these men who are going out to work and it's the women who have to handle life all day long in the estate'.[41] Using this qualitative data Buttimer combined it with quantitative analysis, using activity networks and making centrographic ellipses, drawing on her Washington training; her 1970s' publications on this research were widely read and translated into several languages.

From 1971 Buttimer held various posts at Clark University, USA, culmi-nating in professorship in 1980. In 1976 Buttimer received dispensation from her vows and left Holy Orders, but Dominican values remained an abiding influence in her life and work. In 1978 she married Swedish aca-demic Bertram Broberg. Buttimer spent six years as a research fellow at the University of Lund, Sweden (1982–8) where she worked on the interna-tionally important Dialogue Project with Torsten Hägerstrand. After three years at the Université d'Ottawa (1989–91), she completed her career as professor of geography and head of department at University College, Dublin. Buttimer served on numerous geographical and interdisciplinary committees including those of the AAG, RGS/IBG and Irish National Committee for Geography. Deeply committed to international exchange Buttimer was involved with the IGU throughout her career, she served as vice-president of the IGU from 1996 to 2000 and as president in 2000. She was the first woman to hold this office.

Publications

Buttimer published papers on the French region (1965), French geography in the 1960s (1968a) and social geography (1967, 1968b). The French concepts of *la geographie humaine* and *genres de vie*, derived principally from the work of Vidal de la Blache, have threaded through her work since her doctoral thesis, seen in her AAG monograph *Society and Milieu in the French Geographic Tradition* (1971), which was based on her 1964 thesis. Buttimer reclaimed the French school of geography for the post-1940s' generation of Anglo-American geographers, arguing that the Vidalian tradition was 'a precious ingredient of geographical history, and also an important milestone in the history of ideas' (1971: 1–2). Her 1969 paper on 'Social space in interdisciplinary perspective' published in *The Geographical Review* was widely read and reprinted in several compilation volumes (see Figure 9.5); this was followed in the early 1970s by several papers based on the Glasgow project, addressing planning, social space and welfare, and work on phenomenology and humanistic geography in the 1980s (see Buttimer 1979, 1999).

One of Anne Buttimer's most abiding contributions to the history of geography has been through her work *Values in Geography* (1974), which challenged the limitations of models and quantitative techniques and engaged philosophically with the *raison d'être* of geography as a discipline. This had particular influence on the thinking of Torsten Hägerstrand and led to Buttimer and Hägerstrand undertaking the Dialogue Project at the University of Lund, 1978–88 (see Buttimer 1983).

Margaret Storrie

Born in Paisley on 23 October 1935 to Louisa (née Allison), a former teacher, and Frederick Storrie, a lecturer in chemistry at Glasgow University, Margaret Storrie grew up in a family which placed great store on education and public service. It was the lure of the outdoors which attracted her to geography, despite other talents in mathematics and music (and contrary to parental and school advice to pursue the sciences rather than geography). She also considered studying forestry at Aberdeen but was informed that 'girls' were not eligible.[42] A combination of Scottish tradition and fees paid, courtesy of her father's post in the university, saw Storrie go to Glasgow as an undergraduate. Ronald Miller had recently taken over as head of geography after serving in the army and in Nigeria; he conceived geography as a 'science', with an emphasis on surveying, but Storrie found the old-fashioned course intellectually disappointing. One Miller innovation which was fruitful for Storrie, however, was the introduction of a third-year regional

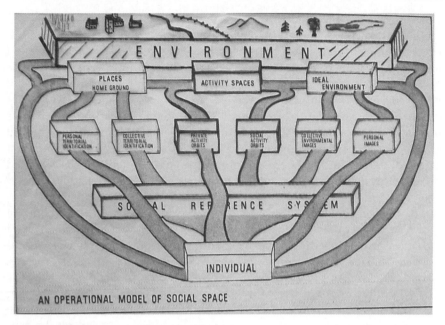

Figure 9.5 Buttimer's model of social space.
Source: Anne Buttimer

dissertation; using a pin in a map, she chose Islay for her study, thus beginning a lifelong engagement with that and other islands. Having rejected teaching, town planning and personnel management as employment possibilities, and with a then fiancé in the Hebrides, Storrie opted to continue with archival and field-based geography. On graduation, she was appointed to a three-year assistant lectureship at Glasgow (1958–60), lecturing on South Africa, taking field and laboratory classes, while working on her PhD on the landscape history of Islay and Ardnamurchan-Sunart. She worked closely with Alan Moisley on the 1957 Outer Hebridean crofting project and he in effect acted as an unofficial PhD supervisor. Interviewed for a post at Hull she decided against working for another military man (Wilkinson), but was offered a two-year tutorial fellowship at Bedford College on the basis of a telephone interview with Gordon Manley. She succeeded Joan Kenworthy to both post and flat in St John's Wood; Smee, Timberlake and Vollans were colleagues. Being in London hugely extended her horizons, especially in the musical worlds, but there were also clouds on the horizon: 'it was only when I went to London I realised the gender bias.'[43]

Her PhD was completed in 1962 (her professorial supervisor never once having discussed any aspect of it!), but Storrie was utterly disappointed by her late evening viva at Miller's home. O'Dell, as external examiner, asked her about one single diagram and then proceeded to discuss gas combustion engines with Miller, the internal examiner: 'It was such a let down: I didn't think they'd even read it'. Storrie recalls a similar experience when interviewed for a post at Queen Mary College the same year, 'they didn't ask a single question about my interests, abilities or research … but Smailes did ask if I had any intention of getting married, and if I did, would I plan to have children' (Storrie obfuscated on this then hypothetical question). QMC was not an all-male preserve, Alice Mutton was a senior colleague and historical geographer and **June Sheppard** had been lecturing there since 1953. Most in the department had been schoolteachers, and the college was focused on undergraduate teaching, with little research culture or encouragement. Storrie recollects that course and other commitments were allocated without discussion and without reference to research interests: a single holiday to North America qualified a person to lecture on it.[44]

In the light of only three advertised posts that year, Manley, whose enthusiasms, intellect and contribution in meteorology Storrie still greatly reveres, advised her to take the lectureship as a first step on the ladder. During the first year of her appointment Storrie then met and married LSE geographer Ian Jackson. On both engagement and marriage she was asked to resign from the department, on the grounds that a married woman on the staff might share departmental secrets with her husband. Furthermore she was not allowed to be involved with the 1964 London IGU congress, which Jackson was helping to organise, and no undergraduate field class expenses were paid because of their dual income. She was the first woman to wear a trouser suit at QMC, which prompted a letter from Smailes telling her trousers were only permissible for women on field courses. (The Registrar's advice was to bring back another couple of trouser suits from her travels.)[44] Storrie applied for other posts but many departments were known never to appoint women; 'Even now I feel quite bitter about such blinkered attitudes … to be held back on those grounds … what it did to one's enthusiasm and energy to fight prejudice like that … it was just so frustrating …. Likewise unwelcome sexual harassment [from colleagues within the discipline]'.[45]

Pregnant in 1971, her resignation was again demanded, but the college Principal overruled, as the Registrar had done previously (Storrie, who sang with the London Symphony Chorus and in the Royal Naval College, collaborated with the Registrar in college musical activities). She led a field trip in Scotland in April 1972, her daughter was born on 1 June and she attended a staff meeting five days later. Jackson had moved to a non-academic post in Canada in 1967 and they 'commuted' until they went separate ways in 1980; in effect Storrie had sole responsibility for their child when in the UK, juggling 5pm meetings and long working hours with the help of non-familial pre- and after-school childcare in term time, and family help during vacation field classes and so on. Storrie published numerous papers on her research in geography journals including *The Geographical Review* (1961), *SGM* (1962), *TIBG* (1962) and *Geografiska Annaler* (1965), followed by research and publications elsewhere on Arran, Islay and others of the Hebrides. She was the first geographer to write on alcohol (Islay being a 'whisky island'), and for decades, Storrie was introduced by male colleagues as 'Whisky Storrie', 'which always got *them* a laugh'.[46]

During the 1980s she was a leading force in the academic development of Canadian Studies, instituting annual conferences and a journal, as well as being active and an office-bearer in national and international associations and policy colloquia. By the 1980s, QMC was one of numerous universities in financial difficulties and in 1988 Storrie took an attractive early retirement package, while agreeing to continue teaching on a part-time basis for 10 years. In 'retirement' her base has become Islay, where she is developing an extensive garden and she retains a bolthole in London. She continues to research, write and broadcast, and has edited *Scottish Archives* for the past 15 years.

Editorial work is a popular option for those taking early retirement. In addition to Storrie and Delano-Smith (discussed above), **Elizabeth Clutton**, another historical geographer (see Clutton 1978 and Clutton and Kenny 1976), similarly took early retirement in the early 1990s from Sheffield and works part time as editor of an academic journal, working from the Isle of Man.[47]

Christine McCulloch

Christine McCulloch (née Sunley) was born in October 1939 and lived in Richmond, Yorkshire with her parents, until they divorced when she was 12 years of age. This resulted in separation from her father, a highway surveyor and country man, when Christine and her mother moved to Scarborough and penury. Her childhood, including the war years with their domestic vegetable plots, pigs and poultry, were a formative influence: 'the country background and open air life in Richmond influenced my interest in geography. The other influence was immense curiosity about the world. Without television, [with] only poorly-illustrated books, no ethnic mix in the local community and very little opportunity to travel, even within Britain, the world was one of imagination'.[48] With two other girls she cycled to the Boys' High School to take A-level classes in chemistry and biology and later, torn between biology and geography, applied to read geography at Cambridge and zoology at King's, University of London. Her experience at Newnham represents a case study in the sociology of the social system of Cambridge entrance procedures in the late 1950s.

The interviews and examinations at Cambridge remain painful in my memory. I don't think they were ready for totally naïve, first-generation university applicants from the North. I had to go to Cambridge by train and stay there for two nights. The first evening we were 'interviewed' by current students at Newnham to see if we would fit in, which I interpreted as a distinct threat in class-ridden Britain. The next day we had a written exam in the morning and then a practical exam in the afternoon. The dons had not taken the trouble to see what syllabus the Northern schools had been following and gave me a formaldehyde-preserved dogfish to dissect. I had never seen a dogfish before! ... After a painful afternoon, poking around in the fish and desperately thinking up names for the parts, the next day we had to go to the Principal's house for interviews. Her grand house with interior court yard was forbidding and then I shudder to think of my answers to the questions posed by two middle-aged ladies perched on a settee a great distance from my interviewee's chair. Questions about what my father did, where my geography teacher had studied, what were my political beliefs were, were fielded hesitantly but when I was asked whether I would be interested in geographical research, I revealed my ignorance by saying 'I did not know that geographers did research!' It was no surprise, and even a relief, to find that I had only made the waiting list.[49]

Sunley went on to King's to read zoology, but found it a disappointing regression to rote learning after the exciting debates about evolution and field study on the foreshore which she had experienced at school and she swapped to geography and geology. 'We did many field trips gratifying my desire to travel and to appreciate the landscape in Britain and, by winning a Fulbright Scholarship, I was able to sail to the USA afterwards. It took 10 days to cross the Atlantic with the reward of coming into New York by the Statue of Liberty'.[50] She had attained a first class honours degree in 1961, one of only four students (two female and two male) out of the University of London cohort of about 150 students. Based at Oberlin College, Ohio, she worked as a graduate assistant for two semesters before travelling for two months and while there received a letter 'offering a lecturing post at Goldsmith's College, without interview. Goldsmith's, then a teacher training college, wanted to introduce an honours degree course in geography in its aspiration to become a mainstream college in the University of London, but candidates for lecturing posts were in short supply'.[51] The post was welcome, but the heavy teaching load left little time for research and after three years Sunley applied for posts advertised at two other University of London colleges in 1965. She was 'outstripped' by a candidate with more research experience at UCL, but was successful at Queen Mary's. However, her gender proved to influence not only the questions asked at interview but also the nature of the job offered: 'I applied for a full-time established lecturing post and after the interview I was told that

I had the best qualifications for the job, but because I was a woman, they would not be able to offer a fixed-term, two-year appointment. Professor Smailes kindly explained that his department of 10 staff already had three women (Alice Mutton, June Sheppard and Margaret Storrie) and he did not want to "unbalance" it any further. At both interviews, I was asked whether I had plans for marriage and had to dissemble'.[52] It was not common for individual women to challenge such sexist practices at that time, especially young women at the beginning of their career and Sunley accepted the job. However, her experience at interview was to prove indicative of attitudes to women within the department. A colleague told her that there was no allowance for maternity leave, and advised her that 'if you want to have a baby you'll have to plan its birth for the beginning of the summer vac!'[53] While marriage was frowned on, single women academics were derided: 'Within the college, the appearance of the few women academics inspired dread of sexless spinsterhood which was so reminiscent of the 1950s' girls' grammar school teachers, who had dedicated their whole lives to their charges. It was a relief at the end of the two years at QMC to join the Natural Environment Research Council [NERC], where explicit equal opportunities reigned and accomplished well-rounded career women flourished.'[54]

Sunley joined the NERC in 1967, only two years after its foundation, which she describes as an exciting time when initiatives were developed to encourage environmental sciences which had previously been dominated by natural sciences. In the face of doubts whether geography was sufficiently scientific to justify NERC support, Sunley invited David Linton, Keith Clayton and others to make a case for geography to be included within the remit of the Council; they put on a 'good show' and geography was admitted to the NERC fold. Sunley's own portfolio included the British Antarctic Survey, hydrology, meteorology and climatology, land use and geomorphology; and while in post she completed an MSc on 'The concept of the graded river: a recurrent problem in geomorphology'. She married Jim McCulloch, Director of the NERC Institute of Hydrology, and resigned just before the birth of their first child. The couple had two children, Richard and Caroline, within 13 months of each other and this combined with live-in elderly relatives in need of care, resulted in a six-year career break for the now Christine McCulloch. Marjorie Sweeting helped McCulloch find a part-time school teaching post, before the fortuitous move of the Economic and Social Research Council (ESRC) to Swindon and McCulloch's appointment there, where she eventually became head of politics, economics and geography research support. After retirement she was recast as a human geographer, completed a doctorate at the University of Oxford on the political ecology of dam building then worked as a research associate at the University of Newcastle. In 2006 McCulloch received the *Area* prize for a new researcher, the same year and at the same age, she noted with irony, that Janet Momsen received a lifetime achievement award.

Experience of sex discrimination in appointments was not unique to the UK, **Janet Momsen** had a similar experience in Canada in the late 1960s. She was born in Cheshire, 10 December 1938, daughter of Eileen (née Quayle) and Fred Henshall. Influenced by her mother, who had been a teacher, and by Miss Leigh, her geography teacher at Manchester High School for Girls (MHSG), Henshall went on to study geography at Lady Margaret Hall, University of Oxford (1958–61). At Oxford Henshall was president of the Junior Common Room and Herbertson Society. In 1960 she was a member of the first Oxford University women's expedition (RGS sponsored) to the Azores and the report on land use in Faial (Henshall & Callender 1968) was published by Stamp. This was followed by a Dip Ed 1961–2 and BLitt 1965. Awarded a Commonwealth Scholarship, she completed an MSc in agricultural economics at McGill University, Canada (1962–4) and was appointed lecturer in geography at King's College, University of London in 1964. While teaching there she worked on her PhD (University of London 1969),[55] moving to the InterAmerican Institute of Agricultural Sciences in Costa Rica in 1967. Her original thesis title was 'A factor analysis of small farming in Barbados and the Windwards', but on Smailes' direction had to change it to 'Land use and population in the Caribbean'. 'It was one of the first quantitative theses but Smailes did not approve and had the power to change the title!'[56] Her next move was to Calgary University, Canada in 1969 where she was offered a job: 'they withdrew the job I had been offered when they discovered that I was pregnant. They then refused to give me a job for the next 10 years because I was a woman.'[57] Staying at Calgary because of her husband's job, 'With great difficulty' Momsen negotiated her two children's upbringing (and care of five visiting step-children) and maintaining academic work. Her list of publications on agriculture and the Caribbean in the 1970s would have been the envy of many a colleague with tenure.[58] An early quantitative paper in TIBG 1966 was followed by a chapter in Chorley and Haggett's (1967) *Models in Geography*, which was particularly influential at this time; Henshall was one of only two women contributors to the volume. In 1979 Momsen returned to the UK, to a post at the University of Newcastle where she worked collaboratively with fellow Oxford graduate **Janet Townsend** (notably the formative collection *Geography of Gender in the Third World* (1987)). During this time Momsen was one of the founders of the IBG Women and Geography Study Group and a contributor to *Geography and Gender* (1984), as well as continuing to write extensively on gender, development, agriculture and tourism. She completed her career at the University of California, Davis. Momsen has recently retired to the UK and continues to be active in research and publication.

Doreen Massey

Doreen Massey was born on 31 January 1944, daughter of Nancy (née Turton) and Jack Massey. She was brought up in Wythenshawe, Manchester and became a pupil at MHSG, having gained a bursary. She went on to read geography at St Hugh's College, Oxford (1963–6). Doreen and her sister Hilary (who studied geography at Southampton) were the first in their family to attend university. While at Oxford Massey was awarded a scholarship (1963) and Special College prize (1966), and attained a first class honours degree. After working in market research (1966–8), she became principal scientific officer in urban and regional research for the government-funded Centre for Environmental Studies in London 1968–80. Massey took a sabbatical to complete an MA in regional science at the University of Pennsylvania (1971–2), which was a critique of industrial location theory.[59] Massey has served on the editorial board of *Environment and Planning A* since 1968 and early publications include 'Towards operational urban development models' (1971) in Chisholm's collection on *Regional Forecasting*, and *Capital and Land* (1978) written with Alex Catalano. She was appointed professor of geography at the Open University in 1982 and has been a huge force in the discipline in the post-1970s' period (see Callard 2004), including recent publications *For Space* (2005) and *World City* (2007). Her post-1970 work merits much fuller treatment elsewhere.

Barbara Kennedy

While studying geography at Newnham College, Kennedy's undergraduate dissertation applied Mark Melton's ideas in France. Having been awarded a double first class honours degree, she studied for her master's with Melton at the competitive graduate centre in the department of geography at the University of British Columbia. This was followed by returning to Cambridge for her PhD under the supervision of Dick Chorley. However, one year into her research she was appointed to a junior teaching post at Cambridge as departmental demonstrator, as the previous incumbent had left because of 'departmental politics'. Despite a large amount of teaching she completed her PhD on schedule and was appointed a fellow of the women's college, New Hall, Cambridge. At this time she was the only woman in the college with a university post.[60] The most significant of Kennedy's publications in the time period studied here was her book co-authored with her supervisor Dick Chorley, *Physical Geography: A Systems Approach*, published in 1971. '[The book was] basically Dick's idea. He provided the structure and I provided the information. I was teaching biogeography and I became more and more aware that if anywhere a systems concept should work, it was the eco-systems, and in fact it didn't, and I got disenchanted with the idea and if you read the introduction you will see two minds at work: one saying this is wonderful and other saying hang on a bit. So Dick and I didn't really see eye to eye'. Although initially pleased to write with an established author and wanting to offer a quantitative approach to physical geography to parallel Haggett's *Location Analysis*, Kennedy became disenchanted with the project and concludes: 'I would say never ever write a book with your PhD supervisor because you don't have the clout to stand up to them'.[61] When a second edition was mooted Kennedy declined on the grounds that it would need complete revision, so she now views it as 'a kind of museum piece'.[62] The book offered an analysis of systematic relations in physical geography, including a new fourfold categorisation of systems types as: morphologic, cascading, process-response and control systems. It concluded with a chapter examining the interaction between socioeconomic and physical systems. Reviewed in the *GJ* by Keith Clayton (1972b), the book was praised as 'a stimulating and worthwhile book' in which the conceptual background to each type of system was helpfully discussed, as well as important discussion of equilibrium and change in systems. Clayton went on, however, to suggest there was too much emphasis on geomorphology at the expense of other aspects of physical geography such as climate, and argued that the reliability of data used was not discussed. Kennedy developed her critique of the over-simplified application of systems approaches, statistical techniques and modelling to geographical issues in a 1979 paper entitled 'A naughty world', where she explicitly challenged the recent publications of Bennett and Chorley and Chapman.

Seeking jobs at the end of her Cambridge fixed-term post, Kennedy was told by a referee that she had not been appointed to one position 'because they didn't want a woman', which demonstrates that informal institutional closure to women could occur regardless of a successful track record. Kennedy subsequently held appointments at Manchester University 1973–9 and St Hugh's College, Oxford from 1979 where she spent the rest of her career. She has continued to research and publish, her most recent book being *Inventing the Earth: Ideas on Landscape Development* since 1740 (2006).

Other women at University of Oxford in the late 1960s and 1970s include **Isabel Thubron** who was a departmental demonstrator when Jean Gottman was head of department (Scargill 1999) and **Judith Pallot** who made her career in Oxford as a fellow of Christ Church College and lecturer in political geography specialising in Russia. **Lucy Caroe** began her career at Aberystwyth when she was appointed to an assistant lectureship in geography in 1963, and later, under her married name Lucy Adrian, became director of studies for geography at Newnham College, where she taught Linda McDowell

1968–71.[63] **Judith Rees** began a near 30-year tenure when she joined the LSE department in 1969 and published *Industrial Demand for Water* the same year (see Balchin 1997: 83 for a graph showing LSE appointments).

Part-Time and Short-Term Posts

One particular problem in examining the production and reception of women's geographical work is that of tracing those women who took short-term or part-time contracts, or worked full time only briefly before undertaking childcare commitments. Male colleagues often take short-term posts in their early career and these can be skipped over in obituaries and histories, but permanent employment usually guarantees a record of their work in some form. A number of women prior to 1970 held temporary or part-time posts, but can only be traced in the form of occasional publications and passing references in conference proceedings or a footnote in departmental histories. Gwyneth Davies's and Nicola Crosbie's biographical accounts provide a window on this negotiation of academic and family life.

Gwyneth Davies

Born on 27 October 1931, daughter of Annie (née Churnside) and Edward Haslam, Gwyneth Davies grew up in Bolton, where her parents worked as a part-time school secretary and accountant, respectively. She gained a scholarship to the Church Institute, a direct grant school in 1949. Too young at 17 to take up a place at Girton College, Cambridge, she accepted an offer at Bedford College instead. Her parents were 'keen on education', but she was only the second family member to go to university. She found Bedford College suffering from post-war fatigue, but worked out that by choosing certain options it was possible to attend a greater variety of lectures within the wider university. Haslam was awarded her BA in geography from London University in 1953, followed by a PGCE in 1954. She then taught at Park School, Preston for three years. Having married Jack Davies in 1957, Gwyneth accompanied him to his post at Khartoum University. Once there she was appointed demonstrator and part-time research assistant by Professor J.H.G. Lebon; she also taught geography at Unity High School, Khartoum and coached the Mahdi's youngest daughter for UK examinations.[64]

The couple returned to live in the UK in 1960 where Jack took up a lectureship in geography at the University of Wales, Swansea. Gwyneth had their first child in the summer of 1961 and was offered a part-time tutorship in the geography department by Professor W.G.V. Balchin in October 1961. Their second child was born in 1964. On principle the couple disapproved of married couples working in the same department, but Gwyneth was able to combine some part-time teaching with family responsibilities.

> I was on annual contracts for 31 years (of which 3.5 were full-time and the rest P/T). During the first couple of years I was paid out of Department's 'petty cash'!! Essentially over the years I was called upon to fill various gaps in the Department's teaching. For some years I was the only female member of the Department's academic staff after Gillian [Groom] left until Rosemary Bromley was appointed. During the 1960s women academic geographers were very few, but personally I did not have problems with men staff.[65]

Encouraged by Professor N. Stephens to take a more formal role in the department, 'I took up organisation of field classes, first year Practical courses and took charge of the Department's tutorial teaching programme … and organised Open Days and the Department's school links programme'. Davies describes herself as writing some 'weak' publications, compared with those of her husband and attributed this in part to the fact that she was better at spoken than written communication, but also because as a part-time academic 'any research was of a limited nature and so produced one-off short articles or book chapters'.[66]

Despite what might be seen through today's viewpoints as at best constrained choices, Davies describes her career in terms of combining positive choices (family and work) and making the most of opportunities: 'Jack, being an Africa specialist was often away – twice for a year, so things were quite a juggle on occasions, but his visits sometimes also provided opportunities. I was able to spend a year in Zaria (Northern Nigeria) 1965–6 and became involved for three months with transport and villages surveys with the Ahmadu Bello University Rural Economy Research Unit to which Jack was attached'. It has been suggested that supportive partners are vital to married women's academic career success (Kirk 2003), and clearly this was a mutually supportive partnership, but one in which the woman whilst having 'complete freedom' to manage her domestic, academic and voluntary work, nonetheless retained the responsibility for that domestic work. Davies highlights the female network which made that possible: 'I was helped to look after family and work through living in a village and always found that I could get help through adopted "grannies", and my mother from the Lake District who came down each year for June to cover the pressure period associated with examinations … [Jack] has always been most supportive, but life was/is hectic.'[67]

Part-time academics not only fall through the gaps in history, they can also fall through administrative gaps and loopholes. Davies reported that the university had not allowed her to transfer from the teaching superannuation scheme to the university one in the 1980s. However, 'Unbeknown to me the AUT [Association of University Teachers] took up my case as an example with the result that 9 years and 10 months after retirement [in 1992] I was awarded a USS pension'. In addition to juggling domestic and academic work, Davies has been deeply embedded in voluntary work. She was chair of the Family Planning Association for South and West Wales until it was taken over by the National Health Service in the mid-1970s, did voluntary work with the Citizens' Advice Bureau, represented the church on the Diocesan Board of Education and became a churchwarden; in 2002 she became president of the Royal Institution of South Wales, the first woman to hold the position since its foundation in 1835. These biographical details give a sense not only of Davies' contribution to the social capital of her locality but also the ways in which academic life relates to personal praxis and how part-time work and retirement accommodate that broader praxis.

Nicola Crosbie

Nicola Crosbie had a similar experience to Davies above. She read geography at Edinburgh in the 1950s, had always intended to pursue a career in cartography, but taught at Bell Baxter School, Fife, for a year, before going to work for the Directorate of Military Survey (DMS) in 1960. 'There were 19 in our year and I think seven girls, one went to Canada and became chief planning officer in part of Ontario, Sally Rankine. All of the others except me went into teaching'.[68] 'In those days the main outlet for all geographers was teaching'.[69] Initially working for the DMS in London, she then took a post in Cyprus, where the base went back to Kitchener's time. The work there was more immediate. Map research officers had to collate information which came in on aerial photographs and from ground surveys. Nicola was involved with producing a New Map Series of East Africa and of Southern Arabia. She spent two years in Cyprus and six months in Aden where she was sent to open a new map library. About six months after she returned home Britain withdrew from Aden. From a personal perspective it was, however, a valuable and interesting experience.

Nicola returned to Scotland in 1964, where she was offered a research fellow in the Air Pollution Survey, headed by Archie MacPherson in the geography department at Edinburgh. It was at this time she met Sandy Crosbie, a lecturer in the department and a widower with two children. They were married in 1965. The findings of the Air Pollution Survey were published in 1968. By this time a second contract to continue research into air pollution in Edinburgh had been agreed between the geography department and Warren Spring Laboratories. MacPherson moved to Vancouver and Crosbie carried on the research, but although this was completed and written up in draft, unfortunately it was never published. Crosbie's life became busy with two more children and increasing commitments to the extended family. Nicola and the children accompanied Sandy to research projects in Southeast Asia and the USA (1973–4), after which Nicola did not return to academic work.

Conclusion

The biographies in this chapter have demonstrated the wide range of teaching, research, fieldwork, committee and editorial work undertaken by women geographers. Several women developed research expertise in physical geography and a number were early users and/or proponents of quantitative methods, contrary to the perception of the Quantitative Revolution as limited to a numerate masculine elite. Many of the women were numerate too and quantitative techniques were a natural progression from earlier methods and calculations used in teaching and research. Women did achieve promotion to chairs in geography at this time of university expansion in the last decade of this period, but there had been a near 20-year gap since Eva Taylor's retirement as professor at Birkbeck. Although promotion came easily to Cole, she was removed as head of department, but kept her research role. By contrast, others such as King were initially sidestepped just as promotion came very late in the careers of Garnett and Campbell. Ironically, oral history accounts suggest evidence that Campbell appears to have blocked other women's promotions, perhaps because her own was so hard won. However, there is evidence of widespread (although not necessarily uniform) institutional sexism, expressed informally through interview questions, modifications to job descriptions and tenure and other working practices, which reflected wider social values. There were social shifts and there was also widespread if patchy support and recognition for women academics from individual heads of department, colleagues, geographical societies and partners. Women lecturers with children developed a range of strategies for negotiating childcare and academic work, ranging from employing au pairs to undertaking part-time work, from research in the school holidays to career breaks. It is quite difficult to identify and assess the career impact of caring for elderly relatives or partners, but several women discussed here and in previous chapters (e.g. Campbell, Garnett, McCulloch and Groom) combined these roles with their geographical work. Those women whose careers began in the 1960s and whose work has had a major impact on the discipline nationally and internationally post 1970 merit more detailed engagement elsewhere (notably Anne Buttimer, Janet Momsen, Janet Townsend and Doreen Massey).

Chapter Ten

Conclusion: Mapping the 'Hidden' Women in British Geography 1900–70

Every geographer, whatever the stage of his [her] study, must at one and the same time be conscious of the immensity of the whole, yet must consider it part by part if he [she] is to examine his [her] material in sufficient detail to understand the complexities involved and see exactly how the parts from the smallest to largest interact and interlock

Jean Mitchell, 1954: 325

In the preceding chapters the production and reception of the geographical work of women belonging to three broad groups of travellers, academics and educationalists have been examined within their own biographical and wider disciplinary and social contexts. Gender has been used in this study in two ways: firstly, and most obviously, as an empirical category to focus on women per se; and secondly, conceptually as a category of analysis in feminist readings of biographies and careers, discursive constructs in the production and reception of these women's work, and in framing a feminist historiography of British geography. This chapter brings together the findings of previous chapters as well as adding some overview data, which combined offers conclusions under three headings: empirical findings; discursive findings; and observations on the historiography of geography and the 'canon'. Inevitably these organisational headings and subdivisions are leaky and there are overlaps and blurring between sections.

I Empirical Findings

The women studied here illustrate a wide variety of experiences in relation to doing 'geographical work' and the analysis that follows attempts to balance a regard for that individuality and making general observations where appropriate, though these rarely if ever apply to all of the women studied within the 120-year period studied here. Most of the women belonged to the 'middle classes', but that category is sufficiently broad to bring in those such as Gertrude Bell who belonged to the moneyed industrial elite and Catherine Raisin whose father was a pannierman at Grey's Inn. For many, their achievements were a reflection of intellectual ability and hard work without the benefit of financial cushioning. While those women born into the socioeconomic elite often lacked access to formal education, e.g. Mary Somerville, others from different social backgrounds had access to private schooling, e.g. Nora MacMunn, Eva Taylor and Sheila Jones (notably several attended the North London Collegiate School). Others benefited from scholarships to grammar schools, e.g. Barbara Kennedy and Alice Coleman, but only a few came through the standard state education system. While a handful of women, such as Cuchlaine King and Jean Grove, followed their parents into university, a significant number of the women who became academics and educationalists were the first in their family to attend university, e.g. Alice Coleman, Gladys Hickman and Gwen Davies. Most of the women needed to earn a living, not least because the majority were single, and even those born into riches at the turn of the century, e.g. Isobel Wylie Hutchison, found socioeconomic conditions after the First World War necessitated income generation and changes in their own lifestyle. Several of the women were 'daughters of the manse' with clergymen fathers, e.g. Hilda Ormsby, Joan Fuller, Kay MacIver and Rachel Fleming; some were explicitly influenced by their faith in their way of life and work, e.g. Kate Marsden, Katy Boswell, Dorothy Sylvester and Grace Dibble; others by their humanist beliefs, e.g. Alice Garnett. Less than 10 percent of the women studied were married. This may reflect their personal choice or sexual preference, or the reduced availability of male partners as a result of male emigration to the Empire in the nineteenth and early twentieth century and male mortality during two world wars. It also reflects the demands of professional work and the frequent characterisation of professional women as 'unfeminine' in the early to mid-twentieth century, especially women teachers and academics. Only a few of the women studied combined having children and continuing a career, e.g. Eva Taylor, Swanzie Agnew, Jean Grove, Margaret Storrie, Gladys Hickman and Christine McCulloch, about half taking career breaks or part-time work while the children were young. Only a handful of the women identified themselves as explicitly 'feminist', e.g. Mary Somerville campaigned for women's suffrage and entrance to universities while maintaining a socially conservative lifestyle, and the biographies of Marion Newbigin and Eva Taylor illustrate feminist lifestyle choices. Others such as Mary Kingsley and Gertrude Bell opposed women's suffrage. Although described as a 'pale northerner', Hilary Chew's family name suggests some Chinese ethnicity otherwise the women discussed are overwhelmingly Caucasian. While many – travel writers, academics and educationalists – wrote within and in support of the colonial discourse, e.g. Joan Reynolds, some were subversive of aspects of this discourse, e.g. Mary Kinglsey and Marion Newbigin. In the 1960s and 1970s while Monica Cole worked in Southern Africa and was an apologist for apartheid, Gladys Hickman and Swanzie Agnew, who also worked in Africa, opposed apartheid through their personal politics, life choices and writing. Some of the women were prolific writers, some acclaimed in academic circles, others within popular literature; some inscribed their work in their institutions or upon the lives of their 'charges' rather than in written texts. Some struggled to get their work published or to get permanent employment or promotion, especially in the university sector; others gained publishers, posts and promotions with apparent ease. Some won critical plaudits, while the publications of others were belittled or ignored. Some received honours from universities and geographical societies (often late in life) at the same time as others of equal standing were omitted.

These few observations give a sense of the diversity of life experience and the varying degrees of women's acceptance within geographical, academic and literary institutions and the rest of this section addresses five strong themes which are threaded through the empirical evidence of the individual biographies outlined in preceding chapters. These themes relate to: (i) the number of women and their presence in geographical societies and higher education; (ii) their subject and methodological groupings within the discipline; (iii) fieldwork; (iv) war work; and (v) the issue of recognition and promotion within geographical institutions.

The number of women and their presence in geographical societies and higher education

The first point to make is that contrary to the impression given by some histories of the discipline, women are not recent arrivals in the production of geographical knowledge. *Many women were producers of geographical knowledge* prior to 1970, and recognition of this fact makes geography not only more feminised, but also more representative of the human race (Monk & Hanson 1982; Domosh 1991a,b; Maddrell 2004a). With the exception of the cohort of 22 admitted 1892–3, women may have been excluded from the RGS until 1913, but women were able to join every other British geographical society from their foundation in the 1880s onwards and had gained access to the BAAS in 1853. It was no small thing for women geographers to be excluded from having the status and benefits offered by the hegemonic RGS, not least the opportunity to be trained in surveying and related expeditionary skills at a time when such methods were at the heart of the subject's epistemology and therefore the definition of what it meant to be a 'geographer'. However, as has been shown in Chapter Two, the RGS was a far from homogeneous group when it came to promoting women's membership and other 'progressive' initiatives such as lobbying universities for geography's academic status. In the meantime, women travellers were speaking at, being published and honoured by other geographical societies as well as gaining popular and critical acclaim for their writings (e.g. Isabella Bird and Mary Kingsley). While women were active members of the non-metropolitan geographical societies from 1884 and the RGS from 1913, it appears to have been the coincidence of women's war work during the First World War, the first phase of the female electoral franchise in 1918, and the growing number of women attaining university qualifications in geography, which acted as a springboard for women's *integration* into geographical institutions. That is to say it was in part a reflection of wider social and cultural discursive shifts as well as developments within the discipline.

Figure 10.1 shows women's tenure as editors of the *SGM* between 1902 and 1967, which if one allows Isobel Wylie Hutchison's period as honorary editor 1944–53, represents continuous editorial work by women for 60 years, omitting the war years 1940–4. However, after 1967 no further women were involved with editing the journal until 1999. Figure 10.2 shows Ella Christie and Marion Newbigin as the first women to be appointed to the RSGS Council in 1923–4 and the continuous presence of women members thereafter, including travellers Ella Christie and Isobel Wylie Hutchison's long service as vice presidents of the Society, a critical mass of women on Council after 1945, and women's leadership in the Dundee branch between 1958 and 1970. Although Gertrude Bell was the first woman fellow to receive a medal from the RGS, almost immediately on her admission in 1913, women's representation on the Society's Council would wait another 17 years until the appointment of Elizabeth Ness in 1930 (Figure 10.3). This was two years after women gained the parliamentary vote as part of the universal suffrage established in 1928, but it is still earlier than many might be supposed for an institution which successfully resisted women's membership for so long. Ness was soon followed by Eva Taylor, the year after her promotion to the chair of geography at Birkbeck and the women members of the RGS Council in the 1930s and 1940s represent a coalition of social elites and academics (though not all of them were in geography departments). This was also reflected in the first women who served as vice presidents of the Society, the first being leading educationalist and public figure Baroness Ravensdale in 1959. The IBG makes for an interesting case as it was founded in 1933 as a conscious attempt by academics to differentiate themselves from the RGS in a professional body which would foster a culture of support and intellectual exchange for the growing number of university lecturers. Women had been involved with initial discussions about setting up the Institute and at its first full membership count in 1935 women represented about 16 percent of the total (Maddrell 2004a). Hilda Ormsby was the first woman to serve on its Council, which she did for the calendar years 1936 and 1937, but as Figure 10.4 shows there was a 12-year hiatus before she was followed by Catherine Snodgrass and although women's representation became more regular, there were gaps in 1959–61 and 1967–70 when there were no women on Council. The last was particularly ironic as it followed Alice Garnett's period as first woman vice president 1964 and 1965 and president in 1966.

Subject and methodological groupings

Those trained within the first phase of establishing geography as a university discipline were schooled in the unified integrative approach to geography and could teach and write on what we would now distinguish as physical and human geography, e.g. Eva Taylor, Hilda Ormsby, Margaret Davies and Jean Mitchell. In the first half of the twentieth century the region dominated geography epistemologically and methodologically, so not surprisingly many women authors wrote regional studies, e.g. Nora MacMunn, Peggy Hobson and Swanzie Agnew. Some had a strong economic content (notably London trained Hilda Ormsby, Hilary Chew and Alice Mutton, but also Catherine Snodgrass), and several worked on Eastern Europe, e.g. Margaret Shackleton, Harriet Steers and Hilda Ormsby; Steers and Ormsby also addressed explicit political themes. Prolific writers Taylor, Newbigin and Coleman were geographical polymaths who covered a wide span of geographical topics in their publications. As the twentieth century progressed, academics increasingly taught systematic rather than regional courses; their related specialist books make it possible to identify some clusters of women authors. Eva Taylor, Eila Campbell, Helen Wallis, Dorothy Sylvester, Joan Fuller, Margaret Davies, Jean Mitchell, Elizabeth Clutton and Catherine Delano-Smith all worked within the sub-discipline of historical geography, with agriculture, the Mediterranean and the history of cartography being significant sub-themes within the group. Eva Taylor is credited with re-establishing and developing the history of cartography within the discipline and certainly drew Campbell into the field. The work of American geographer Ellen Semple and Marion Newbigin on the Mediterranean may have similarly acted as a beacon to other women in the discipline. Beatrice Swainson and Dorothy Sylvester wrote on population and settlement and it has been suggested that women may have been attracted to study these geographical issues as an extension of socialised 'feminine' domestic concerns.[1] Physical geography was also well represented,

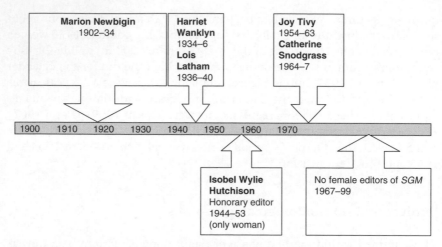

Figure 10.1 Women editors of the *Scottish Geographical Magazine* 1900–70.
Source: Based on details of RSGS roles published annually in the *Scottish Geographical Magazine*

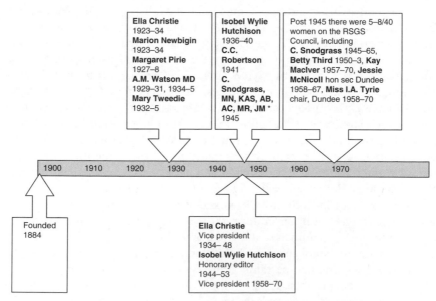

* MN Maude Newbigin, KAS Katherine Addison-Smith, AB Agnes Bruce, AC Annie Cormack, MR Margaret Russell, JM Mrs J. Mathieson

Figure 10.2 Women members of the RSGS Council 1923–70.
Source: RSGS Council membership lists

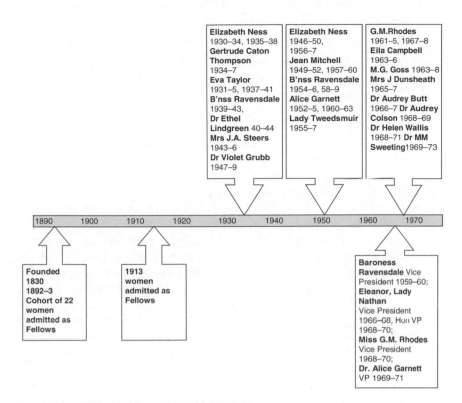

Figure 10.3 Women members of RGS Council 1930–70.
Source: Based on published lists of Council appearing annually in the *GJ*; there may be some discrepancy between appointment to Council and publication of Council lists where these fall either side of years

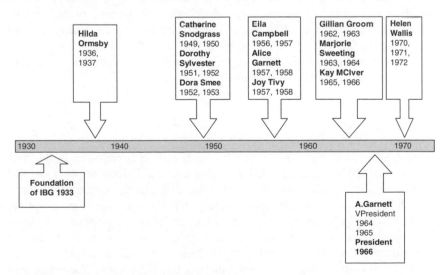

Figure 10.4 Women members of IBG Council 1933–70.
Source: Based on IBG Council lists in Steel (1984). Note IBG years run January to December

especially in the areas of biogeography: Marion Newbigin, Monica Cole, Jean Carter and Joy Tivy; and climatology and related issues: Alice Garnett, Eunice Timberlake and Beatrice Eckstein; and geomorphology: Cuchlaine King, Marjorie Sweeting, Jean Grove, Barbara Kennedy, Christine McCulloch and Gillian Groom (all except the last two being educated at Cambridge). These groupings appear to be borne out in recent findings on female physical geographers, which identified a preponderance of women in biogeography and geomorphology, but this was attributed to the sheer size of these sub-disciplines within geography as a whole rather than female dominance (Dumayne-Peaty & Wellens 1998). However, prior to 1970 biogeography was embryonic and the women associated with the subject were considered pioneers of the field. This may be attributed at least in part to the sustained record of women working in botany since the nineteenth century (Creese & Creese 1998) and its status as a 'soft' science, a subject which was 'unmanly' and suitable for women (McEwan 1998b). The fact that Dumayne-Peaty and Wellens (1998) found the majority of women physical geography academics to be under 40 years of age and only a few in promoted posts, suggests that a generation of women in physical geography had been 'lost' considering their relatively large presence and seniority prior to 1970.

Perhaps because of the number of women involved with physical geography, several were actively concerned with the development and promotion of quantitative techniques in the late 1960s and early 1970s, notably Cuchlaine King, Barbara Kennedy, Gladys Hickman and Sheila Jones. Human geographers such as Catherine Snodgrass, Kay MacIver, Anne Buttimer and Janet Momsen were also highly numerate, using lengthy statistical calculations in the case of the first two and early computerised statistical programs and factor analysis in the case of the latter. Garnett and Cole were among the first cohort of geographers to apply new remote-sensing techniques to their research.

Fieldwork

Powell (2002: 264) has argued that 'more attention is required to the precise material and corporeal history of geographical fieldwork practice, and the role of such performances in enacting disciplinary identity': the evidence of this study shows overwhelmingly that British women geographers have been deeply engaged with fieldwork of many types, human and physical geography, at home and abroad, in extreme and familiar climates and environments. Stoddart's (1991) limited definition of fieldwork as a science of survey may be thought to exclude women, but some Victorian and Edwardian women travellers such as Isabella Bird and Gertrude Bell did undertake survey work and conform to that epistemological and methodological requirement. However, prior to the availability of university degrees in geography, the RGS courses in surveying were the main source of such training and, with the exception of the cohort admitted 1892–3, women were generally excluded from this course until 1913. This constituted a double bind for women whose work was found wanting for failing to employ this methodology, while they were simultaneously excluded from the means of acquiring the said knowledge and skills. Once degree courses in geography were established, surveying was a common element of practical classes, especially for those taking London degrees, which included many of the university colleges across the country before they gained full university status in the interwar years. Many women students were trained in surveying post 1918 and some were taught by women, e.g. Dora Smee and Eunice Timberlake at Bedford College and Cuchlaine King at Nottingham. It has been argued that women were excluded from scientific endeavour in part due to the professionalisation of science in the nineteenth century (McEwan 1998b), and whilst this did occur, women who had access to women's colleges or London University degree courses after 1868 were able to access science degrees, and women's participation in scientific work was more widespread in the late nineteenth century than is often acknowledged (see Creese & Creese 1998). Catherine Raisin and Marion Newbigin, for example, both undertook significant laboratory-based research in the last quarter of the nineteenth century.

However, Domosh (1991a) was right to challenge Stoddart's narrow definition of what constitutes 'fieldwork', and late twentieth-century recognition within human geography of the value of ethnographic field methods makes it easier to recognise the epistemological value of much more varied geographical data collected by women travellers in particular. Although making the gendered assumption that women would not or could not collect the primary phase of exploratory data, Francis Younghusband, addressing the RGS in 1917 stressed that geographical knowledge had to be more than that based on what a tape measure could tell; he suggested that women 'of the type' of Mary Kingsley and Gertrude Bell could make a valuable contribution to geographical work through 'describing any particular locality in such a way as to bring it truthfully and impressively before those of us who have not had a chance of going there ourselves' (1917: 410). At the same time as suggesting that an understanding of beauty should be part of scientific geographical knowledge, he argued that women could make an important contribution to that knowledge. Like their male counterparts, the first generation of women academic geographers adopted the regional paradigm and primarily worked within this framework rather than the 'scientific' modes of exploratory survey or laboratory-based investigation. Vidalian ideas of *genres de vie* and spirit of place imbued much of the regional work written by women geographers 1910–40, which were typically grounded in extensive field knowledge (if not actual surveying) of their study areas, e.g. Ormsby on France or Newbigin on the Mediterranean. At this time many women academics and teachers were involved with serious field study through the Le Play Society and later the Geographical Field Group, using their vacation time and own finances to participate in international field study. Women such as Joan Fuller and Alice Coleman were leaders of GFG, GA and other field courses and numerous women lecturers led fieldwork courses for their own students. Students and staff expressed their pleasure in fieldwork, whether in the opportunity to travel abroad for the first time, as was the case for Gladys Hickman, or the delight in being in demanding and remote environments as locations of personal research, as expressed by Cuchlaine King about the Arctic. Although King had to make a case to be part of Vaughan Lewis' research field team working in Iceland in the early 1950s, as he had simply assumed it would be an all male group, there does not seem to be the same reporting of explicit gendered exclusion from fieldwork in Britain compared to American women during the early to mid-twentieth century (see Monk 2004 and Sack 2004 for examples).

Several British women geographers were individually innovative in their fieldwork, e.g. Garnett's work on alpine insolation and later research on urban pollution, Tivy on ecosystems, Cole on savanna flora and minerals, Taylor's archive work on navigation, Buttimer's social geography in Glasgow, Sweeting's work on karst; others were part of ground-breaking field research teams, e.g. King and Grove's glacial work. These women made fieldwork their own. This is not to dismiss arguments about the gendering of fieldwork, rather it is to illustrate that there is more diversity in women's participation in fieldwork than those arguments might suggest and that it is important to stress women's agency in fieldwork whether in succeeding despite masculinist discourses and practices, or by subverting them. As Bracken and Mawdsley point out, critiques of fieldwork as masculinist can re-inscribe assumptions that all women resent getting dirty or climbing mountains and 'there are other stories to tell around the opportunities and enjoyment that can be part of fieldwork, and the range of methods and places it might involve' (2004: 284). However, there were constraints of funding and time, not least for those women with family responsibilities. King, as a single woman, commented that family commitments could be a 'distraction' from the demands of geomorphological fieldwork (Sack 2004). Although Jean Grove, who has been constructed by Goudie (2001) as a non-feminist heroine who washed her babies' bottoms in meltwater streams while on fieldwork, had a supportive partner and wider support network including paid au pairs, she found it necessary to modify the nature of her glacial fieldwork in order to make it compatible with a family of six children. Women geographers' access to and experience of fieldwork, in common with other forms of geographical work such as writing, can be shaped by lifecycle stage and wider commitments. Grove's family commitments proved a barrier to the type of research she was initially engaged in, but responding to that challenge resulted in innovative work on archive evidence for climate change.

War work

As with fieldwork, it is only through the excavation of individual biographies that a pattern of women geographers' participation in war work emerges. A high proportion of the pre-1970 women geographers studied here were engaged in war work in the First and/or Second World Wars, along with their male colleagues. Their geographical expertise, including experience of travel, surveying, field study, map making and interpretation, and regional research were all utilised by the British state-at-war. Only a few women such as Gertrude Bell, Hilda Ormsby (then Rodwell Jones), Charlotte Simpson and Miss Heath used their geographical skills and knowledge in the First World War. The larger number of women geographers and the greater mobilisation of women for war work during 1939–45 resulted in a larger number of women geographers contributing to the war effort as geographers per se. This included teaching map reading to armed forces cadets and officers in universities (e.g. Taylor and Garnett), contributing to Naval Intelligence Handbooks prepared by editorial teams at Oxford and Cambridge (Garnett, Mutton, Marshall, Mitchell, Latham, Ormsby, Mann and M. Davies) (Maddrell 2008). Travel writers such as Stark and Cressy-Marcks undertook intelligence and/or propaganda work and a cohort of newly qualified graduates fed into the auxiliary women's forces (e.g. King), government ministries (e.g. MacIver and Cole) and map production for the Admiralty (e.g. Jean Carter, Mary Turner and Louise Fell (see Maddrell 2008)). This reinforces the view of geography's close disciplinary links with warfare (see Hudson 1977; Balchin 1987; Heffernan 1996) and demonstrates the blurring of gender boundaries in the war (facilitated by women's geographical qualifications); but all of this needs to be set in the context of the state of total war and female conscription to war work after 1941 (Maddrell 2008).

Recognition and promotion within geographical institutions

Just as women's participation in geography societies (Figure 10.5) and departments has been demonstrated, so too have the number of women being awarded honours by those societies and promoted to higher posts in academia. This includes RSGS awards to Isabella Bird, Alice Lennie, Catherine Snodgrass, Joy Tivy and Isobel Wylie Hutchison and RGS awards to Mary Somerville, Jane Franklin, Gertrude Bell, Freya Stark, Eva Taylor, Eila Campbell, Marion Newbigin, Anne Buttimer and Monica Cole. Within university departments five women were promoted to professorial status 1930–70 (see Figure 10.6). While this creates a feminised hall of fame, which is largely absent from current histories of geography, reclaiming and proclaiming these women's successes can obscure other inequalities. Eva Taylor's remarkable appointment as professor of geography in 1930 is little recorded in recent histories of geography, leaving many to assume that the first women to achieve chairs are to be found in the early period of improved equal opportunities in the 1970s. However, as Figure 10.6 testifies, it was another 32 years before other women followed her into the highest academic ranks, when Alice Garnett was promoted in 1962, Monica Cole in 1964, Cuchlaine King in 1969 and Eila Campbell in 1970. When Taylor was appointed as professor there were only 10 chairs in geography in the country, but when Garnett received her promotion she was one woman among 30 professors of geography. Cole's appointment at Bedford raised the number to two women out of 32 professors and Campbell's appointment at Birkbeck gave a total of three women professors out of 37 professors of geography.[2] It should also be noted that Garnett and Campbell were both promoted only a few years before retirement so had limited opportunity to exert influence through their position and that Kay MacIver was head of department at St Andrews without the recognition of a chair. There are parallels with other countries, for example in Sweden where Gerd Enequist was simultaneously the first woman to be appointed professor of geography and first woman professor at the University of Uppsala when she gained her promotion in 1949 (there was also a long wait before another woman professor in geography was appointed) (Buttimer & Mels 2006). Similarly, in France Jacqueline Beaujeu-Garnier became the first woman Doctor D'Etat in geography in 1960 and was the first female professor at the Sorbonne (Buttimer 1983). In the USA there were quite a large number of women lecturers in geography prior to 1970, but they tended to be overlooked for promoted posts and often only found permanent jobs in teacher education rather than geography departments (Monk 2004, 2007).

It is significant that of the four British women who attained chairs in 1960–70, three were physical geographers. It would appear that it was easier to gain recognition as a physical geographer or it was harder to dispute research findings and thereby block career progression. The epistemological advantage of positivist methods adopted within physical geography paradigms in the post-war years was that they were based on empirical data, which could be charted and analysed statistically, hypotheses proved or

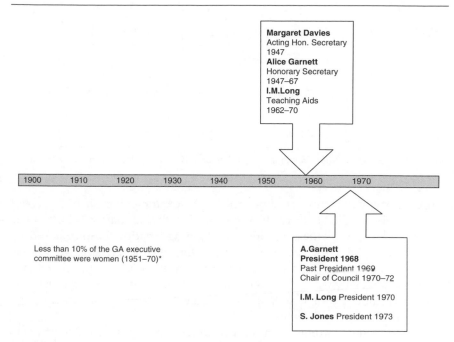

Figure 10.5   Women's roles on the GA Executive Committee 1947–70.
* 1947 Papers indicate as many as five women referred to in relation to the Executive Commitee in the immediate postwar years (1/1/1947, 22/3/1947 28/6/1947, GA papers, Executive Commitee 1945–61).
Source: Data collected from GA Archives, lists of the full GA committee are not extant. GA post holders were appointed at the annual spring conference

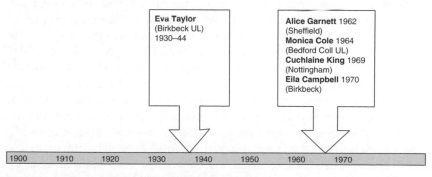

Figure 10.6   Women professors of geography in the UK 1930–70.
Source: data extracted from Balchin 1993: 105–6

disproved. This is not to suggest that physical geography results were uncontested, Marjorie Sweeting's interpretations of Jamaican karst were criticised for example, but quantitative data on climate (Garnett) or glacial deposits (King) or the correlation between flora and mineral deposits (Cole) is much

Table 10.1   Women's academic posts in British geography departments 1963–4

Post	Research Fellow/ Assistant	Demonstrator/ Assistant Lecturer	Lecturer	Senior Lecturer	Reader	Professor
Number	4	6	13	7	2	1
Percentage	12%	18%	39%	21%	6%	3%

Source: extracted from Mike Tanner's (2003) data on UK geography departments 1963–4

more clear cut than historical geography for example, although aspects of that could be studied empirically too. In each case their data was innovative, but it conformed to the scientific paradigm and had to be recognised by those who upheld the same paradigm. Quantifiable physical geography appears to have made for credible academic reputations and associated promotions, especially when promoted posts were in departments sympathetic to women's success such as Bedford College and Nottingham where there was a history of female lecturers.

However, many women lecturers in university geography departments struggled to move off the lowest rungs of the academic hierarchy, e.g. Nora MacMunn and Hilary Chew, where they were often responsible for labour-intensive teaching such as map and practical classes, what has been described as the 'stone floor' rather than the 'glass ceiling' syndrome (Heward & Sinclair Taylor 1995). Accreditation through qualifications gave women access to junior posts but permanent and promoted posts relied on having the right persona and being visible within the discipline through publications, conferences or office in geographical societies, especially the IBG post 1945. Extracting the women from Tanner's (2003) compilation of staffing in geography departments in the academic year 1963–4[3] shows 33 women employed in academic posts in 17 of the 35 British geography departments (see Table 10.1). Just over half of departments had no women at lecturer or above and only six departments had more than one woman at this level: Manchester, Nottingham, Queen Mary's College, Cambridge, Oxford and Bedford College; Bedford was the only department to have three women. Posts ranged from demonstrators and research assistants through to a single professor (Alice Garnett) at a time when promoted posts were strictly rationed in most universities. Four further women worked as map curators or cartographers. This shows a bottom-heavy model of employment status for women academics, with the bulk holding lectureships rather than promoted posts.

Three further points should be made at this juncture. Firstly, Bedford College and other women's colleges at Oxford and Cambridge represented both role models for students who were educated in the belief that women could achieve anything (see Joyce Brown MacPherson in Sack 2004) and significant foci for female employment before most went co-educational in the 1960s and 1970s (see Kirk 2003). Secondly, in more than half of the departments where women were employed as lecturers and above in 1963–4, they were lone female lecturers. This has several implications: women could feel isolated by their gender and 'othered' within the department; they may have felt compelled to conform to masculine cultural norms; 'all women academics' may have been judged by male colleagues based on a single woman lecturer who was not admired, or, a single successful woman could be represented as 'exceptional' (i.e. the exception to the rule). The biographical accounts of Helen Wallis at the British Library, Lois Latham at Hull, and Eva Taylor at Birkbeck demonstrate these experiences. There is also the issue of 'critical mass' of the number of women in departments. Adie (2003) has argued in relation to women in the armed forces that there needs to be a critical mass of women before they can be accepted on equal terms and not judged as the single representative of all women; Monk (2007) has similarly suggested that there needs to be a critical mass of women students 'as precondition for overt resistance' to gendered discrimination, and the same can be applied to members of the faculty. Thirdly, a number of geography departments 1963–4 had no women on the staff, even at assistant lecturer or demonstrator level; excluding departments of two or fewer staff, these were: Aberdeen, Belfast, Birmingham, Bristol, Edinburgh, Leeds, Leicester, Reading, Queen's College Dundee, Southampton and UCL. The Edinburgh department is of particular interest as in the early 1950s it had three women lecturers and in 1963–4 had none. A similar shift occurred in Nottingham in the 1980s where the department, which had consistently employed two or three women from 1945 onwards, reduced this to one after Cuchlaine King retired in 1982, despite a period of rapid growth in student numbers and associated appointments (Cole 2000).

There are parallels between this data and more recent studies of the presence and status of women geographers in British academia. McDowell (1979), writing within a decade of the cut-off date for this study, found that the failure of the large presence of women undergraduates in geography to translate into doctoral students and academics in turn, could be attributed to lack of self-confidence, lack of parental support and the discouraging experience of being taught by all-male staff in some departments. In a follow-up study McDowell and Peake (1990) showed that little had changed in the intervening 10 years and that while only 15 percent of men ended their careers in lecturing posts, this was the case for some 48 percent of women in 1980s' geography departments. This reflected Johnston and Brack's (1983) assertion that while male academics had a one in seven chance of being promoted to Reader, for women it was only one in 14. At the heart of this is a debate about what constitutes academic excellence, including preconceptions that women students tend to do 'solid' work resulting in average rather than high or low degree classifications. McDowell and Peake showed that in 1987 women matched men in the attainment of upper second and first class degrees, but were increasingly underrepresented at graduate and academic appointment level, culminating in approximately only 7 percent of geography academics being women in both 1978 and 1988 and a slightly lower percentage gaining chairs: 6.8 percent in 1978 and 5.3 percent in 1988. Adding these figures to those above relating to chairs in geography held by women 1930–70 shows a continuous decline from approximately 10 percent in 1930 to 5.3 percent in 1988. Driver (1992, 2001) has described the masculine character of 'geography militant' in the nineteenth century[4] and the various data here suggests this applied to academic geography as it became solidly established within British university sector 1930–70, echoing the conjunction between professionalisation and masculinisation in other disciplines/occupations such as medicine (Witz 1992). If Buttimer's (1983, 2005) categories of Phoenix, Faustus and Narcissus are applied to the history of geography, women were appointed to junior posts in the emergent Phoenix stage of the discipline's modern establishment when its status was tenuous and wars necessitated and facilitated women's employment, some of whom gained promoted posts in the period of consolidation. However, the consolidation and professionalisation of geography in the university sector in the Faustus period resulted in attracting more men and the ideal geography academic was discursively constructed in many departments as a bright energetic man, as was recorded in relation to recruitment at Queen's Belfast and in confidential references for Hilary Chew.

Masculinity and the discursive construction of geography as an academic discipline will be returned to in the section below, but it is important to recognise the support of men within geographical institutions alongside identifying individual, symbolic and institutional forms of sexual discrimination against women. As men were the leaders and gatekeepers within most scientific and educational institutions in patriarchal Britain in the nineteenth and early twentieth century, women were dependent on them for admission, book contracts, appointments and promotions. Men such as Douglas Freshfield (RGS honorary secretary) and Robert Needham-Cust actively campaigned for women's admission to the RGS; successive John Murrays published women's travel accounts, e.g. Bird and Stark. Within the university sector, A.J. Herbertson, Halford Mackinder, R.A. Rudmose Brown and Percy Roxby all appointed women to academic posts in early twentieth-century geography departments, which challenges representations of these 'founding fathers' as simplistically masculinist (Maddrell 2004a). Others such as Alan Ogilvie, K.C. Edwards and Frank Debenham appointed women and encouraged them to take higher degrees, publish books and lead fieldwork. Where women married, social convention and employment practices before 1970 often meant withdrawing from paid work on marriage or birth of a child (and attempts were made to impose this on some individuals), and/or submitting their career to their husband's; but as Kirk (2003) points out, a supportive partner could enable a married women's success in academia and professional life. This was evidenced in the biographies of Jean Grove and Margaret Davies, where gendered compromises were made but these compromises were dynamic and evolving and included maintaining the women's professional lives. Not surprisingly, those women who married late, had established careers and did not have children, such as Hilda Ormsby and Alice Garnett, found it the easiest to maintain their careers after marriage. Most single women also lacked the support of a partner which many male academics benefited from, although some lived in residential colleges which gave domestic support or had friends or sisters who acted as 'helpmeets', e.g. Newbigin's sister Florence and Ormsby's sister Dora. Whether married or not, women were frequently defined in the academic arena as stereotypical women even if their academic successes and career choices suggested otherwise. Part of this stereotyping related to their ability to offer administrative or pastoral services to the department or students as an extension of domestic, emotional and 'housekeeping' abilities, e.g. Florence Miller, Alice Garnett and Jean Mitchell. Although publications were not necessarily the priority of academics, whether female or male, in the first half of the twentieth century, when student needs were widely considered as paramount (Buttimer 1983), women lecturers' heavy pastoral and administrative loads were a barrier to research-based publications (Kirk 2003). This clearly resonates with Simone de Beauvoir's (1953/1988) distinction between the immanence of 'women's work' such as housework or emotional work, compared to the transcendence of 'men's work' such as manufacturing or writing, even if this is an oversimplified dichotomy.

II   Discursive Findings

A whole volume could (arguably should) be devoted to discursive constructions within geographical thought and practice, but I would like to address a few points here specifically in relation to women's experience of geography and to the production and reception of their work. My comments come under four headings: (i) definitions of geography and geographical practice in relation to gender; (ii) negotiations of gender in women's writing; (iii) the gendering of the reception of women's geographical work; and (iv) discourses in representing and memorialising women geographers.

Definitions of geography and geographical practice in relation to gender

When J.S. Keltie reported on the state of geographical science for the RGS in 1885, he wrote that despite the RGS' efforts to show geography as an adjunct to the interests of state, commerce and scientific exploration, the subject was not considered a 'manly science', i.e. it was labelled discursively in much the same way as botany was represented as a pursuit suitable for women. It was imperative for those who were seeking to improve the status of geography within the broader intellectual and scientific world, and to establish geography as a university discipline, to shift the understanding of geography away from its perceived base in description and rote learning. This can be seen to have direct bearing on Mackinder's 1887 paper to the RGS 'On the scope and methods of geography' where he articulated geography's utility to statesmen, those in the armed forces, diplomatic and colonial service and commerce; the only category which included women was that of 'teacher'. The crude protectionism of the physical and epistemological space of geography on the part of some fellows in the RGS women's membership debate 1892–3 shows explicit attempts to exclude women from the Society, but other methods of exclusion were less explicit.

The omission of any mention of Hilda Rodwell Jones from correspondence concerning the foundation of a joint school in geography at the University of London exemplifies underlying fears that the presence of women as contributors of geographical knowledge might undermine the credibility of the subject within the university at a sensitive moment in its history. As has been demonstrated, this had implications for the appointment of individual women, the number of women considered 'acceptable' within departments and their access to prestigious promoted posts, as seen in the biographical accounts of Cuchlaine King, Margaret Storrie and Christine McCulloch whose appointments or promotions were explicitly affected by their gender (as was Momsen's experience in Canada). The archive and oral history accounts which substantiate perceptions of preference for 'bright young men' within some university departments go some way to explaining the numerical underrepresentation of women as academics, especially in promoted posts. Autobiographical accounts also testify to the social construction of gender which made it difficult for some women to put themselves forward for promotion prior to the 1970s, as well as departmental leadership having a significant impact on local attitudes to gender.

The language of the history of geography which designated leading figures as 'grand old men' or 'founding fathers' merits some discussion here. While an argument can be made for accepting 'he' as generic in nineteenth- and early- to mid-twentieth-century texts, these are in practice deployed in much more strongly gendered terms within histories. They were never intended to be inclusive and are specific to the masculine gender of the key players in the historical canon. While Beaver (1982) referred to Mackinder as 'the Grand Old Man of British Geography' in the 1930s, there is no comparable phrase for women, 'grand old woman' does not exist, although 'grande dame' might have been applied to Eva Taylor in her later years and Newbigin was described as 'doyenne' by Clout (2003). Similarly, patriarchal linguistic forms and historical categories have designated 'founding fathers' (particularly in US white settler history), but 'founding mother' does not hold the same cultural purchase, despite the attempts of revisionist historians to redress the imbalance. Adams *et al.* (1984) used the term 'founding parents' of the RSGS in order to add Marion Newbigin to the traditional list of 'founding fathers' of British geography, such as Mackinder, Roxby, Herbertson, Chisholm, Ogilvie and Fleure. Mary Somerville has been referred to both as the first (British) woman scientist and the first English (Scottish) geographer; Newbigin was likewise referred to as 'the first scientific geographer', but too often such women were described as 'unique'. They were represented as exceptionally if not uniquely gifted women, meaning they were discursively set apart from other women and could not be equalled. In science this has been called the 'Marie Curie effect', something which can be seen in relation to representations of Mary Somerville in the mid-nineteenth century (Neely 2001) and to Newbigin and Taylor within geography in the twentieth century.

Negotiations of gender in women's writing

There are two key forms of negotiating gender in women's geographical texts: firstly, as an author, and secondly, working with gender as an empirical and conceptual category. The ways in which women represent themselves as authors have been discussed extensively within feminist literary studies and pedagogical writing, autobiography or books based on correspondence have been noted as the dominant discursive forms for women's writing in the nineteenth century (Mills 1991; Pearce 1991). Whilst 'exceptional' individual women such as Mary Somerville transgressed these norms in writing scientific texts with the support of leading male associates, her *Physical Geography* (1848) might be designated as a broadly pedagogic summary of the field. Travel writing by Isabella Bird, Kate Marsden and Mary Kingsley conformed in varying degrees to the autobiographical/diary/correspondence model. However, there were definite shifts in style and voice in Bird's writing after her sister died and Bird joined the RGS, thereby accessing the benefits of training in data collection and photography, and in many ways women's *fin de siècle* travel writing might be seen as transitional in terms of style and content, blending characteristic feminine forms of writing with a desire to collect and communicate empirical and 'scientific' information (i.e. conform to the geographical paradigm). Requirements for income from a popular readership also influenced the character of some women's publications, sometimes necessitating a distinction between what was presented to and through the geographical community compared to mainstream publishers, as seen in the case of Isobel Wylie Hutchison. Kingsley modified part of her account because she thought it would not be deemed credible for a woman travelling in West Africa; and the incredulity of some readers and in response to Kate Marsden's accounts of her travels in Siberia – and by extension to the rest of her work – suggest that Kingsley's presupposition was correct, as the discursive construction of women as unreliable witness was perpetuated into the twentieth century (Mills 1991; Bankey 2001). Academics were perhaps less likely to meet with disbelief of their data, but their interpretations were susceptible to challenge.

There was also the issue of authorship itself, for the first generation of women academics at least. Nineteenth-century women's writings were often prefaced by modest disclaimers for having the temerity to commit their thoughts to public form and academics did not escape this framing. Both of Hilda Ormsby's books include such riders (although they seem to contrary to her confidence in other outlets), echoing 'the anxiety of authorship' (Neeley 2001: 195) of earlier female authors. Eva Taylor's research monographs do not show the same note of apology, but she was already widely published as an author and co-author of non-research based texts and was at the height of her career at the time of their publication, as well as being famously confident in her views. It is notable, however, that Taylor, whether consciously or not, obscured her gender by writing under her initials rather than given name. This may have followed Marianne Evans/George Eliot's lead, or simply echoed academic authorial conventions of the time, including that of her first co-author J.F. Unstead; but coupled with the predominance of male authors in her field it led to the assumption that she was a man, as can be seen in her archived correspondence. This may well have played an important part in gaining a readership for her work beyond her immediate circle, at least until her reputation was established, and other women such as Eila Campbell, Cuchlaine King and Barbara Kennedy adopted the same form of authorial identifier. Ironically this representation of authors as non-gendered and the continuing predominant academic practice of referring to authors principally by their family names (as used here) can perpetuate the impression of at best gender-neutral or at worst universal male authorship.

Another issue arises when women were co-authors with male colleagues. Snodgrass's chapter on Scottish cities was anonymised and marginalised in reviews of Freeman's (1966) *Conurbations*. Smee's chapter on Irish ports and trade in Stamp and Beaver's *The British Isles* (1933) similarly gained little recognition and her name as contributor was dropped from the title page in the rewritten 1943 fourth edition, even though much of her contribution remained. Although co-editor with Darby and author of six chapters in *The Domesday Geography of South-East England* (1962), Campbell's major research and authorial contribution did not receive due credit (Barber 1995), but it may have ultimately helped her gain promotion. This may be the lot of 'junior' authors, whether female or male, but it is worthy of note, especially in terms of the impact of reviewers in marginalising their contributions.

In terms of the content of women's writing, gender did not emerge as a dominant theme, but this is not surprising. Outside of disciplines such as biology and psychology gender was not generally used as a category of analysis in the nineteenth and twentieth century prior to second wave feminism in the 1970s. For the many women who reported identifying themselves in terms of their subject rather than their gender, it may not have entered their minds to consider gender as an issue in their work (also see Sack 2004). For others trying to establish credibility within geographical institutions: (a) it may have been too risky to be associated with the socially avant-garde 'new woman' at the turn of the nineteenth century, to say nothing of militant suffragettes 1900–20; and (b) they may have had more than enough to do to demonstrate they were competent in the established paradigm in the face of possible antipathy or even hostility. Even in the 2000s, women geographers who show an interest in gender issues are at risk of being labelled pejoratively as 'feminist' (Bracken & Mawdsley 2004). However, despite these constraints some women as geographical authors were sensitive to the status or experience of women in their analysis of place and power relations. This can be seen in Bird's argument about the domestic servitude of Korean women required to keep their men dressed in spotless white clothes, but perhaps most notably peppered in Newbigin's writings, e.g. on women's work in Norwegian fishing communities, in Snodgrass's analysis of employment in Scotland, and in Buttimer's awareness of the impact of urban redevelopment in Glasgow on women's daily lives.

The gendering of the reception of women's geographical work: the three 'Rs'

There are several important points to be made about the intersection between gender and the reception of women's geographical work and these are grouped under the three 'Rs': refereeing, reviewing and reputation. As Berg (2001) has pointed out, present-day refereeing of articles for publication in geographical journals is not as 'blind' (objective) as the peer review system purports. In the early twentieth century either no attempt was made to disguise the author's identity or it was identifiable from the research topic or geographical area covered, or from previous presentations at conferences. This familiarity with colleagues' work was marked within the relatively compact pre-1960's British geographical community, so women's work could be identified and evaluated as such, should reviewers have thought in those terms. Archive evidence shows that Marion Newbigin's paper submitted to the *GJ* was criticised for not being sufficiently geographical and that Eila Campbell's paper was criticised by Eva Taylor's disciplinary sparring partners – although both were ultimately published. Documents relating to Taylor's own papers in the *GJ* show both active facilitation of her need to meet institutional requirements for publications in the first half of her career and the rejection of a paper on William Leybourne in 1943 shortly after her retirement. Few papers written by men or women pass through the refereeing process without criticism and women such as Newbigin and Taylor were important gatekeepers themselves as editor and referee. However, referees' reports and other forms of academic critique can be curt and dismissive, and the experience of wounding responses to her work is known to have halted the publication career of Gillian Groom, despite her having original material worthy of publication.

While there is limited data on the reviewing processes for journal papers or publishers' decisions regarding books, the published review is by definition in the public arena. Just as obituaries have been described as the 'first draft of history' for an individual's life, reviews of published work represent the 'first draft' of the critical evaluation of a piece of work. However, they need to be placed within varying discursive conventions and practices which are in part particular to reviewing and in some aspects common to obituaries. In the case of the latter both can be anonymous – depending on the publication's editorial policy at a given time – and can have a tendency to be celebratory. However, the more common practice is that reviews are by named authors and they tend to be critical in more than the literary sense. Reviews are relatively unmediated, they are not peer reviewed and editors often have a 'light touch' in monitoring or changing content. They can also be the means of making a name for the reviewer, especially through 'giant slaying', or may say as much about the reviewer's status, self-representation or disciplinary allegiances and loyalties as the relative merits of the work reviewed.

There are no sweeping generalisations to be made about the critical reception of women's geographical work, but I would like to draw attention to several points. Mary Somerville's work was implicitly deemed acceptable because it was 'unpretending' and 'unassuming in form and pretensions' (see Chapter Two): she worked within boundaries which were acceptable for her gender. The adjective 'suggestive' stands out as having been applied to the work of two late-nineteenth/early-twentieth-century women authors. Mary

Kingsley's work was described as 'valuable and suggestive' in her *GJ* obituary which sought to mark her life while distancing the RGS from association with all her views. It was also used twice in referring to the work of Marion Newbigin, by RGS president Holdich and L.W. Lyde; this appears to have been in the context of being thought provoking, rather than unsubstantiated and the meaning may have changed in usage in the intervening years, but even at best it does not endow a sense of authority. Interestingly, the term was applied to Newbigin again by Freeman (1980) illustrating the process of reiteration once a label has been attached to someone's work. It is also notable that Isabella Bird, who was attributed as the agent provocateur in the 1892–3 women's membership debate by those who opposed the admission of women to the Society, experienced a poor turnout at her first formal paper for the Society and the dinner which followed and a cool reception for her next published work reviewed in *GJ*. On the contrary, Kate Marsden, who had been in the first cohort of women admitted in 1913 when the gender bar was lifted, benefited from a warm review in the *GJ*. Kropotkin used his authority as a Russian geographer and a representative of the RGS to validate her geographical descriptions which, along with her reputation, had been called into question; although the officers of the RGS distanced themselves from Marsden over time as questions about her reputation persisted.

Reviews are only the first stage of critical engagement with a publication and they may not match longer term popularity and usage of texts. In order to assess the reputation of an individual and their work, reviews must be placed in the context of reprints, subsequent editions, and other evidence of longer term engagement through citations and responses to arguments and data presented. Reviews are a useful tool in assessing the initial reception of women's geographical work, but they must be seen in the context of the specificity of time and place and their overall multiplicity (Livingstone 2005; Keighren 2006). Hilda Ormsby's book on France was criticised for its watershed-based regions, but nonetheless it dominated the field for 30 years. Audrey Lambert's book on *The Making of the Dutch Landscape* appears not to have been reviewed in British geographical journals when published in 1971 but went on to be reprinted as a revised second edition in 1985 in the USA. This raises the issue of the reception of work internationally. While publications and reviews in English media may be relatively well circulated and exchanged, (and in turn influence the standing of the author, e.g. Cuchlaine King and Marjorie Sweeting), topics outside the interests of mainstream geographical interests or publications in languages other than English, even if written by British authors, may receive little critical engagement (as was experienced by Catherine Delano-Smith). As Dowson (1995) has argued of women's poetry, women's representation in disciplinary histories has been hampered not by the absence of publications, but by the relative absence of critical engagement with that work.

Futhermore, whether evolutionary or revolutionary, Anglo-American geography has undergone several disciplinary shifts and, as Taylor (1976) has argued of the quantitative turn, an element of this pattern has been the new vanguard's rejection of past work. Historiographers have also played their part in critiquing the Whiggish tendency for new approaches to look for antecedent 'heroic figures'. The result is that past work can be caricatured and dismissed out of hand, as has been argued in the case of the regional method which dominated the epistemology and methodology of British geography for 50 years (Maddrell 2006), and the marginalisation of spatial science in recent work (Johnston 2006).

Discourses in representing and memorialising women geographers

Obituaries have their own discursive structure and have a tendency to be hagiographic. Although multiple obituaries might give the appearance of offering varied perspectives on the deceased, they can be self-referential, repeating earlier versions, or, as in two cases reported in previous chapters, may be written by husband and wife teams who may share the same views and/or sources of information (see Steers and Wanklyn on Margaret Anderson and Viles and Goudie on Jean Grove); alternatively more than one obituary for an individual may be written by the same author (e.g. Potter on Cole). Multiple or not, the obituarist often sets in print a textual memorial which solidifies and fixes the deceased's life story, their personal qualities and professional achievements. Indeed the absence of an obituary in a geographical journal often means there is no coherent overview of a life and career, which in turn can lead to increasing erasure over time from collective memory and history. Several obituaries for women geographers, aware of the trend to measure success in terms of publications, sought to explain their relative lack of publications and to story their lives in terms of service to students, using phrases such as 'she always had time', 'her students are her memorial', e.g. Blanche Hosgood, Florence Miller and Jean Mitchell. Others stressed attributes such as kindliness and generosity, which were not exclusive to, but definitely components of, the feminine ideal type of the nineteenth and twentieth centuries. Although typically celebratory, obituaries can have subtexts found in telling vocabulary and reading between the lines: certain adjectives speak volumes. In obituaries, posthumous or confidential accounts Marion Newbigin, Eva Taylor and Jean Mitchell were all described as 'formidable'; Jane Franklin and Hilary Chew were described as 'determined'; and Jean Carter was categorised as 'difficult'. Notably Monica Cole was described as all three: formidable, determined and difficult. These women were strong characters who did not necessarily conform to gendered norms, they believed in their abilities and their right to be heard whether in or beyond the academy. However, these attributes which were necessary for their success also frequently marked them as undesirable in other ways.

Another factor at play in long-term reputation is women's relative longevity. In some cases this has been rewarded by recognition in the form of awards and honours late in life, e.g. Mary Somerville and Hilda Ormsby. However, this is dependent upon being remembered by the next generation, which in turn relies upon supporters still active within geographical institutions, or, in the case of twentieth-century academics, a school of postgraduates built around one's research. Numerous sources testify to the various student grants and awards given in memory or as a result of bequests by women geographers, e.g. Marion Newbigin, Hilda Ormsby, June Sheppard, Jessie Black, Violet Cressy-Marcks, Alice Garnett, Alice Mutton and Jean Mitchell, but to my knowledge only three women had Festschrifts or memorial compilations written to celebrate their career in textual form, namely Marjorie Sweeting, Eila Campbell and Helen Wallis, all of whom died in the 1990s. Anecdotally, women geographers who had long retirements but continued to be active in geographical institutions and publishing worlds into old age, such as Eila Campbell, have been represented as 'oddballs' who held forth 'uninvited' at conference and geographical society meetings, resulting in parody and mockery, undermining any past scholarly reputation.

III   Reflections on the Historiography of Geography and the Geographical 'Canon'

The sheer volume of women's geographical work discussed in this book challenges those disciplinary histories which omit or minimise the presence of women. It also challenges earlier feminist representations of the production of pre-1970s' geographical knowledge as a masculine preserve, which complicates contemporary feminists' relationship to past geographies. The evidence of innovative acclaimed geographical work by leading figures within their sub-disciplines or the discipline as a whole, such as Eva Taylor, Hilda Ormsby, Alice Garnett, Cuchlaine King, Jean Grove, Marjorie Sweeting and Janet Momsen, feminises the geographical 'halls of fame', counterbalancing the underrepresentation of women. While the many women and their work analysed here can be mapped onto existing approaches to the history of geography as a revisionist project, their presence also calls for a 'rescaling of the map' of the historiography of the discipline. I reiterate the calls for the inclusion of more 'small' stories and 'minor' figures (Livingstone 1992; Lorimer & Spedding 2002), mindful that John Rennie Short's critique of the current RAE regime[5] could equally be applied to the historiography of geography: 'Many academic disciplines at certain times have dominant intellectual figures; there is nothing wrong with that per se, but when power is increasingly channelled into the possession of a few, then it is a legitimate cause for concern' (2002: 324). Linked to this broadening of the cast within the geographical canon is the ongoing need for epistemological and ontological shifts to extend both the definition of what constitutes 'geographical knowledge' and 'geographical work'. In practice this would include not only the writings of the likes of women travellers, but also makes a case for greater recognition of other textual forms such as school texts, guidebooks and other popular forms of writing, as well as non-textual work such as teaching, teacher training, leadership of field visits, public speaking, and committee and editorial work within and beyond geographical institutions. Just as the scientific canon has been shaped by Whewell's narrow definition of scientists as those who make 'original discoveries', geography in its effort to establish its scientific credentials has been too bound up by notions of original exploration (with all its Eurocentric connotations) in the nineteenth century, particular constructs of field-based primary data collection in the early to mid-twentieth century and has arguably been succeeded by theoretical innovation in the late twentieth century. In the case of science and geography these narrow definitions have excluded much of women's work, but also much other valuable secondary phase research, analysis and synthesis, the merits of which can be appreciated outside of prescriptive caricatures. Expanding what is considered to be of historical significance not only increases the visibility of women as active participants in that history and sheds new light on existing questions, it also raises new questions and agendas, making for a different sort of history (Wallach Scott 1996, 1999); the same is true of geography and conceptions of space, place and landscape, as McKittrick's (2006) study of the geographies of black women has shown. While what would be described as internalist overview histories have been most likely to name women's participation in the history of British geography (e.g. Freeman 1960; Stoddart 1983; Steel 1987), these, by their nature, have not engaged in detail with the character of women's (or most men's) geographical work. What is needed is more historical work which is inclusive, contextual and theoretically and analytically sensitive to difference. As the biographical studies in previous chapters demonstrate, individual geographers inhabited frequently complex social, intellectual and institutional positions in their lives as producers of geographical knowledge; an equally nuanced methodology and analysis is needed to begin to understand something of the production and reception of their geographical work, not least in the longer term of the history of geographical thought and practice.

Several aspirations for future historiography arise from this work. Reflecting its aims, this book has strategically focused on women's geographical work and it is hoped that subsequent histories of the discipline will not so easily disregard the geographical work of women in all its complexity and nuance. While there are many interesting stories still to be researched and told about women's geographical work, and individual and collective 'embodied genealogies', it is hoped that a more integrated approach will follow, producing more democratic accounts with an enhanced sensitivity to other forms of underrepresentation within the historiography of geography. However, it is not enough to reference women's geographical work, it is necessary to critically engage with that work, past and present. Gender as an analytical concept needs to be more fully incorporated into the historiography of geography, but this should not depend only upon women undertaking a gendered division of labour: anyone engaging with the history of the discipline should be sensitive to constructions of femininities and masculinities within geographical discourses and practices, as well as other axes of difference. Such an approach can only enhance our appreciation of the intricacies of the detailed tapestry – to use a feminine metaphor – of the history of geography and geographical ideas.

Notes

CHAPTER ONE: PUTTING WOMEN IN THEIR PLACE: WOMEN IN THE HISTORIOGRAPHY OF GEOGRAPHY

1 Driver (1991) also addressed masculinity in relation to late nineteenth-century geographical discourses.
2 The initial cohort admitted in 1892–3 was limited to 22 and no others were allowed to join until 1913.
3 Also see: Holloway (2005) on gender and work and Summerfield (1995) on women's war work; Witz (1992) on women and the professions; David (1987) and Sutherland (1990) on educational reform; Joannou (2005) on women's suffrage and opposition to enfranchisement; Thompson (1990) on social change 1750–1950; and Tosh (1992) on empire and masculinity.

CHAPTER TWO: WOMEN AND BRITISH GEOGRAPHICAL SOCIETIES: MEDALS, MEMBERSHIP, INCLUSION AND EXCLUSION

1 These biographical accounts draw on my entries on Isabella Bird and Mary Somerville previously published in the Thoemmes *Dictionary of Nineteenth-Century Scientists* (Maddrell 2004c, 2004d).
2 M.S.S. Barrow, Additional Papers 2, Royal Geographical Society with the Institute of British Geographers (RGS-IBG) Archives.
3 'Women were eligible to be considered for membership of the AAG from the beginning in 1904, and two of the founding members were women, but since new members were voted on by existing members, very few women were elected' (Personal Communication, Jan Monk). The membership of the American Geographical Society was always open to women and women were more active participants in the early twentieth century. Women were members of the Paris and Vienna geographical societies in the 1850s and 1860s (personal communication Mike Hetternan).
4 Bird Bishop to John Murray, 6 June 1893, cited by Birkett (1989: 223); Maddrell Mander (1995).
5 Letter from Admiral Cave to Duff, 3 December 1892, Additional Papers, Women's Membership, RGS-IBG Archives.

6   Mary Kingsley to J.S. Keltie, 27 November 1899, RGS/IBG Archives. The Geological and Linnaean societies admitted women in 1904 and 1905, respectively (Stoddart 1986).

7   Cited by Birkett (1989), Maddrell Mander (1995), Bell and McEwan (1996).

8   RGS Council Minute Book (1893).

9   For example, Agnes Deans Cameron (1910) after addressing the RGS in February 1910; and Ada Williams, geography teacher in Bangalore (1911) whose application was rejected, marked 'ladies not eligible' (RGS Additional Papers 93 Box 1, Ladies as Fellows 1910–).

10   B. Pullen Burry to RGS, 29 June, RGS Additional Papers 93 Box 1, Ladies as Fellows 1910–).

11   Curzon to RGS Fellowship 1912; cited by Bell and McEwan (1996: 298).

12   Extracted from Record of Lectures and Lecturers, compiled by T.N.L. Brown, no date, handwritten manuscript, Manchester Geographical Society Archives.

13   *Transactions and Report of the Council of the Liverpool Geographical Society January*, 1901; also see Maddrell Mander (1995).

14   A number, such as Martineau and the Carpenters were Unitarians, a denomination which espoused female education (Watts 1981).

15   A. McFarlane to Mary Somerville, 19 October 1835, Somerville Collection, BUS Box II, SC Box 24, University of Oxford Archives.

16   Martha Somerville was also to subsequently receive a civil list pension as her mother's biographer; Letter from Disraeli to Martha Somerville, 19 May 1877, Somerville Collection, Box 20, Celebrities Box II, University of Oxford Archives.

17   Somerville Collection, MSS Box II SC, Box 2, University of Oxford Archives.

18   Darwin to Somerville, 21 January 1869, Somerville Collection Box 20, Celebrities Box II, University of Oxford Archives.

19   *The Quarterly Review* (1848). These comments may be the basis for subsequent claims that this was the first book on physical geography published in Britain.

20   Herschel uses an almost identical phrase, 'the unpretending but most useful treatise', to describe Professor Ansted's account of physical geography (see Baker 1948, footnote 6).

21   Livingstone (2003) makes brief reference to *Physical Geography* in *A Century of British Geography* (R. Johnston & M. Williams (eds)).

22   Also see Letter from Murchison to Somerville, 14 February 1869, Somerville Collection, Box 20, Celebrities Box II, University of Oxford Archives.

23   Bird to J.S. Keltie, 17 April, 1898?, Keltie Correspondence Papers, RGS-IBG Archives.

24   Bird to Murray, 17 November 1899, quoted by Birkett (1989: 228).

CHAPTER THREE: MARION NEWBIGIN AND THE LIMINAL ROLE OF THE GEOGRAPHICAL EDITOR: HIRED HELP OR DISCIPLINARY GATEKEEPER?

1   Public Records Office, Birth Certificate, M.I. Newbigin 1869.

2   Aberystwyth, University of Wales Register, courtesy of Dr Ian Salmon.

3 Hilda Newbigin 1947, 1948 British Association LBR MSS No. 4, copy in RGS-IBG Correspondence Files.
4 Marion Newbigin to E. Heawood, 14 July 1917, RGS-IBG Correspondence Files.
5 Hilda Newbigin 1947, 1948 British Association LBR MSS No. 4, copy in RGS-IBG Correspondence Files.
6 Aberystwyth, University of Wales Register, courtesy of Dr Ian Salmon.
7 Newbigin to Heawood, 14 July 1916, RGS-IBG Correspondence Files 1911–20.
8 Other members of the 1922 Committee included G.G. Chisholm, V. Cornish, L.W. Lyde and Ellen Semple as vice presidents.
9 Semple was also vice president.
10 Correspondence files suggest that Newbigin was notified of the award as early as 1917.
11 Newbigin to Hinks, 28 April 1921 and Hinks to Newbigin, 1 June 1921, RGS-IBG Correspondence Files 1920–30.
12 Newbigin to Heawood, 17 November 1928, RGS Correspondence Files 1921–30.
13 Newbigin to Heawood, 16 January 1917, RGS-IBG Correspondence Files 1911–20.
14 Newbigin to Heawood, 16 April 1917, RGS-IBG Correspondence Files 1911–20.
15 Newbigin to Mill, 8 April 1908, LBR MSS No.7, RGS-IBG Correspondence Files 1911–20.
16 Newbigin to Heawood, 8 December 1915, RGS-IBG Correspondence Files 1911–20.
17 Review Sheet, Douglas Freshfield on Marion Newbigin's paper Race and Nationality, 30 October 1917, RGS Correspondence Files 1911–20.
18 See Monk 2003 and 2004 for photograph.

CHAPTER FOUR: WOMEN TRAVELLERS: INSIDE OR OUTSIDE THE CANON?

1 Kingsley to Lady MacDonald, cited by Gwynn 1933: xx.
2 Kingsley to Roy, 26 August 1893 (transcript), Mary Kingsley Correspondence, LBR MSS No. 52, RGS-IBG Archives.
3 Kingsley to her lecture agent, cited by Blunt 1994: 130.
4 Kingsley to Keltie, 27 November 1899, Correspondence Files, RGS/IBG Archives.
5 Copy of letter to M. Farquharson included in a confidential letter Kingsley to Keltie, 27 November 1899, Correspondence Files, RGS-IBG Archives.
6 Keltie to Kingsley, 16 December 1895, Correspondence Files, RGS-IBG Archives.
7 Kingsley to Keltie, 13 January 1897, Correspondence Files, RGS-IBG Archives.
8 Her court dress photograph held by the RGS gives an impression of a more powerful character than suggested by Figure 4.1.

9 The report was appended to a letter to *The Times* by Kate Marsden, 18 August 1894.

10 Baigent 2008.

11 Cited by Keay 1989: 103.

12 1910, Cited by Wallach Scott 1996: 80.

13 Dick Doughty-Wiley's uncle.

14 Cited by Wallach Scott 1996: 91.

15 See Bell 1995b.

16 Cited in *Geographical* (2007) **79**: 26.

17 Ibid.

18 PRO, L/P&S 10/576, cited by O'Brien 2000.

19 Cited by Wallach Scott 1996: 188.

20 Cited by Wallach Scott 1996: 197.

21 Commander of the British Empire

22 Some of Gardner's letters from this trip are held in the RGS/IBG Archive.

23 Cited by Geniesse 1999: 209 (also see Stark 1951: 195)

24 Traveller Ella Christie was the first woman to be elected to the Council of the RSGS in 1923, and was soon joined by Marion Newbigin. Christie served on Council until 1934 when she was made a vice president of the Society, a position she held until 1948 (see Chapter Eight).

25 Cited by Hoyle 2001: 92.

26 Scottish Screen Archive catalogue.

27 *S2 Weekend. The Scotsman*, 9 February 2002.

28 Cited by Hoyle 2001: 160.

29 *S2 Weekend. The Scotsman*, 9 February 2002.

30 Cited by Hoyle 2001: 208.

31 Marriage certificate Maurice Cressy Marckx and Violet Olivia Rutley, note Marckx was later altered to Marcks.

32 Violet Cressy-Marcks' certificate of fellowship, RGS-IBG Archive.

33 See *Who Was Who* 1961–70.

34 *The Times*, 16 September 1970: 12.

35 Cited by Trollope 1983.

36 *The Times*, 16 September 1970: 5; MI5 would not confirm this in response to my enquiries and I have not as yet been able to find archive evidence in released files from the intelligence services now in the PRO.

37 Will of Violet Fisher, Public Records Office, Probate Registry.

38 *Times Literary Supplement*, 8 December 1932: 491

39 *Times Literary Supplement*, 22 June 1940: 304.

40 E.W. Fletcher to A.R. Hinks, 26 May 1934, E.W. Fletcher Correspondence Files, RGS/IBG Archives.

41 A.R. Hinks to E.W. Fletcher, 5 June 1934, E.W. Fletcher Correspondence Files, RGS/IBG Archives.

42 Letters between Cressy-Marcks and Hinks, Violet Cressy-Marcks Correspondence Files 1931–40, RGS/IBG Archives.

43 Personal communication Stockwell Publishers 2007.

44 See Steel and Lawton 1967, appendix II: 593.

45   I am grateful to Frances Soar, GA manager, who drew my attention to Haslem's papers when I was studying other archive material at the GA; such is the serendipity of archive work.

CHAPTER FIVE: WOMEN IN GEOGRAPHICAL EDUCATION: DEMAND FOR GEOGRAPHY TEACHERS AND TEACHING BY EXAMPLE

1   *Geography* after 1923.
2   Minutes of the meeting at the New Common Room at Christchurch (Oxford), 20 May 1893, GA Archives.
3   First Annual Report 1894, GA Archives.
4   Similar comments were later made about Hilda Fleure when H.J. Fleure served as GA honorary secretary, though it was recognised that 'jointly they rendered an outstanding service' to the Association (*Geography* 1974: 365).
5   Joan Reynolds to Freeman, 25 September 1948, copy in University of Oxford, School of Geography and Environment Archives.
6   Joan Reynolds to Freeman, 25 September 1948, copy in University of Oxford, School of Geography and Environment Archives.
7   The photograph now displayed in the University of Oxford, School of Geography and Environment was provided by Joan Reynolds (copy in University of Oxford, School of Geography and Environment Archives).
8   Joan Reynolds to Freeman, 11 September 1948, copy in University of Oxford, School of Geography and Environment Archives.
9   Ibid.
10   Joan Reynolds to Freeman, 25 September 1948, copy in University of Oxford, School of Geography and Environment Archives.
11   Later still the book would be revised by Ian Scargill and appear in its eighth edition in 1963!
12   Mrs [Estyn] Evans to Bill Mead, courtesy of Bill Mead.
13   Ibid.
14   She also addressed the Anthropological Section in 1926 (Simpson 1930: xi).
15   Simpson (1951) acknowledged that survey was harder – but not impossible – in urban contexts.
16   MS 97, Earliest Impressions, Dorothy Sylvester Papers, National Library of Wales, Aberystwyth.
17   Manchester High School for Girls Staff Register, courtesy of Chris Joy, MHSG archivist.
18   Manchester High School for Girls Staff Register, courtesy of Chris Joy, MHSG archivist; University of Manchester Archives, courtesy of James Peters, University archivist. The MHSG school magazines also refer to her leaving in December 1904 and returning in April 1905, no reason is given.
19   Hugh Perfect, honorary archivist, Moray House School of Education, University of Edinburgh.
20   Coulthard was acknowledged for her assistance by Dorothy Sylvester in the preface to *Maps and Landscape* (1952).
21   Funeral notes for Gladys Hickman, 2006, courtesy of Richard Hickman.
22   Interview with Gladys Hickman, June 2005, Edinburgh.

23   Ibid.

24   Ibid.

25   Bristol University Archives, courtesy of James Webley. The study was on the sequence of man's activities on the flood plain of the Thames, in the vicinity of Deptford Creek, which was close to Goldsmith's College (Richard Hickman).

26   Interview with Gladys Hickman, June 2005, Edinburgh.

27   Interview with Sheila Jones, 2007.

28   Bristol University Archives, courtesy of James Webley.

29   Personal communication, Frances Soar of the GA.

30   Interview with Gladys Hickman, June 2005, Edinburgh.

31   Funeral notes for Gladys Hickman, 2006, courtesy of Richard Hickman.

32   Crewe and Nantwich GA programme cards 1958–80, Dorothy Sylvester Papers 102, National Library of Wales, Aberystwyth.

33   Personal communication from Mike Tanner, 2005; Giles lists both 1953 and 1954 as her leaving dates.

34   Dates extracted from ULIE staff record by Katie Mooney, assistant archivist, ULIE.

35   This dated back to H.R. Mill's 1905 gift of the copyright of his *Hints to Teaching Geography* on the condition a new edition was prepared; J.F. Unstead prepared an updated version for 1909 publication under the title *Guide to Geographical Books and Appliances*. Miss D.M. Forsaith edited the 1932 *Handbook for Geography Teachers* (Balchin 1993).

36   Personal communication from Sheila Jones, 2007.

37   Personal communication from Sheila Jones, 2007.

38   Ibid.

39   Ibid.

40   Interview with Sheila Jones, 2007.

41   Ibid.

42   Jubilee Book, GA Archives, GA, Sheffield; Membership Analysis 1953 map, GA collection 1988/60, Sheffield City Archives.

43   Minute Book of Central London Branch of the GA, GA collection 1988/60, Sheffield City Archives.

44   Crewe and Nantwich GA programme cards 1958–80, Dorothy Sylvester Papers 102, National Library of Wales, Aberystwyth.

45   Figures taken from Lists of the GA Executive Committee 1951–70, GA Collection 1988/60 item 52, Sheffield City Archives.

46   Interview with Sheila Jones, 2007.

CHAPTER SIX: DIPLOMAS, DEGREES AND APPOINTMENTS: THE FIRST GENERATION OF WOMEN GEOGRAPHERS IN ACADEMIA

1   Correspondence from Anna Petre, based on University of Oxford Archives.

2   University of Oxford School of Geography Annual Report 1909, University of Oxford Archives SG/R/1/2.

3   The 1919 Sex Discrimination (Removal) Act compelled many institutions to be open to women.

4   These dates and titles are taken from departmental lecture lists but university records note her as 'Demonstrator and tutor to the women students', 1923–5, 'Lecturer in regional geography' from 1926 and simply 'Lecturer', 1928–35.

5   *Geophil* 1913 MSS, SOGE Library, University of Oxford.

6   See RSGS medal lists at www.geo.ed.ac.uk/rsgs/awards/univmed.htm.

7   C.Withers, 'A short history of geography at Edinburgh. Looking forward, looking back', at www.geos.ed.ac.uk/geography/centenary.

8   Named for Mary Somerville.

9   Birth certificate Hilda Rodwell Jones, PRO; this section draws on my previous entry on Ormsby in the *Oxford Dictionary of National Biography* (2004e) and my 2006 paper in *Environment and Planning A*.

10   Wise undated, unpublished MSS.

11   Ibid.

12   Ormsby undated, unpublished MSS, LSE Archives.

13   The honours degree in geography at LSE was only available from 1921, the first graduates being Elsa Rea and her King's College tutor (and later husband) Dudley Stamp, then lecturer in geology. The first LSE entrant was Alice Peile in 1922. Stamp was appointed reader in geography at LSE in 1926.

14   Ormsby Papers, LSE Archives.

15   Personal communication, Bill Mead 1998.

16   *New York Herald Tribune*, 6 March 1950.

17   Sargent testimonial, 1923, Ormsby Papers, LSE Archives.

18   Ormsby Papers, LSE Archives.

19   See the preface to the second edition.

20   Kate Rutty, 3 October 1905, E.G.R. Taylor Papers, MSS 69467 543A, British Library.

21   *The Times*, 7 July 1966, p. 15.

22   Barclay Barr, 1 September 1935.

23   Cited in Taylor's obituary *TIBG* 1968: 181.

24   W.G. East, 28 September 1960, EGR Taylor Papers, MSS Add 71872, British Library.

25   Eila Campbell to Dr D. Webber of Maldon, Essex, 20 September 1981, EGR Taylor Papers, MSS Add 71872 No. 11. British Library. Campbell points out that Taylor bought the lease for the Oakley Street house from Dunhill in 1920, suggesting either that their relationship had come to an end or that she wished and had the means to be financially independent. The lease ran out during the war years.

26   E.G.R. Taylor to Dr Stopes, 23 March (no year), from 34 Oakley Street, Marie Stopes Papers, MSS Add 58738 (f.24), British Library.

27   Eila Campbell to Dr D. Webber of Maldon, Essex, 20 September 1981, EGR Taylor Papers, MSS Add 71872 No. 11, British Library.

28   E.G.R. Taylor Papers, MSS Add 71872 (various correspondences), British Library.

29   E.G.R. Taylor to Kierwan, 9 October 1946, E.G.R. Taylor Correspondence Block 1941–48, RGS/IBG Archives.

30   Many of her teaching notes as well as research files can be found in the British Library Eva Taylor manuscript collection.

31 *The Times*, 7 July 1966.
32 *Sunday Graphic and Sunday News*, 1 September 1935, p. 9.
33 D.A. Hall-Davies to Taylor, 25 March 1940, E.G.R. Taylor Papers, MSS Add 69466, British Library.
34 RGS Secretary to E.G.R. Taylor, 29 April 1938, E.G.R. Taylor Correspondence Block 1931–40, RGS/IBG Archives.
35 E.G.R. Taylor to Kierwan, 18 March 1947, E.G.R. Taylor Correspondence Block 1941–48, RGS/IBG Archives. Former colleague Andrew O'Dell had moved to Aberdeen to develop the geography department there.
36 Correspondence to Peter de Clerq, cited by de Clerq (2007).
37 E.G.R. Taylor to Crone, 12 March 1945;, secretary to Taylor, 28 December 1945; secretary to Taylor, 1 February 1946, E.G.R. Taylor Correspondence Files 1941–48, RGS/IBG Archives.
38 E.G.R. Taylor to RGS Director, 25 July 1946, 22 October 1946, E.G.R. Taylor Correspondence Files 1941–48, RGS/IBG Archives.
39 News-cutting (1949), *Weekly News*, 'Fish and Chips for the Professor', E.G.R. Taylor Papers, Add. 71872, British Library.
40 *The Times*, 7 July 1966.
41 E. G. R. Taylor Papers, Add 71872/37, British Library.
42 Alphabetical list of subscribers, E.G.R. Taylor Papers, MSS 79 Add 71874, British Library.
43 *Sunday Times*, 20 November 1960.
44 George Philip & Son Publisher to E.G.R. Taylor, 8 January 1965, E.G.R. Taylor Papers, MSS Add 71872, British Library.
45 E.G.R. Taylor to Hinks, Eva Taylor Correspondence Files 1920–1930, RGS-IBG Archives.
46 In a letter from Taylor to Michael Richey, July 1953, cited by de Clerq (2007).
47 E.G.R. Taylor to Eila Campbell, 27 May 1957, E.G.R. Taylor Papers, MSS Add 71872, British Library.
48 Eila Campbell to J.M. Addey, 12 October 1964, E.G.R. Taylor Papers, MSS Add 71872, British Library.
49 *The Times*, 7 July 1966.
50 Francis Chichester to EGR Taylor, 28 March 1963, E.G.R. Taylor Papers, MSS Add 69466, British Library.
51 See Chapter Seven.

CHAPTER SEVEN: FIELDWORK AND WAR WORK: INTERWAR UNIVERSITY GEOGRAPHERS

1 Campbell could be equally well placed in Chapter Eight.
2 Discussion of women geographer's war work draws on my paper, A. Maddrell (2008). The 'map girls'. British women geographers war work 1939–45. *Transactions of the Institute of British Geographers*, **33**: 127–48.
3 Ellis and Hunt suggest it was Garnett's devotion to geography which 'deferred' marriage. Whether by choice or not late marriage reduced the likelihood of parenthood.

4   Alice Garnett curriculum vitae 1962, presented to the committee considering the chair in physical geography, University of Sheffield Archives.

5   RGS-IBG Archives, Certificate of candidature for election: Alice Garnett.

6   Alice Garnett curriculum vitae 1962, University of Sheffield Archives.

7   Ibid.

8   Personal communication, Brian Blouet 2005.

9   Letter Garnett to GA members, 26 January 1959, Sheffield City Archives, GA Collection 1988/87.

10   Seen in a 'Memorandum on geography in the new universities', Sheffield City Archives 1988/60 item 51.

11   Ibid., H.J. Fleure was also known as 'Daddy' Fleure (Ellis & Hunt 1989).

12   Potter (1968) notes the paltry pay Garnett received in her early years at Sheffield.

13   Alice Garnett curriculum vitae 1962, University of Sheffield Archives.

14   RSGS Archives, Honorary Life Members and Fellows. Awards Contd.

15   Swanzie Agnew, interview with Andrew Grout, 10 June 1998, Edinburgh University Geography Department Archives, DG 26/1/3.

16   National Library of Scotland, C.P. Snodgrass Papers Acc. 7861, curriculum vitae, courtesy of Maria Castrillo.

17   Edinburgh University Geography Department Archives, Snodgrass Papers, DG 6/27.

18   Many of the sources used for these papers and a draft of her paper on employment, as well as other research projects, can be found in the Edinburgh University Geography Department Archives, DG 6/27.

19   Edinburgh University Geography Department Archives, Snodgrass Papers, DG 6/27. Handwritten draft of Notes of the Geographical Distribution of Employment in Scotland, p. 16.

20   National Library of Scotland, C.P. Snodgrass Papers Acc. 7861, 4. (ii), courtesy of Maria Castrillo.

21   The other was Mrs Isobel Robertson for her 1949 paper on the 'Head dyke: a fundamental line in Scottish geography'.

22   Swanzie Agnew interview, 1998, Edinburgh University Geography Department Archives.

23   Snodgrass was one of two women to present papers in Section E that year, the other being Mrs D. Portway Dobson who spoke on 'The Bristol district in the Prehistoric period'.

24   Newnham College Register Volume I; also see Coppock 1990.

25   Although always known as Harriet Grace Wanklyn/Steers, she was only registered as Harriet (Public Record Office birth certificate).

26   RSGS Archives, *Scottish Geographical Magazine*, Harriet Wanklyn references and curriculum vitae.

27   Ibid.

28   Ibid.

29   Shackleton was the only female contributor, Stamp wrote three volumes and S.W. Wooldridge the remaining volume on geomorphology.

30   Public Record Office, birth certificate Dora Kate Smee; her obituary gives her birth date as 1899.

31   Letter (2 June 1943) Borough of Cambridge, Northamptonshire Record Office ZB 291/291.
32   Northamptonshire Record Office ZB 281/238.
33   Northamptonshire Record Office ZB 291, Smee's diary.
34   Northamptonshire Record Office ZB 291/294–5.
35   Letter from Richard Wertall (?), Tally Ho House, Hasselbech, 9 July 1976, Northamptonshire Record Office ZB 291/471.
36   Letter from Desmond Newby, Brankley Farm, Haselbech, 12 July 1976, Northamptonshire Record Office ZB 291/471.
37   Letter, 28 May 1935, from College of Arms.
38   Smee inherited Haselbech Cottage from Constance Ismay.
39   Northamptonshire Record Office ZB 291/291: a letter acknowledges receipt of the second of seven annual payments to endow the Bedford travelling grant at the University of Cambridge (19 May 1949); another letter also informs of the recipient, Alan Maley of St John's College, gaining the award.
40   Smee memorandum 1950, Northamptonshire Record Office ZB 291/470.
41   RHUL Archives, Staff Card D1275.
42   Timberlake was assistant staff representative on the Academic Board in 1947, RHUL Archives, Staff Card.
43   RHUL Archives Staff Card.
44   BAAS Annual Report 1926.
45   Letter B. Tunstall to Mr Smith, 17 May 1933, Staff Files, QMUL Archives.
46   Eric Rawstron in Sheppard (1994: 92–3).
47   Alice Mutton curriculum vitae, Department of Geography Papers, QMUL Archives.
48   Ibid.
49   Ibid.
50   Ibid.
51   She had to fight to get the permanent contract she had understood she was being appointed to in 1933; see Mutton's QMC Staff File.
52   Letter H.W. Melville to M.A. Baatz, 30 November 1971, Academic Registrar, University of London.
53   Personal communication James Peters, University archivist: Staff records, University of Manchester Archives.
54   MS 97, Earliest Impressions, Dorothy Sylvester Papers, National Library of Wales, Aberystwyth.
55   Ibid.
56   Ibid. (university years).
57   Personal communication James Peters, University Archivist: Staff records, University of Manchester Archives; also Dorothy Sylvester Papers, National Library of Wales, Aberystwyth.
58   See M. Mifflin (ed.) (1980) *A Book of Meditational Verse by the Wistaston Poets*, and hymns by Dorothy Taylor in *The Methodist Hymn Book*.
59   Geoffrey North personal communication; MS 102 Geographical Association, Dorothy Sylvester Papers, National Library of Wales, Aberystwyth.
60   Crewe and Nantwich GA branch programme cards, *op. cit.*

61   Papers at the NLW indicate that she studied basic Welsh; see MS Earliest Impressions op. cit., NLW.

62   Her papers, deposited by her brother Professor Peter Campbell, are lodged in the collection of the British Museum Map Room. Eila Campbell Papers, I/I/1 and I/I/5, British Library.

63   British Library, Map Room Archives Eila M.J. Campbell Papers, I/I/3 testimonials, EGR Taylor, E. Willatts.

64   British Library Map Room Archives, Eila M.J. Campbell Papers, I/I/3/10 brief curriculum vitae.

65   British Library Map Room Archives, Eila M.J. Campbell Papers, I/I/3 testimonials, E.G.R. Taylor, 9/9/1941 and March 1944.

66   British Library Map Room Archives, Eila M.J. Campbell Papers, I/I/3 testimonials, 21. Vernon testimonial 7 December 1945

67   British Library Map Room Archives, Eila M.J. Campbell Papers, I/I/3 testimonials, 43. Letter (3 March 1947), Academic Registrar University of London to Eila Campbell. Note: Taylor had retired in 1940.

68   British Library Map Room Archives, Eila M.J. Campbell Papers, I/I/2 testimonials, 47. Letter Wooldridge to Campbell.

69   Op. cit. I/3/19 23 and 61.

70   Ibid., 63.

71   British Library Map Room Archives, Eila M.J. Campbell Papers, I/I/2 (46). Draft Letter to Wooldridge (8 August 1947).

72   British Library Map Room Archives, Eila M.J. Campbell Papers, I/3/9.

73   British Library Map Room Archives, Eila M.J. Campbell Papers I/4/19.

74   British Library Map Room Archives, Eila M.J. Campbell Papers I/2/I.

75   Letter Eila Campbell to Peter Campbell, 9 August 1949, British Library Map Room Archives, Eila M.J. Campbell Papers, I/2/4.

76   Letter Eila Campbell to Peter Campbell, 21 August 1949, British Library Map Room Archives, Eila M.J. Campbell Papers.

77   Typed note Eila Campbell to Don Thompson in response to his letter re hers of 14 October 1949, British Library Map Room Archives, Eila M.J. Campbell Papers I/2/4: 57, I/2/6.

78   Letter Eila Campbell to Peter and Mummy, 29 August 1950, British Library Map Room Archives, Eila M.J. Campbell Papers I/2/2 101.

79   British Library Map Room Archives, Eila M.J. Campbell Papers I/4.

80   Letter Eva Taylor to Eila Campbell, no date (annotated before 1939), Eva Taylor Papers Add. 71872, No. 21, British Library.

81   Letter Eva Taylor to Eila Campbell, Wednesday 7 (no month or year), Eva Taylor Papers Add. 71872, No. 28, British Library.

82   Cited by de Clerq 2007.

83   Interview Catherine Delano-Smith 2006.

84   Interview Anne Buttimer 2006.

85   Letter Eila Campbell to Mother, 12 June 1950, British Library Map Room Archives, Eila M.J. Campbell Papers, I/2/1. 43. Note sugar rationing continued in the UK until 1953.

86   British Library Map Room Archives, Eila M.J. Campbell Papers, I/3/8.7.

87   Eva Taylor to Eila Campbell 7/6/51, Eva Taylor Papers MSS Add 71872, British Library.
88   Personal communication with Crone at the RGS. Campbell bemoans the effort expended in rewriting and correcting two papers for Bagrov, only receiving minimal postage expenses in return! Campbell to Crone (20 April 1953), Eila Campbell, RGS Personal Communication, Block CB11 1941–50.

CHAPTER EIGHT: THE WAR YEARS AND IMMEDIATE POST-WAR PERIOD

1   MS N6 Margaret Davies (Biographical Notes), Elwyn and Margaret Davies Papers, NLW, Aberystwyth.
2   MSS B14–16, Elwyn and Margaret Davies Papers, NLW, Aberystwyth.
3   GA Executive Committee minutes 1/1/1946, 1/1/1947, GA Archives, Sheffield.
4   Elwyn became secretary to the Council of the University of Wales and after 1963 became permanent secretary to the Welsh Department of the Ministry of Education.
5   See presidential address on 'The food of primitive peoples', MS G5, E. Elwyn and Margaret Davies Papers, NLW, Aberystwyth.
6   *The Times*, 16 October 1982, p. 10.
7   Archdeacon Ivor Phillips, speaking at her funeral at Parc Gwyn, Narbeth (*Tenby Observer*, 15 October 1982).
8   Ibid.
9   See MSS G5, *op. cit.* for list and copy of lectures; several were used for two or three audiences, carefully annotated on the script. The lecture to the Federation of University Women in Swansea on 'Conservation in rural Wales' was given in March 1971.
10   See MSS B119 *op. cit.* including Davies's field diary, photographs and diagrams of the site. Elwyn Davies also undertook research in the Isle of Man on land divisions, see E. Davies (1956) *TIBG*.
11   She also drew many illustrations for history pamphlets published by the Schools Service of BBC Wales.
12   *The Times*, 16 October 1982.
13   MS N6, op. cit., NLW.
14   *The Times*, 16 October 1982, p. 10; *The Tenby Observer*, 15 October 1982; *The Western Telegraph*, 14 October 1982.
15   Graduates of the University of Sheffield (1932), Sheffield University Archives.
16   Personal communication Jay Appleton 2007.
17   Ibid.
18   Personal communication Roy Ward 2007.
19   Personal communication Derek Spooner 2007.
20   Personal communication Roy Ward 2007.
21   Personal communication Derek Spooner 2007.
22   Ibid.
23   One colleague kept some books and slides relating to the Hull region and the Welsh borderlands.

24   MSS Miss M.S. Williamson, University of Hull Archives.
25   Interview Cuchlaine King 2007.
26   Ibid.
27   Margaret King (later Ritchie) served as geographical assistant in the Geographical Section of the General Staff War Office 1942–5 (see Maddrell 2008).
28   Interview Cuchlaine King 2007.
29   Ibid.
30   Ibid.
31   University of Nottingham Department of Geography Newsletter, January 1966, 1968.
32   Interview Cuchlaine King 2007.
33   University of Nottingham Department of Geography Newsletter, January 1965.
34   Interview Cuchlaine King 2007.
35   Ibid.
36   Ibid
37   Interview Cuchlaine King 2007.
38   Matriculation form, Mary Marshall, University of Oxford Archives.
39   According to Buxton (1984) she also contributed to the volume on Palestine and Trans-Jordan.
40   Interview C. Delano-Smith 2006.
41   A small collection of work-related papers and personal communication, Mary Marshall Papers, School of Geography and Environment Library, University of Oxford.
42   Interview Delano-Smith 2006.
43   Michaelmas 1965, Mary Marshall Papers, School of Geography and Environment Library, University of Oxford.
44   Delano-Smith op.cit.; personal communication from D.I. Scargill 2007.
45   Delano-Smith op.cit.
46   D.I. Scargill op.cit.
47   Personal communication Kay MacIver 2007.
48   Interview Kay MacIver 2005.
49   Ibid.
50   Ibid.
51   Ibid.
52   Ibid.
53   Ibid.
54   Ibid.
55   Ibid.
56   Ibid.
57   Ibid.
58   Ibid.
59   Ibid.
60   Ibid.
61   Ibid.
62   Dates courtesy of Rachel Hart, Muniments Archivist, Special Collections, University of St Andrews Library.

63 From the Dutch 'Zwaantje' meaning 'Little Swan'.

64 Swanzie Agnew, interview with Andrew Grout (10 June 1998), Edinburgh University Geography Department Archive DG 26/1/3.

65 *The Telegraph Online*, Swanzie Lady Agnew of Lochnaw, last updated 23 August 2001.

66 Ibid.

67 See *Scotsman* obituary for photograph of Agnew.

68 Personal communication Cosmo Ngongondo 2007.

69 Cole (1998–2000) has 1908 as Fuller's birth date, and Wheeler puts her birth-place as Worthing, but her birth certificate states 1909 and Swindon, respec-tively. The place confusion may have arisen from the peripatetic nature of the Methodist ministry necessitating regular moves for the family. Her father's last post was in Ticehurst and he retired in 1937 to Worthing.

70 *Hill's Arrangements, Minister's and Probationers with Circuits*, 1936.

71 Interview C.A.M. King 2007. They subsequently married and the couple went to Glasgow before northern Nigeria; they published a joint paper on tran-shumance in Fjaerland in the *SGM* in 1949.

72 Obituary G. Joan Fuller 1909–92, *University of Nottingham Gazette*, autumn 1992.

73 Interview Cuchlaine King 2007.

74 Wheeler (1992) also notes her pastoral role in the first women's hall of resi-dence Florence Boot Hall.

75 *Nottingham University Department of Geography Newsletter*, No. 11, 1969.

76 Personal communication David McEvoy 2007, former student at the University of Manchester geography department.

77 Manchester University Department of Geography Minute Book 1966–72, 20 November 1968.

78 Personal communication Geoffrey North 2005.

79 Ibid. and university records courtesy of James Peters, University Archivist, Manchester University.

80 Personal communication Alice Coleman 2007.

81 Interview Alice Coleman 2005.

82 Ibid.

83 Ibid.

84 Ibid.

85 Ibid.

86 Ibid.

87 Ibid.

88 Ibid.

89 The award is now given to outstanding business women and includes Anita Roddick amongst its holders.

90 Interview Alice Coleman 2005.

91 Ibid.

92 Short interview with Alice Coleman, 31 August 2005.

93 Interview with Alice Coleman, 8 November 2005 and Coleman 1997c.

94 Personal communication Roy Walsh 2007.

95 Personal communication Gwyneth Davies 2007; death date Swansea University Archives 2007.

96   Marjorie Sweeting to Eila Campbell, 17 July 1957, Campbell Papers I/3/19.22, British Library.
97   Personal communication Gwyneth Davies 2007.
98   Personal communication Jack and Gwyneth Davies 2007. Dave Herbert also remembered the negative impact of critical comments by referees on Groom's research; Personal communication D.T. Herbert 2007 (Herbert was both a student and colleague of Groom's).
99   Personal communication Rosemary Bromley 2007. Jack Davies and Dave Herbert echoed the view that she was an excellent teacher.
100 Senate minutes of November 1947, November 1950, May 1960, Queen's University Belfast Archives, courtesy of Ursula Mitchel.
101 She resigned in 1965, Senate minutes, 8 June 1965, Queen's University Belfast Archives, courtesy of Ursula Mitchel.

CHAPTER NINE: UNIVERSITY EXPANSION, SPECIALISATION AND QUANTIFICATION: 1950–70

1   Monica Mary Cole, birth certificate, Public Record Office.
2   For example Michael Chisholm, Dudley Stamp and Peggie Hobson (for the latter see Bartholomew 1994).
3   'The North West Primorje of Jugoslavia. A Report of the Geographical Investigations made during August 1952', The Geographical Field Group, December 1952 (Nottingham University Archives); also see Moodie *et al.* (1955).
4   Personal communication Jacky Tivers 2005.
5   Personal communication Jacky Tivers 2007.
6   Bedford College University of London Council Reserved Area Business Vol LVIX, Confidential annexe 19/7/73; Confidential annexe 21/2/74, Archives, Royal Holloway, University of London.
7   Personal communication Jacky Tivers 2007.
8   Newnham College Register Vol. 1: 45–6.
9   Newnham College Register Vol. 1: 45–6.
10   Interview Barbara Kennedy 2006.
11   Ibid.; *GJ* 1955: 237; 1980: 487.
12   *Tropical Geomorphology Newsletter*, volume 18–19, http://www.zikzak.net.tgn (accessed 21 February 2006).
13   The University of Oxford holds 71 of Sweeting's field notebooks but none are titled 'Spitsbergen'.
14   Yuan Daoxian, The results of 15 years of intimate scientific exchange, http://www.karst.edu.cn (accessed 21 February 2006).
15   Letter University of Hull Registrar to H.I. Loeten, 13 July 1954; letter of application, 15 June 1954, H. Chew curriculum vitae, Helen Chew papers, University of Hull; R.W. Steel and R. Lawton (1967), appendix I.
16   Letter Manley to King, 6 July 1954; letter Bryan to King, 6 July 1954, University of Hull Archives.
17   H. Chew curriculum vitae, University of Hull Archives.
18   Personal communication Dick Grove 2007.

19   Ibid.
20   Ibid.
21   Ibid.
22   Interview Barbara Kennedy 2006.
23   Personal communication Dick Grove 2007.
24   Ibid.
25   PRO, birth certificate. Also see T. Campbell 2004.
26   Although this was also the practice at Bedford College (and is the convention in academic writing).
27   Interview 2006.
28   The Map Room moved from the British Museum to the British Library in 1973.
29   The British Library holds a recording of this service held on 5 May 1995 and an interview with Wallis after her retirement.
30   Interview Margaret Wilkes 2005; personal communication Matthew Zawadzki, Sheffield University Calendars, Sheffield University Archives 2007; personal communication Margaret Wilkes 2008.
31   Interview Margaret Wilkes 2005; Leicester Alumni 2000.
32   Interview Delano-Smith 2006.
33   Ibid.
34   Personal communication Delano-Smith.
35   Ibid.
36   Ibid.
37   Ibid.
38   Ibid.
39   Ibid.
40   Interview Anne Buttimer 2006.
41   Ibid.
42   Telephone interview Margaret Storrie 2008.
43   Ibid.
44   Ibid.
45   Ibid.
46   Personal communication Margaret Storrie 2008.
47   Interview Elizabeth Clutton 2005.
48   Personal communication Christine McCulloch 2007.
49   Ibid.
50   Ibid.
51   Ibid.
52   Ibid.
53   Ibid.
54   Ibid.
55   She drafted the conclusion to her thesis on the labour ward while giving birth to her first child and had her viva three weeks later.
56   Personal communication Janet Momsen 2008; interview 2008.
57   These were published under her maiden name Henshall until the mid-1970s; Janet Momsen curriculum vitae.
58   Doreen Massey curriculum vitae and personal communication Doreen Massey 2008.

59   Interview Barbara Kennedy 2006.
60   Ibid.
61   Ibid.
62   Personal communication Linda McDowell 2007.
63   Personal communication Gwyneth Davies 2007.
64   Ibid.
65   Ibid.
66   Ibid.
67   Nicola Crosbie, interview Nicola and Sandy Crosbie 2005.
68   Ibid.

CHAPTER TEN: CONCLUSION: MAPPING THE 'HIDDEN' WOMEN IN BRITISH GEOGRAPHY 1900–70

1   Interview Margaret Wilkes 2005.
2   Figures based on the list and dates of chairs in geography in Balchin (1993), Appendix L.
3   I am grateful to Mike Tanner for generously sharing this data and allowing me to use it.
4   Halford Mackinder felt compelled to be the first European to climb Mt. Kenya in 1891, in order to chalk up a 'first', thereby conforming to the masculine paradigm of exploration to avoid being called an armchair geographer (see Parker 1982).
5   The Research Assessment Exercise used to allocate government funding for research in higher education.

Bibliography

A.C. (1955). Review: *Map and Landscape* by Dorothy Sylvester. *The Geographical Journal*, **121**: 368–9.

A.G.O. (1931). Review: *The Geographical Interpretation of Topographical* Maps by Alice Garnett. *The Geographical Journal*, **77**: 188–9.

A.J.W. (1938). Review: Isobel Wylie Hutchison, *Stepping Stones from Alaska to Asia. The Geographical Journal*, **91**: 184–5.

A.M. (1940). Review: V. Cressy-Marcks, *Journey into China. The Geographical Journal*, **96**: 214–5.

Adams, J.H, Crosbie, A.J. & Gordon, G. (1984). *The Making of Scottish Geography. 100 Years of the Royal Scottish Geographical Society.* Edinburgh: RSGS.

Adie, K. (2003). *Corsets to Camouflage. Women and War.* London: Coronet.

Adrian, L. (1990) Jean Brown Mitchell 1904–1990. *The Geographical Journal*, **156**: 242–3.

Agnew, S. (1946a). The vine in Bas Languedoc. *Geographical Review*, **36**: 67–79.

Agnew, S. (1946b). Brig o' Turk – a Highland aspect. *Scottish Geographical Magazine*, **62**: 61–8.

Agnew, S. (1946c). Montpellier: giver of names. *Scottish Geographical Magazine*, 85–8.

Agnew, S. (1978). Review: *Southern Africa* by A.J. Christopher. *The Geographical Journal*, **144**: 502.

Alberti, J. (2002). *Gender and the Historian.* London: Longman.

Allen, H.D. (2003). Jean Grove 1927–2001. Cambridge Quaternary Research Group, www.quaternary.group.ac.uk/history/others/Grove, accessed 23 March 2007.

Allison, J.E. (1955). Miss Edith Marjorie Ward. *Geography*, **40**: 279.

Anon. [Bell, G.] (1918). *The Arabs of Mesopotamia.* Basra: Government Press.

Anon. [Bird, I.] (1856). *An Englishwoman in America.* London: John Murray.

Anon. [Cubbitt] (1919) War work of the society. *The Geographical Journal*, **53**: 336–9.

Ashton, R. (1986). *Little Germany. Exile and Asylum in Victorian England.* Oxford: Oxford University Press.

Baigent, E. (2004). The geography of biography, the biography of geography: rewriting the *Dictionary of National Biography. Journal of Historical Geography*, **30**: 531–51.

Baigent, E. (2008). Kate Marsden. *Geographers Biobibliographical Studies*, **27**: 63–92.

Baker, A.R.H. (2003). *Geography and History. Bridging the Divide*. Cambridge: Cambridge University Press.

Baker, A. *et al.* (1970). *Geographical Interpretations of Historical Sources*. London: David & Charles.

Baker, J.N.L. (1934). Review: E.G.R. Taylor, *Late Tudor and Early Stuart Geography. The Geographical Journal*, **84**: 171–2.

Baker, J.N.L. (1948). Mary Somerville and geography in England. *The Geographical Journal*, **111**: 207–22.

Baker, J.N.L. & Gilbert, E.W. (1944). The doctrine of the axial belt of industry in England., *The Geographical Journal*, **103**: 49–63.

Balchin, W.G.V. (1960). Review: C.A.M. King, *Beaches and Coasts. The Geographical Journal*, **126**: 231.

Balchin, W.G.V. (1987). United Kingdom geographers in the Second World War. *The Geographical Journal*, **153**: 159–80.

Balchin, W.G.V. (1993). *The Geographical Association. The First One Hundred Years*. Sheffield: Geographical Association.

Balchin, W.G.V. (1997). The immediate post-war years. In W.G.V. Balchin (ed.), *The Joint School Story. The Seventy-Fifth Anniversary of the Establishment of the Joint School of Geography for Geographical Cooperation between the London School of Economics and King's College London 1922–1997* (pp. 18–20). London: The Joint School Society.

Bankey, R. (2001). *La Donna é mobile*: constructing the irrational woman. *Gender, Place and Culture*, **8**: 37–54.

Barber, P. (1995). The Eila Campbell Papers. *Imago Mundi*, **47**: 10–11.

Barnard, H.C. (1961). *A History of English Education*. London: University of London Press.

Barnes, F. & King, C.A.M. (1955). Beach changes in Lincolnshire since the 1953 storm surge. *East Midland Geographer*, **4**: 18–28.

Barnes, F. & King, C.A.M. (1957). The spit at Gibraltar Point, Lincolnshire. *East Midland Geographer*, **8**: 22–31.

Barnes, F. & King, C.A.M. (1961). Salt marsh development at Gibraltar Point, Lincolnshire. *East Midland Geographer*, **15**: 20–31.

Barnes, T.J. (2001). Lives lived and lives told: biographies of geographies quantitative revolution. *Environment and Planning D. Society and Space*, **19**: 409–29.

Barnett, C. (1995). Awakening the dead: who needs the history of geography? *Transactions of the Institute of British Geographers*, **20**: 417–19.

Barnett, C. (1998). Impure and worldly geography: the Africanist discourse of the Royal Geographical Society, 1831–73. *Transactions of the Institute of British Geographers*, **23**: 239–51.

Barr, P. (1985). *A Curious Life for a Lady. The Story of Isabella Bird*. London: Macmillan.

Barratt, J. & Hickman, G.M. (1997). *Exploring China*. London: Hodder & Stoughton.

Bartholomew, J.C. (1989). Obituary: Peggy Hobson. *Scottish Geographical Magazine*, **105**: x.

Beaver, S.H. (1962). Charlotte A. Simpson. *Geography*, **47**: 314.

Beaver, S.H. (1982). Geography in the British Association for the Advancement of Science. *Geographical Journal*, **148**: 173–81.

Beckinsale, R.P. (1978). Review: C.A.M. King (ed.), *Landforms and Geomorphology*. *The Geographical Journal*, **144**: 323–4.

Bedini, S.A. (1967). Review: E.G.R. Taylor, *The Mathematical Practitioners of Hanoverian England*. *Isis*, **58**: 120–1.

Bee, A., Madge, C. & Wellens, J. (1998). Women, gender feminisms: visiting physical geography. *Area*, **30**: 195–6.

Bell, G. (1907). *The Desert and the Sown*. London: E.P. Dutton & Co.

Bell, G. (1911). *Amurath to Amurath*. London: Heinemann.

Bell, M. (1993). 'The pestilence that walketh in darkness'. Imperial health, gender and images of South Africa 1880–1910. *Transactions of the Institute of British Geographers*, NS18: 327–41.

Bell, M. (1995a). Edinburgh and empire. Geographical science and citizenship for a 'new' age, ca. 1900. *Scottish Geographical Magazine*, **111**: 139–49.

Bell, M. (1995b). 'Citizenship not charity': Violet Markham on nature, society and the state in Britain and South Africa. In M. Bell *et al.* (eds), *Geography and Imperialism 1820–1940* (pp. 189–220). Manchester: Manchester University Press.

Bell, M. *et al.* (eds.) (1995). *Geography and Imperialism 1820–1940*. Manchester: Manchester University Press.

Bell, M. & McEwan, C. (1996). The admission of women fellows to the Royal Geographical Society, 1892–1914: the controversy and the outcome. *The Geographical Journal*, **162**: 295–312.

Benest, E.E. & Timberlake, E.M. (1945). *Astro-Navigation Tables for Common Tangent Method*. Cambridge: Cambridge University Press.

Bentley, L. (1991) *Educating Women. A Pictorial History of Bedford College, University of London 1849–1985*. Surrey: Alma Publishers.

Berg, L.D. (2001). Masculinism, emplacement, and positionality in peer review. *The Professional Geographer*, **53**: 511–21.

Bird, I. (1856). *An Englishwoman in America*. London: John Murray.

Bird, I. (1875). *The Hawaiian Archipelago: Six Months among Palm Groves, Coral reefs and Volcanoes of the Sandwich Islands*. London: John Murray.

Bird, I. (1879). *A Lady's Life in the Rocky Mountains*. London: John Murray.

Bird, I. (1880). *Unbeaten Tracks in Japan: An Account of Travels on Horseback in the Interior Including Visits to the Aborigines of Yezo and the Shrines of Nikko and Ise*. London: John Murray.

Bird, I. (1883). *The Golden Chersonese and the Way Thither*. London: John Murray.

Bird, I. (1891). *Journeys in Persia and Kurdistan, Including a summer in the Upper Karun Region and a visit to the Nestorian Rajahs*. London: John Murray.

Bird, I. (1894). *Among the Tibetans*. London: SPCK.

Bird, I. (1898). *Korea and Her Neighbours*. London: John Murray.

Bird, I. (1899). *The Yangtze Valley and Beyond: an Account of Journeys in China, Chiefly in the Province of Sze Chuan and among the Man-tze of the Somo Territory*. London: John Murray.

Bird, J.H. (1968). Obituary: Florence Clark Miller. *Geography*, **LIII**: 328.

Birkett, D. (1989). *Spinsters Abroad*. Oxford: Blackwell.

Birkett, D. (2004a). *Off the Beaten Track. Three Centuries of Women Travellers.* London: National Portrait Gallery.

Birkett, D. (2004b). Mary Kingsley. *Oxford Dictionary of National Biography.* Oxford: Oxford University Press.

Blanchard, R. (1933). Review: *France: A Regional and Economic Geography. Geographical Review,* **23**: 156–7.

Blumen, O. & Bar-Gal, Y. (2006). The Academic Conference and the Status of Women: The Annual Meetings of the Israeli Geographical Society. *Annals of the Association of American Geographers,* **58**: 341–55.

Blunt, A. (1994). *Travel, Gender and Imperialism. Mary Kingsley and West Africa.* New York: Guilford.

Blunt, A. & Willis, J. (2000). *Dissident Geographies. An Introduction to Radical Ideas and Practice.* London: Pearson Education.

Blunt, A. (2005). *Domicile and Diaspora.* Oxford: RGS-IBG Series/Blackwell.

Blunt, A. & Rose, G. (eds.) (1994). *Writing Women and Space: Colonial and Postcolonial Geographies.* New York: Guilford.

Board, C. (1962). Review: *South Africa* by Monica Cole. *The Geographical Journal,* **128**: 222–3.

Bondi, L. (1990). Feminism, postmodernism and geography: space for women? *Antipode,* **22**: 156–67.

Boog Watson, W.N. (1967–8). The first eight ladies. *University of Edinburgh Journal,* **23**: 227–34.

Bordo, S. (1990). Feminism, postmodernism, and gender scepticism. In L.J. Nicholson (ed.), *Feminism/Postmodernism* (pp. 133–56). London: Routledge.

Bowden, M.J. (1972). Review: *Three Generations of British Historical Geography. Economic Geography,* **48**: 214–16.

Bowen, E.G. (1970). Obituary. Herbert John Fleure. *Geographical Review,* **60**: 443–5.

Bracken L.J. & Mawdsley, E. (2004). 'Muddy glee': Rounding out the picture of women and physical geography fieldwork. *Area,* **36**: 280–6.

Bremner, C. (1987). *The Education of Girls and Women in Britain.* London: Swan Sonnenschein.

Brunskill, I. (2005). *Great Lives. A Century of Obituaries.* London: HarperCollins.

Buchanan, R.O. (1954). The IBG: retrospect and prospect. *Transactions of the Institute of British Geographers,* **20**: 1–14.

Burstall, S. (1911). *The Story of Manchester High School for Girls.* Manchester: Manchester University Press.

Burton, A. (2003). *Dwelling in the Archive. Women Writing House, Home and History in Late Colonial India.* Oxford: Oxford University Press.

Burton, A. (2005). Introduction: archive fever. In A. Burton (ed.), *Archive Stories. Facts, Fictions and the Writing of History* (pp. 1–24). Durham: Duke University Press.

Butler, J. (1990). *Gender Trouble. Feminism and the Subversion of Identity.* New York: Routledge.

Buttimer, A. (1965). The changing French region. *Professional Geographer,* **XVII**: 1–5.

Buttimer, A. (1967). Réflections sur la géographie sociale. *Bulletin de la Société Géographique de Liége,* **III**: 27–49.

Buttimer, A. (1968a). French geography in the Sixties. *Professional Geographer,* **XX**: 134–45.

Buttimer, A. (1968b). Social geography. In *Macmillan's Revised International Encyclopaedia of the Social Sciences*. New York: Free Press.

Buttimer, A. (1969). Social space in interdisciplinary perspective. *The Geographical Review*, **LIV**: 417–26.

Buttimer, A. (1971). *Society and Milieu in the French Geographic Tradition*. Chicago: Rand McNally/AAG.

Buttimer, A. (1974). *Values in Geography*. Commission on College Geography, Resource Paper 24, Washington DC.

Buttimer, A. (1979). Uniqueness, universality, and the challenge of *la géographie humaine*. In D. Ley & M. Samuels (eds), *Humanistic Geography* (pp. 58–76). Chicago: Maroufa Press.

Buttimer, A. (1983). *The Practice of Geography*. London: Longman.

Buttimer, A. (1987). A social topography of home and reach. *Journal of Environmental Psychology*, **7**: 307–19.

Buttimer, A. (1993). *Geography and the Human Spirit*. Baltimore: Johns Hopkins University Press.

Buttimer, A. (1999). Humanism and relevance in geography. *Scottish Geographical Journal*, **115**: 2, 103–16.

Buttimer, A. (2001). Home-reach-journey. In P. Moss (ed.), *Placing Autobiography in Geography* (pp. 22–41). Syracuse: Syracuse University Press.

Buttimer, A. & Mels, T. (2006). *By Northern Lights. On the Making of Geography in Sweden*. Aldershot: Ashgate.

Caine, N. (1976). Review: C.A.M. King, *Periglacial Processes. Journal of Arctic and Alpine Research*, **8**: 413.

Caird, J. (1996). Obituary: Emeritus Professor Joy Tivy. *Scottish Geographical Magazine*, **111**: 64.

Callard, F. (2004). Doreen Massey. In P. Hubbard, R. Kitchin & G. Valentine (eds), *Key Thinkers on Space and Place* (pp. 219–25). London: Sage.

Calloway, H. (1987). *Gender, Culture and Empire. European women in Colonial Nigeria*. London: Macmillan.

Calvert, H.R. (1966). Review: E.G.R. Taylor, *The Mathematical Practitioners of Hanoverian England, 1714–1840. The Geographical Journal*, **132**: 520–1.

Campbell, E.M.J. (1949). An English philosophico- chronological chart. *Imago Mundi*, **6**: 79–84.

Campbell, E.M.J. (1949a). The early development of the atlas. *Geography*, **34**: 187–95.

Campbell, E.M.J. (1951). Land and population problems in Fiji. *The Geographical Journal*, **118**: 477–82.

Campbell, E.M.J. (1962a). Bedfordshire. In H.C. Darby & E.M.J. Campbell (eds), *The Domesday Geography of South-East England* (pp. 1–47). Cambridge: Cambridge University Press.

Campbell, E.M.J. (1962b). Hertfordshire. In H.C. Darby & E.M.J. Campbell (eds), *The Domesday Geography of South-East England* (pp. 48–96). Cambridge: Cambridge University Press.

Campbell, E.M.J. (1962c). Middlesex. In H.C. Darby & E.M.J. Campbell (eds), *The Domesday Geography of South-East England* (pp. 97–137). Cambridge: Cambridge University Press.

Campbell, E.M.J. (1962d). Buckinghamshire. In H.C. Darby & E.M.J. Campbell (eds), *The Domesday Geography of South-East England* (pp. 138–185). Cambridge: Cambridge University Press.

Campbell, E.M.J. (1962e). Berkshire., In H.C. Darby & E.M.J. Campbell (eds), *The Domesday Geography of South-East England* (pp. 239–86). Cambridge: Cambridge University Press.

Campbell, E.M.J. (1962f). Kent. In H.C. Darby & E.M.J. Campbell (eds), *The Domesday Geography of South-East England* (pp. 483–562). Cambridge: Cambridge University Press.

Campbell, E.M.J. (1968). Eva Germaine Rimington Taylor. *Transactions of the Institute of British Geographers*, **45**: 181–6.

Campbell, E.M.J. (1972). Eva Taylor. *Dictionary of National Biography*. Oxford: Oxford University Press.

Campbell, E.M.J. (1973). A.E.F. Moodie. *Transactions of the Institute of British Geographers*, **59**: 159–61.

Campbell, E.M.J. (1992). Obituary: J. Brian Harley 1932–1991. *The Geographical Journal*, **158**: 252–3.

Campbell, E.M.J. & Donnelly, U. (1947). Peasant life in the glens of Antrim. *Economic Geography*, **23**: 10–14.

Campbell, J.A. (1978). The Queens University of Belfast Department of Geography Jubilee 1928–1978: Geography at Queen's, an historical survey. Departmental Research Papers 2, Queens University of Belfast.

Campbell, T. (2004). Helen Margaret Wallis. *Oxford Dictionary of National Biography*. Oxford: Oxford University Press.

Carlquist, S. (1978). Review of *Biogeography: a Study of Plants in the Ecosphere* by Joy Tivy. *The Quarterly Review of Biology*, **53**: 188.

Carter, D. (1943). Review: V. Cressy Marcks, *Journey into China. Journal of Modern History*, **15**: 71–2.

Castree, N., Fuller, D. & Lambert, D. (2007). Geography without borders. *Transactions of the Institute of British Geographers*, **32**: 129–32.

Chance, J. (2005). Introduction. 'What has a woman to do with learning?'. In J. Chance (ed.), *Women Medievalists*. Wisconsin: University of Wisconsin Press.

Chew, H.C. (1953). The post-war land use pattern of the former grasslands of eastern Leicestershire. *Geography*, **XXXVIII**: 286–95.

Chew, H.C. (1956). Changes in land use and stock over England and Wales 1939 to 1951. *The Geographical Journal*, **122**: 466–70.

Chisholm, G.G. (1898). Mrs. Bishop on Korea and the Koreans. *The Geographical Journal*, **XI**: 288–9.

Chorley, R.J. & Kennedy, B.A. (1971). *Physical Geography: A Systems Approach*. London: Edward Arnold.

Claridge, E. (1982). Introduction to Mary H. Kingsley, *Travels in West Africa*. London: Virago.

Clark, A. (1965). Review: *An Historical Geography of South Africa. The Geographical Review*, **55**: 133–5.

Clay, J. (2001). *Machonochie's Experiment*. London: John Murray.

Clayton, K. (1972a). Review: J.C. Doornkamp and C.A.M. King, *Numerical Analysis in Geomorphology. The Geographical Journal*, **138**: 86.

Clayton, K. (1972b). Review: R.J. Chorley and B.A. Kennedy, *Physical Geography. A Systems Approach. The Geographical Journal*, **138**: 246–7.

Cloke, P., Philo, C. & Sadler, D. (1991). *Approaching Human Geography. An Introduction to Contemporary Theoretical Debates*. London: Paul Chapman.

Clout, H. (2003). Place description, regional geography and area studies: the chorographic inheritance. In R. Johnston & M. Williams (eds), *A Century of British Geography* (pp. 247–74). Oxford: British Academy/Oxford University Press.

Clout, H. (2003b). *Geography at University College, London. A Brief History*. London: UCL Geography Department.

Clout, H. & Gosme, C. (2003). The naval intelligence handbooks: a monument in geographical writing. *Progress in Human Geography*, **27**: 153–73.

Clutton, E. (1978). Political conflict and military strategy: the case of Crete as exemplified by Basilicata's Relatione of 1630. *Transactions of the Institute of British Geographers*, NS3: 274–84.

Clutton, E. & Kenny, A. (1976). *Crete*. Newton Abbott: David Charles.

Cole, J. (2000). *Geography at Nottingham 1922–70. A Record*. University of Nottingham. Geography Cole, J. & King, C.A.M. (1968). *Quantitative Geography. Techniques and Theories in Geography*. London: John Wiley & Sons, Ltd.

Cole, M. (1949). The Elgin basin: a land utilisation survey. *South African Geographical Journal*, **XXXI**: 1–41.

Cole, M. (1956). Vegetation studies of South Africa. *Geography*, **41**: 114–22.

Cole, M. (1957). The Witwatersrand conurbation: a watershed mining and industrial region. *Transactions of the Institute of British Geographers*, **23**: 249–65.

Cole, M. (1960). Cerrado, caatinga and pantal: distribution and origin of the savanna vegetation of Brazil. *The Geographical Journal*, **126**: 168–79.

Cole, M. (1961). *South Africa*. London: Methuen.

Cole, M. (1963). Vegetation nomenclature and classification with particular reference to the savannas. *South African Geographical Journal*, **55**: 3–14.

Cole, M. (1966). *South Africa*, 2nd edition. London: Methuen.

Cole, M. (1986). *The Savannas. Biogeography and Geobotany*. London: Academic Press.

Cole, M., Owen Jones, E.S. & Custance, N.D.E. (1974). Remote sensing in mineral exploration. In E.C. Barrett & L. Curtis (eds), Environmental Remote Sensing – Applications and Achievements (pp. 51–66). London: Edward Arnold.

Cole, M. & Smith, R.F. (1984). Vegetation as indicator of environmental pollution. *Transactions of the Institute of British Geographers*, NS9: 447–93.

Coleman, A. & Beaver, S.H. (1955). Changing geographical values in Dale-i-Sunnfjord. *The Geographical Journal*, **1**: 51–63.

Department, Nottingham.

Coleman, A. (1955). Land reclamation at a Kentish Colliery. *Transactions of the Institute of British Geographers*, **4**: xx.

Coleman, A. (1958). The terraces and antecedence on part of the River Salzach *Transactions of the Institute of British Geographers*.

Coleman, A. (1965). Cartographic methods for use in land-use memoirs. *Transactions of the Institute of British Geographers*.

Coleman, A. (1970). The conservation of wildscape. A quest for facts. *The Geographical Journal*.

Coleman, A. (1985). *Utopia on Trial*. London: Hilary Shipman.

Coleman, A. (1997a). Fieldwork. In W.G.V. Balchin (ed.), *The Joint School Story. The Seventy-Fifth Anniversary of the Establishment of the Joint School of Geography for Geographical Cooperation between the London School of Economics and King's College London 1922–1997* (pp. 23–5). London: The Joint School Society.

Coleman, A. (1997b). The Second Land Utilisation Survey of Britain. In W.G.V. Balchin (ed.), *The Joint School Story. The Seventy-Fifth Anniversary of the Establishment of the Joint School of Geography for Geographical Cooperation between the London School of Economics and King's College London 1922–1997* (pp. 37–40). London: The Joint School Society.

Coleman, A. (1997c). DICE design. In W.G.V. Balchin (ed.), *The Joint School Story. The Seventy-Fifth Anniversary of the Establishment of the Joint School of Geography for Geographical Cooperation between the London School of Economics and King's College London 1922–1997* (pp. 73–5). London: The Joint School Society.

Coleman, A. *et al.* (1988). *Altered Estates*. London: Adam Smith Institute.

Collins, L. (1991). James Wreford Watson 1915–1990. *Transactions of the Institute of British Geographers*, **16**: 227–32.

Cooke, R.U. & Robson, B.T. (1976). Geography in the UK, 1972–6. *The Geographical Journal*, **142**: 81–100.

Coppock, J.T. (1964). Post-war studies in British agriculture. *Geographical Review*, **54**: 409–26.

Coppock, J.T. (1990). Obituary: Jean Brown Mitchell 1904–1990. *Scottish Geographical Magazine*, **106**: 64.

Cornwell, T. (2004). Isabella Bird, I presume. *The Scotsman*, S2, 18 March.

Coulthard, E.M. (1942). Experiments in teaching current affairs. *Geography*, **27**: 69–71.

Coulthard, E.M. (1943). A school experiment. *Geography*, **27**: 117–18.

Coulthard, E.M. (1946). A spring term experiment. *Geography*, **31**: 134–8.

Court, A. (1971). Review: J.P. Cole and C.A.M. King, *Quantitative Geography*. *Geographical Review*, **61**: 157–8.

Creese, M. (2004a). Mary Somerville. *Oxford Dictionary of National Biography*. Oxford: Oxford University Press.

Creese, M. (2004b). Marion Newbigin. *Oxford Dictionary of National Biography*. Oxford: Oxford University Press.

Creese, M. (2004c). Catherine Alice Raisin. *Oxford Dictionary of National Biography*. Oxford: Oxford University Press.

Creese, M. & Creese, T. (1998) *Ladies in the Laboratory*. Lanham, MA: Scarecrow Press.

Cressy-Marcks, V. (1932). *Up the Amazon and Over the Andes*. London: Hodder & Stoughton.

Cressy-Marcks, V. (1940). *Journey into China*. London: Hodder & Stoughton.

Crone, G.R. (1931). Review: E.G.R. Taylor, *Tudor Geography*. *The Geographical Journal*, **77**: 289–90.

Crone, G.R. (1955). Review: E.G.R. Taylor, *The Mathematical Practitioners of Tudor and Stuart England*. *The Geographical Journal*, **121**: 104–5.

Crone, G.R. (1962). Review: *Friedrich Ratzel* by Harriet Wanklyn. *The Geographical Journal*, **128**: 104.

Crone, G.R. (1964). British geography in the twentieth century. *The Geographical Journal*, **130**: 191–215.

Crone, G.R. (1966). Obituary: Professor E.G.R. Taylor, D.Sc. *The Geographical Journal*, **132**: 594–6.

Crone, G.R., Campbell, E.M.J. & Skelton, R.A. (1962). Landmarks in British cartography. *The Geographical Journal*, **128**: 406–26.

Crosbie, A.J. (1991). Professor James Wreford Watson. *Scottish Geographical Magazine*, **107**: 72.

Crouse, N.M. (1934). *The Search for the Northwest Passage*. New York: Columbia University Press.

Curtis, S.J. & Boultwood, M.E.A. (1970). *An Introductory History of Education in Britain since 1800*. London: University Tutorial Press.

Curzon, G.N. (1892). *Persia and the Persian Question*. London: Longman, Green & Co.

D.C.S.-T. (1933). Review: V. Cressy-Marcks, *Up the Amazon and Over the Andes. The Geographical Journal*, **81**: 555.

D.G.H. (1926). Obituary: Gertrude Lowthian Bell. *The Geographical Journal*, **68**: 363–8.

Daniels S. & Nash, C. (2004). Lifepaths: geography and biography. *Journal of Historical Geography*, **30**: 449–58.

Darby, H.C. (ed.) (1936). *An Historical Geography of England before AD 1800*. Cambridge: Cambridge University Press.

Darby, H.C. (1960). *An Historical Geography of England*: twenty years after. *The Geographical Journal*, **126**: 147–59.

Darby, H.C. (1983a). Historical geography in Britain, 1920–1980: continuity and change. *Transactions of the Institute of British Geographers*, **4**: 421–8.

Darby, H.C. (1983b). Academic geography in Britain: 1918–1946. *Transactions of the Institute of British Geographers*, **8**: 14–26.

Darby H.C. & Campbell, E.M.J. (eds) (1962) *The Domesday Geography of South-East England*. Cambridge: Cambridge University Press.

Darling, E. & Whitworth, L. (2007). *Women and the Making of Built Space in England, 1870–1950*. Aldershot: Ashgate.

David, D. (1987). *Intellectual Women and Patriarchy*. London: Macmillan.

Davies, M. (1947). Megalithic correlations with the Lusitanian flora. *The Geographical Journal*, **110**: 262.

Davies, M. (1951). *Wales in Maps*. Cardiff: University of Wales Press.

Davies, M. (1955). The open fields of Laugharne. *Geography*, **XL**: xx.

Davies, M. (1956). Rhossili open field and related South Wales filed patterns. *Agricultural History Review*, **IV**: 80–96.

Davies, M. (ed.) (1967). *Brecon Beacons National Park Guide*. London: HMSO.

Davies, M. (1979). *The Story of Tenby*. Tenby: Tenby Museum.

Davies, M. (1983). *Pembrokeshire Children in History*. Llandysul: Gomer Press.

De Clerq, P. (2007). The life and work of E.G.R. Taylor (1879–1966), author of *The Mathematical Practitioners of Tudor and Stuart England* and *The Mathematical Practitioners of Hanoverian England. Journal of the Hakluyt Society*.

Delano-Smith, C. (1967). Ancient landscapes of the Tavoliere, Apulia, *Transactions of the Institute of British Geographers*, **41**: 203–8.

Delano-Smith, C. (1979). *Western Mediterranean Europe. A Historical Geography of Italy, Spain and Southern France since the Neolithic.* London: Routledge.

Delano-Smith, C. & Kain, R.J.P. (1999). *English Maps: a History.* London: The British Library and Toronto University Press.

Deleuze G. & Guattari, F. (1988). New mappings in politics, philosophy, and culture. In E. Kaufman (ed.), *New Mappings in Politics, Philosophy, and Culture.* University of Minnesota Press, Minneapolis.

Diamond, D. (1995). Obituary: Joy Tivy 1924–1995. *The Geographical Journal*, **161**: 355–6.

Dibble, L.G. (1993). *Return Tickets in Pictures for Armchair Globetrotters.* Ilfracombe: Arthur H. Stockwell.

Dickinson, R.E. (1969). *The Makers of Modern Geography.* London: Routledge & Kegan Paul.

Dickinson, R.E. (ed.) (1976). *The Regional Concept. The Anglo-American Leaders.* London: Routledge & Kegan Paul.

Domosh, M. (1991a). Towards a feminist historiography of geography. *Transactions of the Institute of British Geographers*, **16**: 95–104.

Domosh, M. (1991b). Beyond the frontiers of geographical knowledge. *Transactions of the Institute of British Geographers*, **16**: 488–90.

Doornkamp, J.C. & King, C.A.M. (1971). *Numerical Analysis in Geomorphology: an Introduction.* London: Edward Arnold.

Dowson, J. (1995). Women's poetry: the thirties and the nineties. *Women: A Cultural Review*, **6**: 216–303.

Driver, F. (1991). Henry Morton Stanley and his critics: geography, exploration and empire. *Past and Present*, **133**: 134–66.

Driver, F. (1992). Geography's empire: histories of geographical knowledge. *Environment and Planning D: Society and Space*, **10**: 23–40.

Driver, F. (ed.) (1995). Geographical traditions: rethinking the history of geography. *Transactions of the Institute of British Geographers*, NS20: 403–22.

Driver, F. (2001). *Geography Militant. Cultures of Exploration and Empire.* Oxford: Blackwell.

Driver, F. & Rose, G. (1992). Introduction. Towards new histories of geographical knowledge. In F. Driver & G. Rose (eds), *Nature and Science: Essays in the History of Geographical Knowledge, Historical Geography Research Series 28* (pp. 1–7).

Dumayne-Peaty, L. & Wellens, J. (1998). Gender and physical geography in the United Kingdom. *Area*, **30**: 197–205.

Dunlop, M. (1939). Lines of cultural communication in Bronze Age France. *Geographical Review*, **29**: 274–90.

E.E.E. (1935). Review: *Europe. A Regional Geography. Geography*, **20**: 230.

Eaglesham, E.J.R. (1967). *The Foundation of Twentieth-Century Education in England.* London: Routledge & Kegan Paul.

Ellis, R. & Hunt, A. (1989). Obituary: Alice Garnett. *Geography*, **74**: 274–5.

Embleton, C. & King, C.A.M. (1968). *Glacial Geomorphology.* London: Edward Arnold.

Evans, D. (1986). Le Play House and the regional survey movement in British sociology 1920-1955, unpublished M.Phil. thesis. City of Birmingham Polytechnic.

F.S.C. (1935). Review: Isobel Wylie Hutchison, *North to the Rime-Ringed Sun: Being the Record of an Alaskan-Canadian Journey Made in 1933–34*, The Geographical *Journal*, **85**: 289.

Fairchild, W.B. (1976). Gladys Mary Wrigley 1885–1975. *Geographical Review*, **66**: 331–3.

Fara, P. (2004). *Pandora's Breeches: Women, Science and Power in the Enlightenment*. London: Pimlico.

Feaver, I. (1997). The Joint School Society. In W.G.V. Balchin (ed.), *The Joint School Story. The Seventy-Fifth Anniversary of the Establishment of the Joint School of Geography for Geographical Cooperation between the London School of Economics and Kings College London 1922–1997*. London: The Joint School Society.

Flemming, F. (1998) *Barrow's Boys*. London: Granta.

Fleure, H.J. (1932). Review: *France: A Regional and Economic Geography. The Geographical Journal*, **79**: 511–12.

Fleure, H.J. (1968). Obituary: Rachel Mary Fleming. *Geography*, **53**: 327–8.

Fleure, H.J. & Davies, M. (1970). *A Natural History of Britain* (revised edition). London: Fontana/Collins.

Fleure, H.J. & Davies, M. (1989) *A Natural History of Britain* (revised edition). London: Bloomsbury/ Collins.

Fleure, H.J. & Dunlop, M. (1942). Glendarragh circle and alignments. *Antiquities Journal*, **XXII**: xx.

Fleure, H.J. & Fleure, H.M. (1968). Alice Garnett. Honorary Secretary 1947–67. *Geography*, **LIII**: 93–5.

Foucault, M. (1972/1992). *The Archaeology of Knowledge*. London: Routledge.

Fox Keller, E. (1982). Feminism and science. In Keohane N. *et al.* (eds), *Feminist Theory. A Critique of Ideology*. Chicago: University of Chicago Press.

Fox Keller, E. (1984). *Reflections on Gender and Science*. New Haven: Yale University Press.

Frank, K. (1986). *A Voyager Out. The life of Mary Kingsley*. Boston: Houghton Mifflin.

Franklin Rawnsley, W. (ed.) (1923). *The Life, Diaries and Correspondence of Jane, Lady Franklin, 1792–1875*. London: Erskine MacDonald.

Fraser, N. & Nicholson, L.J. (1990). Social criticism without philosophy: an encounter between feminism and postmodernism. In L.J. Nicholson (ed.), *Feminism/ Postmodernism* (pp. 19–38). London: Routledge.

Freeman, T.W. (1954). Early developments in geography at Manchester University. *The Geographical Journal*, **CXX**: 118–19.

Freeman, T.W. (1961). *A Hundred Years of Geography*. London: Duckworth.

Freeman, T.W. (1966). *The Conurbations of Great Britain*. Manchester: Manchester University Press.

Freeman, T.W. (1976a). The *Scottish Geographical Magazine*: its first thirty years. *Scottish Geographical Magazine*, **92**: 92–100.

Freeman, T.W. (1976b). Two Ladies. *The Geographical Magazine*, **49**: 208.

Freeman, T.W. (1980). *A History of Modern British Geography*. London: Longman.

Freeman, T.W. (1984). The Manchester and Royal Scottish Geographical Societies. *The Manchester Geographical Journal*, **150**: 55–62.

Fuller, G.J. (1954). Geographical aspects of the development of Boston (Lincolnshire) between 1700 and 1900 AD. *East Midland Geographer*, 1: 3–13.

Fuller, G.J. (1955a). Settlement in Northamptonshire between 500 AD and Domesday. *East Midland Geographer*, 1: 25–36.

Fuller, G.J. (1955b). The Trient Valley: an introductory study. *Geography*, **XL**: 28–39.

Fuller, G.J. (1957). The development of drainage, agriculture and settlement in the Fens of south-east Lincolnshire during the nineteenth century. *East Midland Geographer*, 1: 3–15.

Fuller, G.J. (1959). Communications in the port of New York. *Geography*, **XLIV**: 128–30.

Fuller, G.J. (1965). Lead mining in Derbyshire in the mid-nineteenth century. *East Midland Geographer*, 3: 373–93.

Fuller, G.J. (1966). The Peak lead industry. In K.C. Edwards (ed.), *Nottingham and its Region*. Nottingham: British Association for the Advancement of Science (Local Executive Committee).

Fuller, G.J. (1970). Early lead smelting in the Peak District: another look at the evidence. *East Midland Geographer*, 5: 1–18.

Furley, P.A. (1989). Review: *The Savannas* by Monica Cole. *The Geographical Journal*, 155: 122–3.

Gardiner, J. (2004). *Wartime Britain 1939–45*. London: Review.

Garnett, A. (1928). The capitals of Morocco. *Scottish Geographical Magazine*, 44: 31–41.

Garnett, A. (1930). *The Geographical Interpretation of Topographical Maps*. London: George Harrap.

Garnett, A. (1935). Insolation, topography and settlement in the Alps. *The Geographical Review*, 25: 601–17.

Garnett, A. (1937). Insolation and relief: their bearing on the human geography of Alpine regions. *Transactions of the Institute of British Geographers*, 4/5: 1–71.

Garnett, A. (1939). Diffused light and sunlight in relation to relief and settlements in high latitudes. *Scottish Geographical Magazine*, 55: 271–84.

Garnett, A. (1940). An unusual fall of ice crystals. *Quarterly Journal of the Royal Meteorological Society*.

Garnett, A. (1945a). Air pollution: geographical factors. *Air Pollution*, 73–113.

Garnett, A. (1945b). Climate, relief and atmospheric pollution in the Sheffield region. *Advancement of Science*, 13: 331–41.

Garnett, A. (1945c). The Loess regions of Central Europe in Prehistoric times. *The Geographical Journal*, 106: 132–43.

Garnett, A. (1945d). Nauczanie geografi w wielkiej Brytanii. *GEografi w Szkole* Warsaw, 6: 314–21.

Garnett, A. (1956). Climate. In *Sheffield and its Region: a Scientific and Historical Survey*. Sheffield: BAAS.

Garnett, A. (1962a). Obituary: Elsa Stamp. *Geography*, 47: 423.

Garnett, A. (1962b). From Pennine high peak to the Humber. In J.B. Mitchell (ed.), *Great Britain: Geographical Essays* (pp. 330–53). Cambridge: Cambridge University Press.

Garnett, A. (1963). Survey of air pollution under characteristic winter and anti-cyclonic conditions. *International Journal of Air and Water Pollution*, 7: 963–8.

Garnett, A. (1967). Some climatological problems in urban geography with reference to air pollution. *Transactions of the Institute of British Geographers*, 42: 21–43.

Garnett, A. (1969). Teaching geography: some reflections. *Geography*, 54: 385–400.

Garnett, A. (1983). The IBG. The formative years – some reflections. *Transactions of the Institute of British Geographers*, NS 8: 27–35.

Garnett, A. (1987). The pioneers: some reminiscences. *Transactions of the Institute of British Geographers*, NS 12: 240–1.

Garnett, A., Pead, P. & Finch, D. (1976). The use of conversion factors in air pollution studies. *Atmospheric Environment*, 10: 325–78.

Garnett, J. (1995). Women and gender in the *New DNB*. *Oxford DNB Newsletter*, 1: 2–3.

Garnett, O. (1934, 1965). *Fundamentals in School Geography*. London: Harrap.

Garnett, O. (1966). Obituary: J.F. Unstead. *Geography*, 51: 151–3.

Gedalof, I. (1996). Can nomads learn to count to four? Rosi Braidotti and the Space for Difference in Feminist Theory. *Women: A Cultural Review*, 7(2): 189–201.

Gedalof, I. (2000). Identity in transit: nomads, cyborgs and women. *European Journal of Women's Studies*, 7: 337–54.

Geniesse, J.F. (2000). *Freya Stark. Passionate Nomad*. London: Pimlico.

George, F. (1968). Review: C. Embleton and C.A.M. King. *The Geographical Journal*, 134: 589–90.

Gergits, J. (1996). Isabella Lucy Bird. In B. Brothers & J. Gergits (eds), *Dictionary of Literary Biography: British Travel Writers 1837–1875* (pp 29–49). London: Brucoli Clark Layman.

Gibb, L. (2005). *Lady Hester. Queen of the East*. London: Faber & Faber.

Gilbert, E.W. (1965). Andrew John Herbertson 1865–1915. An appreciation of his life and work. *Geography*, 50: 313–31.

Gilbert, E.W. (1970). Review: *The Rural Landscape of the Welsh Borderland: a Study in Historical Geography. The Geographical Journal*, 136: 260–1.

Giles, J. (1987). Obituary. Beatrice Saward. *Geography*.

Glacken, C.J. (1962). Review: *Friedrich Ratzel* by Harriet Wanklyn. *The Geographical Review*, 52: 467–8.

Glick, T. (1984). History and philosophy of geography. *Progress in Human Geography*, 8: 275–83.

Goddard, J. (1970). Review: J.P. Cole and C.A.M. King, *Quantitative Geography. The Geographical Journal*, 136: 141–3.

Godlewska, A. (1999). *Geography Unbound. French Geographic Science from Cassini to Humboldt*. Chicago: Chicago University Press.

Godlewska, A. & Smith, N. (eds) (1994) *Geography and Empire*. Oxford: Blackwell.

Goffman, E. (1959). *The Presentation of Self in Everyday Life*. New York: Anchor.

Goodman, S. (1985). *Gertrude Bell*. Leamington Spa: Berg.

Goudie, A.S. (1995). Obituary. Marjorie Sweeting. *The Geographical Journal*, 161: 239–40.

Goudie, A.S. (2001). Obituary: Jean Grove. *The Independent*, 12 March 2001.

Grace, J. (1974). Review: *Organic Resources of Scotland. Journal of Applied Ecology*, 11: 821–2.

Graham, E. (1995). Postmodernism and the possibility of a new human geography. *Scottish Geographical Magazine*, **111**: 175–8.

Graves, N. (1997). Geographical education in the 1990s. In D. Tilbury & M. Williams (eds), *Teaching and Learning in Geography*. London: Routledge.

Gregory, K.J. (1985). *The Nature of Physical Geography*. London: Edward Arnold.

Grenville, J.A.S. & Fuller, G.J. (1962). *The Coming of the Europeans*. London: Longman.

Groom, G.E. (1959). Niche glaciers in Bünsow Land Vestspitsbergen. *Journal of Glaciology*, **3**.

Groom, G.E. & Williams, V.H. (1965). The solution of limestone in South Wales. *The Geographical Journal*, **131**: 37–41.

Grove, J. (1966). The Little Ice Age in the Massif of Mont Blanc. *Transactions of the Institute of British Geographers*, **40**: 129–43.

Grove, J. (1972). The incidence of landslides, avalanches and floods in western Norway during the Little Ice Age. *Arctic and Alpine Research*, **4**: 131–8.

Grove, J. (1979). The glacial history of the Holocene. *Progress in Physical Geography*, **3**: 1–54.

Grove, J. (1988). *The Little Ice Age*. London: Routledge Keegan & Paul.

Grove, J. (2004). *Little Ice Ages: Ancient and Modern*. London: Routledge.

Guelke, J.K. & Morin, K. (2001). Gender, nature, empire: women naturalists in nineteenth-century British travel literature. *Transactions of the Institute of British Geographers*, **26**: 306–26.

Gwynn, S. (1933). *The Life of Mary Kingsley*. London: Macmillan.

H.C.K.H. (1933). Review: Southern Europe. *Geographical Journal*, **81**: 70.

H.R.W. (1953). Review: *The Geographical Interpretation of Topological Maps* by Alice Garnett. *The Geographical Journal*, **119**: 491–2.

H.T. (1963). Review: Darby and Smee. *Geography*.

Hamilton (1848). *Journal of the Royal Geographical Society*, **18**: xl–xli.

Hansen, P.H. (2004). Freya Stark. *Oxford Dictionary of National Biography*. Oxford: Oxford University Press.

Haraway, D. (1989). *Primate Visions. Gender, Race and Nature in the World of Modern Science*. London: Routledge.

Haraway, D. (1991). *Simians, Cyborgs and Women: The Reinvention of Nature*. London: Routledge.

Harding, S. (1986). *The Science Question in Feminism*. Milton Keynes: Open University Press.

Harding, S. (ed.) (1987). *Feminism and Methodology: Social Science Issues*. Milton Keynes: Open University Press.

Harding, S. (1991). *Whose Science? Whose Knowledge? Thinking from Women's Lives*. Milton Keynes: Open University Press.

Harris, A. (1978). Geography at Hull: the early years. In D Symes (ed.), *Northumberside – Introductory Themes*. Hull: University of Hull Department of Geography.

Harrison Church, R.J. (1981). Hilda Ormsby 1877–1973. In T.W. Freeman (ed.), *Geographers' Bio-bibliographical Studies* (pp. 95–7). London: International Geographical Union.

Harrison Church R.J. *et al.* (1967). *An Advanced Geography of Northern and Western Europe*. London: Hulton.

Heffernan, M. (1996). Geography, cartography and military intelligence: The Royal Geographical Society and the First World War. *Transactions of the Institute of British Geographers*, 21, 504–33.

Heffernan, M. (2003). Histories of geography. In S. Holloway, S.P. Rice & G. Valentine (eds), *Key Concepts in Geography* (pp. 3–22). London: Sage.

Henshall, J.D. (1966). The demographic factor in the structure of agriculture in Barbados. *Transactions of the Institute of British Geographers*, 38: 183–95.

Henshall, J.D. (1967). Models of agricultural activity. In P. Haggett & R.J. Chorley (eds), *Models in Geography* (pp. 425–58). London: Methuen.

Herbertson, A.J. & Herbertson, F.D. (1899). *Man and His Work. An Introduction to Human Geography*. London: Adam Black.

Herbertson A.J., MacMunn, N.E. & Unstead, F.J. (eds) (1909). *Guide to Geographical Books and Appliances*. London: Geographical Association.

Herbertson, F.D. (1914). *The Three Southern Continents*. Oxford: Clarendon Press.

Herbertson, F.D. (1950). *The Life of Frederic Le Play*. Ledbury: Le Play Society.

Herries Davies, G.L. (2004). Our first sixty years: one editor remembers, *Irish Geography*, 37: 6–14.

Oxford University Press.

Heward, C. & Sinclair Taylor, A. (1995). Women into management workshops: shattering the glass ceiling or raising the stone floor? *Journal of Teacher Development*, 4(2).

Hickman, G.M. (1978). *The New Africa*, 2nd edition. London: London University Press.

Hickman, G.M. (1983). *Introducing the New China*. London: Hodder & Stoughton.

Hickman, G.M. & Dickins, W. (1960). *Lands and Peoples of East Africa*. Harlow: Longman.

Hickman, G.M. & Mayo, R.E. (1961). *Adventuring Abroad. Pilgrim Way Geographies Book 2*. London: Blackie.

Hickman, G.M., Reynolds, J. & Tolley, H. (1973). *A New Professionalism for a Changing Geography*. Bristol: Schools Council Geography 14–18 Project.

Hilling, D. (1994). Geography. In J. Morduant Crook (ed.), *Bedford College, University of London, Memories of 150 Years* (pp. 131–47). Egham: Royal Holloway and Bedford New College.

Holloway, G. (2005). *Women and Work since 1840*. London: Routledge.

Holloway, S., Rice, S.P. & Valentine, G. (eds) (2003). *Key Concepts in Geography*. London: Sage.

Holt, D.N. (1973). Review *Malawi in Maps. The Geographical Journal*, 139: 181–2.

Hoyle, G. (2001). *Flowers in the Snow. The Life of Isobel Wylie Hutchison*. Lincoln: University of Nebraska Press.

Hubbard, P. *et al.* (2004). *Key Thinkers in Geography*. London: Sage.

Hudson, B. (1977). The new geography and the new imperialism. *Antipode*, 9: 12–19.

Ihse, M. (1990). Review: *Agricultural Ecology* by Joy Tivy. *Geografiska Annaler*, 72**B**: 37–8.

Ives, J. (1977). Review: C. Embleton and C.A.M. King, *Glacial Geomorphology. Annals of the Association of American Geographers*, 67: 151–2.

J.A.S. (1938). Review: 'Insolation and relief: their bearing on the human geography of Alpine Regions' by Alice Garnett. *The Geographical Journal*, 91: 63–4.

J.G.T. (1952). Review: *Wales in Maps. The Geographical Journal*, 118: 215–6.

Jackson, S. (1994). Obituary: Monica Mary Cole 1922–1994. *The Geographical Journal*, **160**: 244–5.

Jacobus, M.E. *et al.* (1990). *Body/Politics. Women and the Discourse of Science.* New York: Routledge.

Jay, L.J. (1979). Andrew John Herbertson, 1865–1915. In T.W. Freeman & P. Pinchemel (eds), *Geographers: Biobibliographical Studies*, **3**: 85–92.

Joannou, M. (2005). Mary Augusta Ward (Mrs Humphry) and the opposition to women's suffrage. *Women's History Review*, **14**: 561–80.

Johnson F.R. (1935). Review: E.G.R. Taylor, *Tudor Geography*, and *Late Tudor and Early Stuart Geography. Isis*, **23**: 289–94.

Johnson, F.R. (1957). Review: E.G.R. Taylor, *The Mathematical Practitioners of Tudor and Stuart England. Isis*, **48**: 377–8.

Johnson, H. (1895). *The Life of Kate Marsden*. London.

Johnston, R.J. (2006). The politics of changing human geography's agenda: textbooks and the representation of increasing diversity. *Transactions of the Institute of British Geographers*, NS31: 286–303.

Johnston, R.J. & Brack, E.V. (1983). Academic appointment and promotion in the academic labour market: a preliminary survey of British University Departments of Geography, 1933–1982. *Transactions of the Institute of British Geographers*, NS8: 100–11.

Johnston, R.J. & Sidaway, J.D. (2004). *Geography and Geographers. Anglo-American Geography since 1945*, 6th edition. London: Arnold.

Johnston, R.J. & Williams, M. (eds) (2003). *A Century of British Geography*. Oxford: The British Academy/Oxford University Press.

Jordan, E. (1999). *The Women's Movement and Women's Employment in Nineteenth-Century Britain*. London: Routledge.

Kain, R. & Delano-Smith, C. (2003). Geography displayed. Maps and mapping. In R. Johnston & M. Williams (eds), *A Century of British Geography* (pp. 371–428). Oxford: The British Academy/Oxford University Press.

Kanner, B.P. *et al.* (1997). *Two Hundred Years of British Women's Autobiographies. A Reference Guide and Reader*. London: Prentice-Hall.

Kaye, E. (2002). Review: Gwyneth Hoyle, *Flowers in the Snow: The Life of Isobel Wylie Hutchison. Arctic, Antarctic, and Alpine Research*, **34**: 114–15.

Kearns, G. (1997). The imperial subject: geography and travel on the work of Mary Kingsley and Halford Mackinder. *Transactions of the Institute of British Geographers*, **22**: 450–72.

Keay, J. (1989). *With Passport and Parasol*. London: BBC.

Keighren, I.M. (2006). Bringing geography to the book: charting the reception of influences of geographic environment. *Transactions of the Institute of British Geographers*, **31**: 525–40.

Keltie, J.S. (1885). *Geographical Education. Report to the Council of the Royal Geographical Society*. London: John Murray.

Kennedy, B.A. (1979). A naughty world. *Transactions of the Institute of British Geographers*, NS4: 550–8.

Kennedy, B.A. (1995). Marjorie Sweeting. *The Independent*, 18 January 1995.

Kennedy, B.A. (2006). *Inventing the Earth: Ideas on Landscape Development since 1740*. Oxford: Blackwell.

King, C.A.M. (1959). *Beaches and Coasts*. London: Edward Arnold.

King, C.A.M. (1960). The Yorkshire Dales. *British Landscapes through Maps* 7. The Geographical Association.

King, C.A.M. (1962). *Oceanography for Geographers*. London: Edward Arnold.

King, C.A.M. (1964). The character of the off-shore zone and its relationship to the foreshore near Gibraltar Point, Lincolnshire. *East Midland Geographer*, 3: 230–43.

King, C.A.M. (1966). *Techniques in Geomorphology*. London: Edward Arnold.

King, C.A.M. (1968). Beach measurements at Gibraltar Point, Lincolnshire. *East Midland Geographer*, 4: 295–300.

King, C.A.M. (1969). Moraine types on Henry Kater Peninsula, East Baffin Island, N.W.T., Canada. *Arctic and Alpine Research*, 1: 289–94.

King, C.A.M. (1975a). *Introduction to Marine Geology and Geomorphology*. London: Edward Arnold.

King, C.A.M. (1975b). *Introduction to Physical and Biological Oceanography*. London: Edward Arnold.

King, C.A.M. (1976). *Northern England*. London: Methuen.

King, C.A.M. (ed.) (1976a). *Landforms and Geomorphology: Concepts and History*, Benchmark Papers in Geology No. 28. Stroudsburg, PN: Dowden, Hutchinson & Ross, Inc.

King, C.A.M. (ed.) (1976b). *Periglacial Processes*, Benchmark Papers in Geology No. 27. Stroudsburg, PN: Dowden, Hutchinson & Ross, Inc.

King, C.A.M. (1978). Changes on the foreshore and the spit between 1972 and 1978 near Gibraltar Point, Lincolnshire. *East Midland Geographer*, 7: 73–82.

King, C.A.M. & Buckley, J.T. (1969). Geomorphological investigations in West-Central Baffin Island, N.W.T., Canada. *Arctic and Alpine Research*, 1: 105–19.

Kingsley, M.H. (1897). West Africa from an ethnologist's point of view. *Transactions of the Liverpool Geographical Society*, xx.

Kingsley, M.H. (1897/1982). *Travels in West Africa, Congo Français, Corisco and Cameroons*. London: Virago.

Kingsley, M.H. (1899). *The Story of West Africa*. London: Horace Marshall & Son.

Kirk, E. (2003). Women academics at Royal Holloway and Bedford Colleges, 1939–69. *Historical Research*, 76: 128–50.

Kirwan, L.P. (2004). Gertrude Caton-Thompson. *Oxford Dictionary of National Biography*. Oxford: Oxford University Press.

Kleinberg, S.J. (1992). Introduction. In S.J. Kleinberg (ed.), *Retrieving Women's History. Changing the Perceptions of the Role of Women in Politics and Society* (pp. ix–xi). Oxford: Berg/UNESCO.

Knox, J.C. (1973). Review: R.J. Chorley and B.A. Kennedy, *Physical Geography. A Systems Approach. Annals of the Association of American Geographers*, 63: 556–7.

Kobayashi, A. (1995). Review: *The Geographical Tradition* by David Livingstone. *Annals of the Association of American Geographers*, 85: 192–4.

Kuchler, A.W. (1973). Review of *Biogeography: a Study of Plants in the Ecosphere* by Joy Tivy. *Geographical Review*, 63: 595–7.

Kwan, M.-P. (2004). Beyond difference: from canonical geography to hybrid geographies. *Annals of the Association of American Geographers*, **94**, 756–63.

L.W.L. (1921). Review: *Aftermath* by M.I. Newbigin. *The Geographical Journal*, **58**: 137–8.

Lambert, A. (1971). *The Making of the Dutch Landscape. An Historical Geography of the Netherlands*. London: Seminar Press.

Lewis, V. & Sweeting, M.M. (1959). Correspondence. The karstlands of Jamaica: cockpits or rounded hills? *The Geographical Journal*, **125**: 289–91.

Light, R.U. (1950). Gladys Mary Wrigley. *Geographical Review*, **40**: 4–6.

Linton, D.L. (1968). Professor Alice Garnett. *University of Sheffield Gazette*, **48**: 67–8.

Liu, T. (1991). Teaching the differences among women from a historical perspective. Rethinking race and gender as social categories. *Women's Studies International Forum*, **4**: 265–76.

Livingstone, D.N. (1979). Some methodological problems in the history of geographical thought. *TESG*, **70**: 226–31.

Livingstone, D.N. (1990). Geography, tradition and the Scientific Revolution: an interpretative essay. *Transactions of the Institute of British Geographers*, NS15: 359–73.

Livingstone, D.N. (1991). The moral discourse of climate: historical consideration on race, place and virtue. *Journal of Historical Geography*, **17**: 413–34.

Livingstone, D.N. (1992). *The Geographical Tradition*. Oxford: Blackwell.

Livingstone, D.N. (1995). Conversing with my critics. *Scottish Geographical Magazine*, **111**: 196–8.

Livingstone, D.N. (2005). Science, text and space: thoughts on the geography of reading. *Transactions of the Institute of British Geographers*, **30**: 391–401.

Lloyd, A. (ed.) (1992). *Manchester Geographical Society Library Catalogue*. Manchester: Manchester Geographical Society.

Lochhead, E. (1981). Scotland as the cradle of modern academic geography in Britain. *Scottish Geographical Magazine*, **97**: 98–109.

Lochhead, E. (1984). The Royal Scottish Geographical Society: the setting and sources of its success. *Scottish Geographical Magazine*, **100**: 12–22.

Long, I.L.M. (1953). Children's reactions to geographical pictures. *Geography*, **38**: 100–7.

Long, I.L.M. (1961). Research in picture study: the reaction of grammar school pupils to geographical pictures. *Geography*, **46**: 322–37.

Long, M. (ed.) (1964). *Handbook for Geography Teachers*, revised 5th edition. London: Methuen.

Long, M. & Roberson, B.S. (1966). *Teaching Geography*. London: Methuen.

Lorimer, H. (2003). Talking small stories: spaces of knowledge and the practice of geography. *Transactions of the Institute of British Geographers*, **28**: 197–217.

Lorimer, H. & Spedding, N. (2002) Excavating geography's hidden spaces. *Area*, 294–302.

Lukitz, L. (2004). Gertrude Lowthian Bell. *Oxford Dictionary of National Biography*. Oxford: Oxford University Press.

Lyde, L.W. (1915). Balkan geography and politics. *Geographical Journal*, **46**: 386.

MacGreggor, D.R. (1975). Catherine Park Snodgrass. *Scottish Geographical Magazine*, **91**: 128–9.

MacKenzie, J. (ed.) (1990). *Imperialism and the Natural World*. Manchester: Manchester University Press.

MacKenzie, J. (1992). Geography and imperialism: British provincial geographical societies. *Historical Geography Research Papers*, 28: 49–62.

MacKenzie, J. (1995). The provincial geographical societies in Britain, 1884–1918. In M. Bell *et al.* (eds.), *Geography and Imperialism 1820–1940* (pp. 93–124). Manchester: Manchester University Press.

Mackinder, H.J. (1887). On the scope and methods of geography. *Proceedings of the Royal Geographical Society*, 9: 141–60.

MacMunn, N.E. (1906). The areas of the orographical regions of England and Wales. *The Geographical Journal*, 27: 288–91.

MacMunn, N.E. (1907–8). The economic geography of a county. Illustrated from Essex and Cumberland. *The Geographical Teacher*, 4: 29–38.

MacMunn, N.E. (1913). *The Upper Thames Country and the Severn-Avon Plain*. Oxford: Clarendon Press.

MacMunn, N.E. (1915–16). The French and German borderlands. Geographical aspects of the political history. *The Geographical Teacher*, 8: 312–17.

MacMunn, N.E. & Coster, G. (1922). *Europe. A Regional Geography*. Oxford: Clarendon Press.

Maddrell, A. (1996). Empire, emigration and school geography: changing discourses of Imperial citizenship, 1880–1925. *Journal of Historical Geography*, 22: 373–87.

Maddrell, A. (1997). Marion Newbigin and the scientific discourse. *Scottish Geographical Magazine*, 113: 33–41.

Maddrell, A. (1998). Discourses of race and gender in school geography texts 1830–1918. *Environment and Planning D. Society and Space*, 16: 81–103.

Maddrell, A. (2004a). Complex locations. Historiography, feminism and difference. In J. Sharp & D. Thien (eds), *Geography and Gender Reconsidered*. London: RGS-IBG Women and Geography Study Group.

Maddrell, A. (2004b). Marion Newbigin. *Dictionary of Nineteenth-Century British Scientists*. Bristol: Thoemmes Press.

Maddrell, A. (2004c). Mary Somerville. *Dictionary of Nineteenth-Century British Scientists*. Bristol: Thoemmes Press.

Maddrell, A. (2004d). Isabella Bird. *Dictionary of Nineteenth-Century British Scientists*. Bristol: Thoemmes Press.

Maddrell, A. (2004e). Hilda Ormsby. *Oxford Dictionary of National Biography*. Oxford: Oxford University Press.

Maddrell, A. (2004f). Alice Garnett. *Oxford Dictionary of National Biography*. Oxford: Oxford University Press.

Maddrell, A. (2004g). Charlotte Cameron. *Oxford Dictionary of National Biography*. Oxford: Oxford University Press.

Maddrell, A. (2004h). Violet Cressy-Marcks. *Oxford Dictionary of National Biography*. Oxford: Oxford University Press.

Maddrell, A. (2006). Revisiting the region: 'ordinary' and 'exceptional' regions in the work of Hilda Ormsby 1917–1940. *Environment and Planning A*, 38: 1739–52.

Maddrell, A. (2007). Teaching a contextual history of geography through role play: women's membership of the Royal Geographical Society (1892–3). *Journal of Geography in Higher Education*, 31: 393–412.

Maddrell, A. (2008). The 'Map Girls'. Women geographers' war work 1939–45: shifting boundaries of gender and reflections on the history of geography. *Transactions of the Institute of British Geographers*, **33**: 127–48.

Maddrell, A. (forthcoming 2009). The history and philosophy of geography. In R. Kitchen & N. Thrift (eds), *The International Dictionary of Human Geography*. Elsevier.

Maddrell Mander, A. (1995). Geography, gender and the state. A critical evaluation of the development of geography 1830–1918. Unpublished DPhil thesis, University of Oxford.

Maguire, S. (1998). Gender differences in attitudes to undergraduate fieldwork. *Area*, **30**: 207–14.

Marcus, M.G. (1969) Review: C. Embleton and C.A.M. King, *Glacial Geomorphology*. *Geographical Review*, **59**: 635–36.

Margary, H. (1995). *Imago Mundi* saved by Eila Campbell. *Imago Mundi*, **47**: 8–9.

Markham, C. (1881). *Fifty Years' Work of the Royal Geographical Society*. London: John Murray.

Marlow, J. (ed.) (2001). *Votes for Women. The Virago Book of Suffragettes*. London: Virago.

Marsden, K. (1893). *On Sledge and Horseback to Outcast Siberian Lepers*. London: Record Press.

Marsden, K. (1921). *My Mission to Siberia: a Vindication*. London: Edward Stanford.

Marsden, W. (1999). A British historical perspective on geographical fieldwork from the 1820s to the 1970s. In R. Gerber & G.K. Chuan (eds), *Fieldwork in Geography. Reflections, Perspectives and Actions* (pp. 15–36). Springer.

Martin, G.J. (2005). *All Possible Worlds*. Oxford: Oxford University Press.

Massey, D. (1971). Towards operational urban development models. In R. Chisholm, A.E. Frey & P. Haggett (eds), *Regional Forecasting*. London: Butterworths.

Massey, D. (2005). *For Space*. London: Sage.

Massey, D. (2007). *World City*. Cambridge: Polity.

Massey, D. & Catalano, A. (1978). *Capital and Land: Landownership by Capital in Great Britain*. London: Edward Arnold.

Matless, D. (1992). An occasion for geography: landscape, representation and Foucault's corpus. *Environment and Planning D. Society and Space*, **10**: 41–56.

Matless, D. (1998). *Landscape and Englishness*. London: Reaktion Books.

Matless, D. (1999). Editorial. *East Midland Geographer*, **22**: 3–4, 71–2.

Matless, D. & Cameron, L. (2006) Experiment in landscape: the Norfolk excavations of Marietta Pallis. *Journal of Historical Geography*, **32**: 96–126.

Mayall, L. (2005). Creating the 'suffragette spirit': British feminism and the historical imagination. In A. Burton (ed.), *Archive Stories. Facts, Fictions and the Writing of History* (pp. 232–50). Durham: Duke University Press.

McConnell, A. (2004). Eila Campbell. *Oxford Dictionary of National Biography*. Oxford: Oxford University Press.

McDowell, L. (1979). Women in British geography. *Area*, **11**: 151–4.

McDowell, L. (1993a). Space, place and gender relations: Part I. Feminist empiricism and the geography of social relations. *Progress in Human Geography*, **17**: 157–79.

McDowell, L. (1993b). Space, place and gender relations: Part II. Identity, difference, feminist geometries and geographies. *Progress in Human Geography*, 17: 305–18.

McDowell, L. & Peake, L. (1990). Women in British geography revisited – or the same old story. *Journal of Geography in Higher Education*, 14: 19–30.

McEwan, C. (1995). 'The Mother of all the peoples': geographical knowledge and the empowering of Mary Slessorin. In M. Bell *et al.* (eds), *Geography and Imperialism 1820–1940* (pp. 125–50). Manchester: Manchester University Press.

McEwan, C. (1998a). Cutting power lines within the palace? Countering paternity and eurocentrism in the 'geographical tradition'. *Transactions of the Institute of British Geographers*, 23: 371–84.

McEwan, C. (1998b). Gender, science and physical geography in nineteenth-century Britain. *Area*, 30: 215–24.

McKittrick, K. (2006). *Demonic Grounds: Black Women and the Cartographies of Struggle*. Minneapolis: University of Minnesota Press.

Mead, W.R. (1990). Obituary: Harriet Grace Steers 1906–1990. *Transactions of the Institute of British Geographers*, 156: 353.

Mead, W.R. (1995). Eila as geographer. *Imago Mundi*, 47: 7–8.

Meller, H. (1990). *Patrick Geddes. Social Evolutionist and City Planner*. London: Routledge.

Middleton, D. (1965). *Victorian Lady Travellers*. London: Routledge & Kegan Paul.

Middleton, D. (1980). Travel literature in the Victorian and Edwardian eras. *Hakluyt Society Annual Report*, London.

Middleton, D. (1993). Obituary: Freya Stark. *The Geographical Journal*, 159: 368–70.

Middleton, D. (2004a). Jane Franklin. *Oxford Dictionary of National Biography*. Oxford: Oxford University Press.

Middleton, D. (2004b). Isabella Bishop (née Bird). *Oxford Dictionary of National Biography*. Oxford: Oxford University Press.

Middleton, D. (2004c). Kate Marsden. *Oxford Dictionary of National Biography*. Oxford: Oxford University Press.

Mill, H.R. (1930). *Record of the Royal Geographical Society 1830–1930*. London: Royal Geographical Society.

Mill, H.R. (1951). *Hugh Robert Mill. An Autobiography*. London: Longman.

Miller, F.C. (1921–2). Early maps of China and of the Mediterranean. *The Geographical Teacher*, 11: 296–300.

Miller, F.C. (1952). Obituary: Miss K.C. Boswell. *Geography*, **XXXVII**: 236.

Mills, S. (1991). *Discourses of Difference. An Analysis of Women's Travel Writing and Colonialism*. London: Routledge.

Mitchell, J.B. (1933). The Matthew Paris maps. *The Geographical Journal*, 81: 27–34.

Mitchell, J.B. (1938). The growth of Cambridge. In H.C. Darby (ed.), *The Cambridge Region*. Cambridge: Cambridge University Press.

Mitchell, J.B. (1954). *Historical Geography*. London: English Universities Press.

Mitchell, J.B. (1962). Population. In J.B. Mitchell (ed.), *Great Britain: Geographical Essays* (pp. 33–54). Cambridge: Cambridge University Press.

Mitchell, J.B. (ed.) (1962). *Great Britain: Geographical Essays*. Cambridge: Cambridge University Press.

Mitchell, R. (1995). A monstrous regiment of women: adding nineteenth-century women to the *New DNB*. *Oxford DNB Newsletter*, 1: 3–4.

Mohanty, C. (1987). 'Under western eyes': feminist scholarship and colonial discourses. *Feminist Review*, 30: 61–88.

Moi, T. (1985). *Sexual/Textual Politics: Feminist Literary Theory*. London: Methuen.

Momsen, J. & Townsend, J. (eds) (1987). *Geography of Gender in the Third World*. London: Hutchinson.

Monk, J. (1998). The women were always welcome at Clark. *Economic Geography*, 74: 14–30.

Monk, J. (2003). Women's worlds at the American Geographical Society. *Geographical Review*, 93: 237–57.

Monk, J. (2004). Women, gender and the histories of American geography. *Annals of the Association of American Geographers*, 94: 1–22.

Monk, J. (2007). Changing expectations and institutions: American women geographers in the 1970s. *Geographical Review*, 96: 259–77.

Monk, J. & Hanson, S. (1982). On not excluding half of the human race in human geography. *Professional Geographer*, 34: 11–23.

Moodie, A.E., Fuller, G.J., Cole, M.M. & Butland, G.J. (1955). The Upper Soča Valley. *Geographical Studies*, 2.

Morantz-Sanchez, R. (1995). *Gender and History*. Oxford: Blackwell.

Morison, S.E. (1958). Review: E.G.R. Taylor, *The Haven-Finding Art*, *Isis*, 49: 352–3.

Morris, M. & O'Connor, L. (eds) (1996). *The Virago Book of Women Travellers*. London: Virago.

Morris, R.J. (1990). Clubs, societies and associations. In F.M.L. Thompson (ed.), *The Cambridge Social History of Britain 1750–1950. Volume 3 Social Agencies and Institutions* (pp. 395–444). Cambridge: Cambridge University Press.

Murchison, R.I. (1897). Address to Royal Geographical Society. *Journal of the Royal Geographical Society*, 27: xciv–cxviii.

Mutton, A. (1938). The Black Forest: its human geography. *Economic Geography*, 14: 131–53.

Mutton, A. (1951). Hydroelectric power in Western Europe. *The Geographical Journal*, 117: 328–42.

Mutton, A. (1953). Hydroelectric power in Norway. *Transactions of the Institute of British Geographers*, 19: 123–30.

Mutton, A. (1961). *Central Europe. A Regional and Human Geography*. London: Longman.

Naish, G.P.B. (1957). Review: E.G.R. Taylor, *The Haven-Finding Art*. *The Geographical Journal*, 123: 85–6.

Neeley, K.A. (2001). *Mary Somerville: Science, Illumination and the Female Mind*. Cambridge: Cambridge University Press.

Newbigin, M.I. (1898). *Colour in Nature. A Study in Biology*. London: John Murray.

Newbigin, M.I. (1901). *Life by the Sea-shore*. London: Allen and Unwin.

Newbigin, M.I. (1905). The value of geography. *Scottish Geographical Magazine*, XXI: 1–4.

Newbigin, M.I. (1906). The Kingussie district: a geographical study. *Scottish Geographcial Magazine*, **XXII**: 285–315.

Newbigin, M.I. (1907). The study of weather as a branch of nature knowledge. *Scottish Geographical Magazine*, **XXIII**: 627–48.

Newbigin M.I. (1911a) *Modern Geography*. London: Williams & Norgate.

Newbigin, M.I. (1911b). *Ordnance Survey Maps. Their Meaning and Use*. London.

Newbigin, M.I. (1911c). *Tillers of the Ground*. London: Macmillan .

Newbigin, M.I. (1912). *Man and His Conquest of Naure*. London: A&C Black.

Newbigin, M.I. (1913). *Animal Geography. The Faunas of the Natural Regions of the World*. Oxford: Clarendon Press.

Newbigin, M.I. (1914). *The British Empire Beyond the Seas*. London: Bell.

Newbigin, M.I. (1915a). *Geographical Aspects of Balkan Problems*. London: Constable.

Newbigin, M.I. (1915b). The Balkan peninsula: its peoples and its problems. *Scottish Geographical Magazine*, **XXXI**: 281–303.

Newbigin, M.I. (1916). The geographical treatment of rivers. *Scottish Geographical Magazine*, **XXXII**: 57–69.

Newbigin, M.I. (1917). Race and nationality [with transcript of following discussion]. *The Geographical Journal*, **L**: 313–35.

Newbigin, M.I. (1918). The origin and maintenance of diversity in man. *Geographical Review*, **VI**: 411–20.

Newbigin, M.I. (1920a). *Aftermath. A Geographical Study of the Peace Terms*. London: Johnston.

Newbigin, M.I. (1920b). *A New Geography of Scotland*. London: H. Russell.

Newbigin, M.I. (1922). *Frequented Ways*. London: Constable William & Norgate.

Newbigin, M.I. (1924). *Commercial Geography*. London.

Newbigin, M.I. (1924a). *The Mediterranean Lands. An Introductory Study in Human and Historical Geography*. London: Christophers.

Newbigin, M.I. (1927). *Canada: the Great River, the Lands and the Men*. London: Christophers.

Newbigin, M.I. (1929). *A New Regional Geography of the World*. London: Christopers.

Newbigin, M.I. (1932). *Southern Europe. A Regional and Economic Geography of Mediterranean Lands*. London: Methuen.

Newbigin, M.I. (1936). *Plant and Animal Geography*, London: Metheun. [posthumous].

Nichols, P. (1997). A student's eye view. In W.G.V. Balchin (ed.), *The Joint School Story. The Seventy-Fifth Anniversary of the Establishment of the Joint School of Geography for Geographical Cooperation between the London School of Economics and King's College London 1922–1997* (pp. 63–5). London: The Joint School Society.

Nigel, T. & Brown, L. (1971). *The History of the Manchester Geographical Society 1884–1950*. Manchester: Manchester University Press.

O'Brien, R. (2000). *Gertrude Bell. The Arabian Diaries 1913–1914*. New York: Syracuse University Press.

Ogilvie, A.G. (ed.) (1928). *Great Britain: Essays in Regional Geography by Twenty-Six Authors*. Cambridge: Cambridge University Press.

Ormsby, H. (1923). The Danube as a waterway. *Scottish Geographical Magazine*, **34**: 103–11.

Ormsby, H. (1924). *London on the Thames*. London: Sifton-Praed.

Ormsby, H. (1927). Regional survey in a large city. *Geography*, **14**: 40–5.

Ormsby, H. (1931). *France. A Regional and Economic Geography*. London: Methuen.

Ormsby, H. (1932). The limestones of France and their influence on human geography. *Geography*, **17**: 11–21.

Ormsby, H. (1935). The definition of *Mitteleuropa* and its relation to the conception of Deutschland in the writings of modern German geographers. *Scottish Geographical Magazine*, **51**: 337–46.

Oughton, M. (1963). Ellen Jennifer Rickard. *Geography*, **48**: 418.

Our Obituary Record (1872). Mary Somerville. *Our Obituary Record*, 14 February 1872.

P.K. (1893). Review: Kate Marsden, *On Sledge and Horseback to Outcast Siberian Lepers. The Geographical Journal*, **LXX**: 376.

Parker, W.H. (1982). *Geography as an Aid to Statecraft*. Oxford: Clarendon Press.

Parks, G.B. (1931). Review: E.G.R. Taylor, *Tudor Geography. The Geographical Review*, **21**: 686–7.

Parks, G.B. (1934). Review: E.G.R. Taylor, *Late Tudor and Early Stuart Geography. Geographical Review*, **24**: 693–4.

Paterson, K. & Sweeting, M.M. (eds) (1986). *New Directions in Karst*. Norwich: GeoBooks.

Patmore, J.A. (1987). British geography, 1918–1945: a personal perspective. In R.W. Steel (ed.), *British Geography 1918–1945*. Cambridge: Cambridge University Press.

Patterson, E.C. (1969). Mary Somerville. *British Journal for the History of Science*, **IV**: 311–36.

Patterson, E.C. (1983). *Mary Somerville and the Cultivation of Science, 1815–1840*. The Hague.

Peake, L., Staeheli, L. & Koffman, E. (eds) (2004). *Mapping Women, Making Politics: Feminist Perspectives on Political Geography*. London: Routledge.

Pearce, L. (1991). *Woman/Image/Text*. London: Harvester Wheatsheaf.

Pemberton, P.H. (ed.) (1970). *Geography in Primary Schools*. Sheffield: The Geographical Association.

Perry, A. (2005). The colonial archive on trial: possession, dispossession, and history in *Delgamuukw v. British Columbia*. In A. Burton (ed.), *Archive Stories. Facts, Fictions and the Writing of History* (pp. 325–50). Durham: Duke University Press.

Petellier, M. (1995). Eila and the international conferences on the history of cartography. *Imago Mundi*, **47**: 9.

conferences on the history of cartography, *Imago Mundi* 47: 9.

Phillips, P. (1990). *The Scientific Lady: a Social History of Women's Scientific Interests 1520–1918*. London: Weidenfeld & Nicolson.

Philo, C. (1998). Reading *Drumlin*: academic geography and a student geographical magazine. *Progress in Human Geography*, **22**: 344–67.

Ploszajska, T. (1996). Constructing the subject. Geographical models in English schools 1870–1944. *Journal of Historical Geography*, **22**: 288–98.

Ploszajska, T. (1999). *Geographical Education, Empire and Citizenship: Geographical Teaching and Learning in English Schools, 1870–1944*, Vol. 35, Historical Geography Research Series. Cambridge: Cambridge University Press.

Polhandt-McCormick, H. (2005). In good hands: researching the 1976 Soweto uprising in the state archives of South Africa. In A. Burton (ed.), *Archive Stories. Facts, Fictions and the Writing of History* (pp. 299–324). Durham: Duke University Press.

Pollock, N.C. (1957). Review: *Land Use Studies in the Transvaal Lowveld* by Monica Cole. *The Geographical Journal*, **123**: 399–400.

Pollock N.C. & Agnew, S. (1964). *A Historical Geography of South Africa*. London: Longman, Green & Co.

Potter, G. (1968). Professor Alice Garnett. *University of Sheffield Gazette*, **48**: 66–7.

Potter, R. (1994a). Obituary: Professor Monica Mary Cole. *The Independent*, 22 January 1994.

Potter, R. (1994b). Obituary: Monica Cole. *Royal Holloway and Bedford Association College Journal*, **8**: 15–16.

Potter, R. & Catt, P. (1994). Obituary: Monica Mary Cole 1922–1994. *Transactions of the Institute of British Geographers*, NS19: 373–7.

Powell, R.C. (2002). The Sirens' voices? Field practices and dialogue in geography. *Area*, **34**: 261–72.

Power Cobbe, F. (1872). 'Blessed Old Age', Obituary Mary Somerville. *The Echo*, 3 December.

Pratt, M.L. (1992). *Imperial Eyes. Travel Writing and Transculturation*. London: Rouledge.

Pratt, M.L. & Hanson, S. (1994). Geography and the construction of difference. *Gender, Place and Culture*, **1**: 5–30.

Pruitt, E. (1961). Review: C.A.M. King, *Beaches and Coasts. Geographical Review*, **51**: 458–60.

Ravenhill, W. (1996). Obituary: Helen Margaret Wallis 1924–1995. *Transactions of the Institute of British Geographers*, NS21: 299–301.

Rees, J. (1969). *Industrial Demand for Water. A Study of South East England*. London: Weidenfeld & Nicolson.

Reynolds, J.B. (1901). Class excursions in Wales and England. *The Geographical Teacher*, **I**: 32–6.

Reynolds, J.B. (1904a). The regional method of teaching geography. *The Geographical Teacher*, **II**: 224–8.

Reynolds, J.B. (1904b). Official geographical literature on the colonies. *The Geographical Teacher*, **II**: 30–2, 70–6.

Reynolds, J.B. (1910). Map reading and imagination. *The Geographical Teacher*, **V**: 81–5.

Reynolds, J.B. (ed.) (1915a). *Scotland. Quotation and Picture Series*. London: A&C Black.

Reynolds, J.B. (1915b). *Regional Geography: The World*. London: A&C Black.

Reynolds, J.B. (1918). *Europe and the Mediterranean Region*, 8th edition. London: A&C Black.

Rickard, E. & Michaelis, M. (1912). The use of statistics in the teaching of geography: a reply. *The Geographical Teacher*, **6**: 147–53.

Robinson, G.M. (1991). James Wreford Watson 1915–1990. *The Geographical Journal*, **157**: 244.

Robinson, J. (1990). *Wayward Women. A Guide to Women Travellers*. Oxford: Oxford University Press.

Robinson, J. (1994). *Unsuitable for Ladies. An Anthology of Women Travellers*. Oxford: Oxford University Press.

Rojek, C. (2000). Leisure and the rich today: Veblen's thesis after a century. In J.T. Haworth (ed.), *Work, Leisure and Wellbeing* (pp. 117–30). London: Routledge.

Roque Ramirez, H. (2005). A living archive of desire: Teresita la Campesina and the embodiment of Queer Latino community histories. In A. Burton (ed.), *Archive Stories* (pp. 111–35). Durham: Duke University Press.

Rose, G. (1993). *Feminism and Geography: The Limits of Geographical Knowledge*. Cambridge: Polity.

Rose, G. (1995). Tradition and paternity: same difference? *Transactions of the Institute of British Geographers*, **20**: 414–16.

Rose, H. (1994). *Love, Power and Knowledge: Towards a Feminist Transformation of the Sciences*. Cambridge: Polity Press.

Royal Scottish Geographical Society (1904). Obituary. Mrs. Isabella Bishop. *Scottish Geographical Magazine*, **XX**: 595–7.

Royal Scottish Geographical Society (1934). Obituary. Marion Isabel Newbigin. *Scottish Geographical Magazine*, **L**: 331–3.

Russell, M. (1986). *The Blessings of a Good Thick Skirt. Women Travellers and their World*. London: Flamingo.

Russell, M. (1994). *The Blessings of a Good Thick Skirt. Women Travellers and their World*. London: Flamingo.

Ryan, J.R. (2006). 'Our home on the ocean': Lady Brassey and the voyages of the *Sunbeam*, 1874–1887. *Journal of Historical Geography*, **32**: 579–604.

Sack, D. (2004). Experiences and viewpoints of selected women geomorphologists from the mid-twentieth century. *Physical Geography*, **25**: 438–52.

Said, E. (1978). *Orientalism*. London: Penguin.

Samers, M. (2001). What is the point of economic geography? *Antipode*, **33**: 183–93.

Sanderson, M. (1974). Mary Somerville. Her work in physical geography. *Geographical Review*, **LXIV**: 410–20.

Scargill, I. (1999). *The Oxford School of Geography 1899–1999*. Oxford: University of Oxford School of Geography and Environment Research Paper Series.

Secor, A. (1999). Orientalism, gender and class in Lady Mary Wortley Montagu's Turkish Embassy Letters: to persons of distinction, men of letters etc. *Ecumene*, **6**: 375–98.

Shackleton, M.R. (1939). *Europe. A Regional Geography*, third edition. London: Longman.

Shackleton, M.R. (1959). *Europe. A Regional Geography*, sixth edition. London: Longman.

Sheppard, J.A. (1953). Horticultural developments in east Yorkshire. *Transactions of the Institute of British Geographers*, **19**: 73–80.

Sheppard, J.A. (1994). *A Century of Geography. A History of the Geography Department of the East London College (to 1935), Queen Mary College (1935–1989), Queen Mary and Westfield College (1989–)*. London: Queen Mary and Westfield College.

Sheppard, J.A. & Rawstron (1980). Alice Mutton 1908–1979. *Transactions of the Institute of British Geographers*, NS5: 123–4.

Shimwell, D.W. (1982). Review: *Biogeography: a Study of Plants in the Ecosphere* (3rd edition) by Joy Tivy. *Journal of Applied Ecology*, **19**: 992–3.

Sidaway, J. (1997). The production of British geography. *Transactions of the Institute of British Geographers*, **22**: 488–504.

Simpson, C. (1930). *Rediscovering England*. London: Ernest Benn.

Simpson, C. (1934). *The Study of Local Geography – A Handbook for Teachers*. London: Methuen.

Simpson, C. (1945). A venture in field geography. *Geography*, **30**: 35–44.

Simpson, C. (1951). *Making Local Surveys. An Eye for Country*. London: Pitman.

Simpson, H. (1950). *A Woman among Savages*. London: Puffin.

Sinnhuber, K. (1954). Central Europe – Mitteleuropa – Europe Centrale: an analysis of a geographical term. *Transactions of the Institute of British Geographers*, **20**: 15–39.

Skelton, R.A. (1968). Professor E.G.R. Taylor. *Imago Mundi*, **22**: 114–16.

Skinner, Q. (1969). Meaning and understanding in the history of ideas. *History and Theory*, **8**: 3–53.

Smee, D.K. (1953). Obituary: Miss Blanche Hoogood. *Geography*, **XXXVIII**: 320.

Smith, B.G. (1986). The contribution of women to modern historiography in Great Britain, France and the United States, 1750–1940. *American Historical Review*, **89**: 709–32.

Smith, B.G. (1996). Historiography, objectivity and the case of the abusive widow. In J. Wallach Scott (ed.), *Feminism and History* (pp. 547–9). Oxford: Oxford University Press.

Snodgrass, C. (1941). The density of agricultural population in Scotland with English and European comparisons. *The Geographical Journal*, **4**: 236–45.

Snodgrass, C. (1944a). Recent population changes in Scotland. *Scottish Geographical Magazine*, **60**: 33–8.

Snodgrass, C. (1944b). Notes on the geographical distribution of employment in Scotland. *Scottish Geographical Magazine*, **60**: 69–75.

Snodgrass, C. (1946). Land utilisation in the Lothians. *Scottish Geographical Magazine*, **62**: 28–31.

Snodgrass, C. (1953). County of East Lothian. *Third Statistical Account of Scotland*. Edinburgh: Oliver Boyd.

Snodgrass, C. (1966). Scottish conurbations. In T.W. Freeman, *The Conurbations of Great Britain*. Manchester: Manchester University Press.

Somerville, M. (1826). On the magnetizing power of the more refrangible solar rays. *Philosophical Transactions of the Royal Society of London*, **116**: 132–9.

Somerville, M. (1831). *The Mechanism of the Heavens*. London: John Murray.

Somerville, M. (1832). *A Preliminary Dissertation on the Mechanism of the Heavens*. Philadelphia.

Somerville, M. (1834). *On the Connexion of the Physical Sciences*. London: John Murray.

Somerville, M. (1845). On the action of the rays of the spectrum on the vegetable juices. Extract from a letter from Mrs. M. Somerville to Sir J.F.W. Herschel, Bart., dated Rome, September 20, 1845. Communicated by Sir J. Herschel. *Philosophical Transactions of the Royal Society of London*, **136**: 111–20.

Somerville, M. (1848). *Physical Geography*. London: John Murray.

Somerville, M. (1869). *On Molecular and Microscopic Science.* London: John Murray.

Somerville, M. (ed.) (1873/1996). *Personal Recollections from Early Life to Old Age of Mary Somerville, with Selections from her Correspondence. by her Daughter, Martha Somerville.* London: John Murray.

Somerville, M. (1935). The Comet. *Quarterly Review,* **55**: 195–223.

Sparke, M. (1996). Displacing the field in fieldwork: masculinity, metaphor and space. In N. Duncan (eds), *BodySpace* (pp. 212– 33). London: Routledge.

Sparks, B. (1966). Review: C.A.M. King, *Techniques in Geomorphology. The Geographical Journal,* **132**: 555–6.

Stafford, R.A. (1989). *Scientist of Empire. Sir Roderick Murchison, Scientific Exploration and Victorian Imperialism.* Cambridge: Cambridge University Press.

Stamp, D., Beaver, S. & Smee, D. (1933). *The British Isles. A Geographical and Economic Survey.* London: Longmans, Green & Co.

Stark, F. (1934). *The Valleys of the Assassins.* London: John Murray.

Stark, F. (1936). *The Southern Gates of Arabia.* London: John Murray.

Stark, F. (1940). *A Winter in Arabia.* London: John Murray.

Stark, F. (1951). *Beyond Euphrates.* London: John Murray.

Steel, R.W. (1984). *The Institute of British Geographers: the First Fifty Years.* London: Institute of British Geographers.

Steel, R.W. (ed.) (1987). *British Geography 1918–1945.* Cambridge: Cambridge University Press.

Steel, R.W. & Lawton, R. (eds) (1967). *Liverpool Essays in Geography.* London: Longman.

Steers, H.G. (1962). The Grampians. In J.B. Mitchell (ed.), *Great Britain: Geographical Essays* (pp. 527–42). Cambridge: Cambridge University Press.

Steers, H.G. (1962a). Review: *Central Europe: A Regional and Human Geography. The Geographical Journal,* **128**: 78–9.

Steers, J.A. (1973). Review: *Karst Landforms* by Marjorie Sweeting. *The Geographical Journal,* **139**: 142–4.

Steers, J.A. & Mitchell, J.B. (1962). In J.B. Mitchell (ed.), *Great Britain: Geographical Essays* (pp. 86–103). Cambridge: Cambridge University Press.

Steers, J.A. & Woodbridge, S.W. (1976). Interwar trends. In R.E. Dickinson (ed.), *The Regional Concept: the Anglo-American Leaders* (pp. 135–62). London: Routledge & Kegan Paul.

Steven, A. (2000). Lady Agnew, *The Scotsman,* 9 October: 14.

Stoddart, D.R. (ed.) (1981). *Geography, Ideology and Social Concern.* Oxford: Blackwell.

Stoddart, D.R. (1983). Progress in geography: the record of the IBG. *Transactions of the Institute of British Geographers,* NS8: 1–13.

Stoddart, D.R. (1986). *On Geography.* Oxford: Blackwell.

Stoddart, D.R. (1988). James Alfred Steers, 1899–1987. *Transactions of the Institute of British Geographers,* NS 13: 109–15.

Stoddart, D.R. (1991). Do we need a feminist historiography of geography? And if we do, what should it be? *Transactions of the Institute of British Geographers,* **16**: 484–7.

Storrie, M.C. (1961). Islay: a Hebridean exception. *The Geographical Review,* **51**: 87–108.

Storrie, M.C. (1962a). The census of Scotland as a source in the historical geography of Islay. *Scottish Geographical Magazine*, 78: 152–65.

Storrie, M.C. (1962b). The Scotch whisky industry. *Transactions of the Institute of British Geographers*, NS31: 97–114.

Storrie, M.C. (1965). Landholdings and settlement evolution in West Highland and Scotland. *Geografiska Annaler*, 47: 138–61.

Storrie, M. (1981). *Islay. Biography of an Island*. Islay: The OA Press.

Summerfield, P. (1995). Women and war in the twentieth century. In J. Purvis (ed.), *Women's History: Britain 1850–1945. An Introduction* (pp. 307–32). London: UCL Press.

Sutherland, G. (1990). Education. In F.M.L. Thompson (ed.), *The Cambridge Social History of Britain. Volume 3 Social Agencies and Institutions* (pp. 119–70). Cambridge: Cambridge University Press.

Swallow, M. (1963). Review C.A.M. King, *Oceanography for Geographers. The Geographical Journal*, 129: 548.

Sweeting, M.M. (1943). Wave trough experiments on beach profiles. *The Geographical Journal*, 101: 163–72.

Sweeting, M.M. (1950). Erosion cycles and limestone caverns in the Ingleborough District. *The Geographical Journal*, 115: 63–78.

Sweeting, M.M. (1958). The karstlands of Jamaica. *The Geographical Journal*, 124: 184–99.

Sweeting, M.M. (1964). Some factors in the absolute denudation of limestone terrains. *Erdkunde*, 18: 92–5.

Sweeting, M.M. (1965). Denudation in limestone regions. A symposium. Introduction. *The Geographical Journal*, 131: 34–7.

Sweeting, M.M. (1966). The weathering of limestones. In G.H. Dury (ed.), *Essays in Geomorphology* (pp. 177–210). London: Heinemann.

Sweeting, M.M. (1972). *Karst Landforms*. London: Macmillan.

Sweeting, M.M. (1978). Karst water and karst hydrology. *Progress in Physical Geography*, 2: 99–106.

Sweeting, M.M. (1979) Karst morphology and limestone petrology. *Progress in Physical Geography*, 3: 102–10.

Sweeting, M.M. (ed.) (1980). *Karst Geomorphology*, Benchmark Series in Geomorphology. Pennsylvania: Dowden, Hutchison & Ross.

Sweeting, M.M. (1991). Jean Brown Mitchell. *Newnham College Roll Letter*, Cambridge, pp. 120–4.

Sweeting, M.M. (1995). *Karst in China: its Geomorphology and Environment*. Berlin: Springer-Verlag.

Sweeting, M.M. Daoxian, Y., Hu, M. & Bull, P.A. (1985). Aspects of the Quaternary deposits around Guilin, Guangxi, S China: a scanning electron microscope study. *Carsologica Sinica*, 3: 267–71 (English) and 272–8 (Chinese).

Sweeting, M.M. & Groom, G.E. (1956). Notes on the glacier fluctuations in Bünsow Land, Central Vestspitsbergen. *Journal of Glaciology*, 2: 640–1.

Sweeting, M.M. & Groom, G.E. (1958). Valleys and raised beaches in Bünsow Land, Central Vestspitsbergen. *Norsk Polar-institutt Skrifter*, 115: 1–18.

Sweeting, M.M. & Sweeting, G.S. (1969). Some aspects of the carboniferous lime-stone in relation to its landforms with particular reference to NW Yorkshire and County Clare. *Mediterranée*, 7: 201–9.

Sylvester, D. (1939). Obituary: Dorothy Mary Preece. *Geography*, **XXIV**: 136.

Sylvester, D. (1943). A method of panorama from contoured maps. *Geography*, **28**: 12–18.

Sylvester, D. (1947). The hill villages of England and Wales. *The Geographical Journal*, **110**: 76–93.

Sylvester, D. (1952). *Maps and Landscape*. London: George Philip & Son.

Sylvester, D. (1969). *The Rural Landscape of the Welsh Borderland: a Study of Historical Geography*. London: Macmillan.

Sylvester, D. (1980). *A History of Cheshire*, Darwen County History Series (2nd edition). Chichester: Phillimore.

Sylvester, D. (1983). *A History of Gwynedd*, Darwen County History Series. Chichester: Phillimore.

Sylvester, D. & Nulty, D. (eds) (1958). *The Historical Atlas of Cheshire*. Chester: Cheshire County Council.

Tanner, M. (2003). List of staff in university geography departments 1963–4, personal communication.

Taylor, D. (1994). Review: *Biogeography* (3rd edition). *Journal of Biogeography*, **6**: 671.

Taylor, E.G.R. (1921). *A Sketch-map Geography*. London: Methuen.

Taylor, E.G.R. (1927). The earliest account of triangulation. *Scottish Geographical Magazine*, **43**: 421–5.

Taylor, E.G.R. (1928a). *Oceans and Rivers*, Benn's Sixpenny Library No. 31. London: Ernest Benn.

Taylor, E.G.R. (1928b). William Bourne: a chapter in Tudor geography. *The Geographical Journal*, **72**: 329–39.

Taylor, E.G.R. (1929a). Roger Barlow: a new chapter in early Tudor geography. *The Geographical Journal*, **74**: 157–66.

Taylor, E.G.R. (1929b). Jean Rotz: his neglected treatise on nautical science. *The Geographical Journal*, **73**: 455–9.

Taylor, E.G.R. (1930). *Tudor Geography, 1485–1583*. London: Methuen.

Taylor, E.G.R. (1930a). *Production and Trade. A Geographical Survey of all the Countries of the World*. London: George Philip.

Taylor, E.G.R. (1934a). Obituary. Dr Marion I. Newbigin. *The Geographical Journal*, **LXXXIV**: 367.

Taylor, E.G.R. (1934b). *Late Tudor and Early Stuart Geography, 1583–1650*. London: Methuen.

Taylor, E.G.R. (1936a). Leland's England. In H.C. Darby (ed.) *An Historical Geography of England before AD 1800*. Cambridge: Cambridge University Press.

Taylor, E.G.R. (1936b). Cambden's England, in Darby H.C. (ed.) In H.C. Darby (ed.) *An Historical Geography of England before AD 1800*. Cambridge: Cambridge University Press.

Taylor, E.G.R. (1947a). Geography in war and peace. *Scottish Geographical Magazine*, **63**: 97–108.

Taylor, E.G.R. (1947b). 'They rejoiced in things stark naughty'. *Scottish Geographical Magazine*, **20**: 304–9.

Taylor, E.G.R. (1954). *The Mathematical Practitioners of Tudor and Stuart England.* Cambridge: Cambridge University Press (for Institute of Navigation).

Taylor, E.G.R. (1956). *The Haven-Finding Art.* London: Hollis & Carter.

Taylor, E.G.R. (1966). *The Mathematical Practitioners of Hanoverian England, 1714–1840.* Cambridge: Cambridge University Press.

Taylor, E.G.R. & Richey, M.W. (1962). *The Geometrical Seaman: A Book of Early Nautical Instruments.* London: Hollis & Carter.

Taylor, G. (ed.) (1957). *Geography in the Twentieth Century.* London: Metheun.

Taylor, P. (1976). An interpretation of the quantification debate in British geography. *Transactions of the Institute of British Geographers,* NS1: 129–42.

Thane, P. (1990). Government and society in England and Wales, 1750–1914. In F.M.L. Thompson (ed.), *The Cambridge Social History of Britain. Volume 3 Social Agencies and Institutions* (pp. 1–62). Cambridge: Cambridge University Press.

The Geographical Journal (1900). Obituary: Miss Mary H. Kingsley. *The Geographical Journal,* 16: 114–15.

The Geographical Journal (1924). *London on the Thames.* H. Ormsby. *The Geographical Journal,* LXIII: 534.

The Geographical Journal (1985). Obituary: Gertrude Caton-Thompson. *The Geographical Journal,* 151: 437–8.

The Geographical Journal (1987). Obituary: James Alfred Steers, CBE, 1899–1987. *The Geographical Journal,* 153: 436–8.

The Geographical Journal (1990). Obituary: Jean Mitchell. *The Geographical Journal,* 156: 242–3.

The Geographical Journal (1936) Obituary: May French Sheldon. *The Geographical Journal,* LXXXVII: 288.

The Geographical Teacher (1916). Editorial. The late Mrs Herbertson. *The Geographical Teacher,* 44: 211.

The Geographical Teacher (1924). *London on the Thames.* H. Ormsby. *The Geographical Teacher,* XII: 324.

The Morning Post (1872). Obituary Mary Somerville. *The Morning Post,* 2 December 1872.

The Times (1872). Obituary Mary Somerville. *The Times,* 2 December 1872.

Third, B.W. (1955). Changing landscape and social structures in Scottish lowlands as revealed by eighteenth century estate plans. *Scottish Geographical Magazine,* 71: 83–93.

Thomas, N. (2004). Exploring the boundaries of biography: the family and friendship networks of Lady Curzon, Vicerene of India 1898–1905. *Journal of Historical Geography,* 30: 496–515.

Thompson, D.W. (1924). The origin of London: a review. *Scottish Geographical Magazine,* XL: 97–9.

Thompson, F.M.L. (1990). *The Cambridge Social History of Britain 1750–1950.* Cambridge: Cambridge University Press.

TIBG (1968). Obituary: Florence Clark Miller. *Transactions of the Institute of British Geographers,* 44: 186–7.

TIBG (1968). Obituary: Professor Eva Germaine Rimington Taylor. *Transactions of the Institute of British Geographers,* 45: 181–6.

Tiltman, M. (1935). *Women in Modern Adventure.* London: George Harrap.

Timberlake, E. (1983). Dora K. Smee 1899–1982. *Transactions of the Institute of British Geographers*, NS8: 120–1.

Tivy, J. (1962a). An investigation of certain slope deposits in the Lowther Hills, Southern Uplands of Scotland. *Transactions of the Institute of British Geographers*, 30: 59–73.

Tivy, J. (1962b). Some thoughts on field work. *The Drumlin*, 24–6.

Tivy, J. (1965). Some thoughts on biogeography. *The Drumlin*, 32–6.

Tivy, J. (1968). The teaching of geography. *The Drumlin*, 39–42.

Tivy, J. (1971). *Biogeography: a Study of Plants in the Ecosphere*. Edinburgh: Oliver & Boyd.

Tivy, J. (1972). The concept and determination of carrying capacity of recreational land in the USA. Occasional Paper No. 3, Countryside Commission of Scotland, Buttleby, Redgorton.

Tivy, J. (ed.) (1973). *The Organic Resources of Scotland. Their Nature and Evaluation*. Edinburgh: Oliver & Boyd.

Tivy, J. (1982). *Biogeography: a Study of Plants in the Ecosphere*, 2nd edition. London: Longman.

Tivy, J. (1990) *Agricultural Ecology*. Harlow: Longman.

Tivy, J. (1993). *Biogeography: a Study of Plants in the Ecosphere*, 3rd edition. London: Longman.

Tivy, J. & O'Hare, G. (1981). *The Human Impact on the Ecosystem*. Edinburgh: Oliver & Boyd.

Topographical Antiquities (1924). Review: *London on the Thames* by H. Ormsby. *Topographical Antiquities*, 551–2.

Tosh, J. (1992). The flight from domesticity and imperial masculinity in Britain, 1880–1914. Paper presented at the Gender and Colonialism Conference, University College, Galway.

Trollope, J. (1983). *Britannia's Daughters. Women of the British Empire*. London: Pimlico.

Unstead, J.F. & Taylor, E.G.R. (1910/1926). *General and Regional Geography for Students*. London: George Philip & Son.

Unstead, J.F. & Taylor, E.G.R. (1912a). Correspondence. *The Geographical Journal* 39: 293–4.

Unstead, J.F. & Taylor, E.G.R. (1912b). Correspondence. *The Geographical Journal* 40: 94–5 and 223–4.

Van Tilburg, J. (2003). *Among Stone Giants. The Life of Katherine Routledge and her Remarkable Expedition to Easter Island*. New York: Scribner.

Vickery, A. (1993). Golden age to separate spheres? A review of the categories and chronology of English women's history. *The Historical Journal*, 36: 383–414.

Viles, H.A. (1996). Obituary. Marjorie Sweeting 1920–1994., *Transactions of the Institute of British Geographers*, NS21: 429–32.

Viles, H.A. (1997). A lifetime of landform: Marjorie Sweeting's work on tropical and sub-tropical karst. In P.W. Williams (ed.), Tropical and Sub-tropical Karst. Essays Dedicated to the Memory of Dr. Marjorie Sweeting. *Zeitschrift fur Geomorphologie, Supplementbände*, Band 108.

Vincent, P. (1995). *Limestone Pavements in the British Isles*: a review. *The Geographical Journal*, 161: 265–74.

Vollans, E. (1949). Derby, a railway town and regional centre. *Transactions of the Institute of British Geographers*, **15**: 91–112.

Vollans, E. (1992). Obituary. Eunice M. Timberlake. *Transactions of the Institute of British Geographers*, NS17: 247–8.

W.G.E. (1950). Obituary: Margaret Shackleton. *The Geographical Journal*, **115**: 267–8.

Wagstaff, M. (1996). Geography. *The First Seventy-Five Years at the University of Southampton, Department of Geography.* Southampton: Southampton University.

Walford, R. (2001). *Geography in British Schools 1850–2000. Making a World of Difference.* London: Woburn Press.

Wallach Scott, J. (1988). *Gender and the Politics of History.* New York: Columbia University Press.

Wallach Scott, J. (1992). The problem of invisibility. In S.J. Kleinberg (ed.), *Retrieving Women's History. Changing the Perceptions of the Role of Women in Politics and Society* (pp. 5–18). Oxford: Berg/UNESCO.

Wallach Scott, J. (1996). *Desert Queen. The Extraordinary Life of Gertrude Bell: Adventurer, Adviser to Kings, Ally of Lawrence of Arabia.* London: Weidenfeld & Nicholson.

Wallach Scott, J. (1999). Some reflections on gender and politics. In M.Marx Ferree, J. Lorber & B.B. Hess (eds), *Revisioning Gender.* London: Sage.

Wallis, H. (1951). The first English globe. A recent discovery. *The Geographical Journal*, **117**: 275–90.

Wallis, H. (1994). Eila Muriel Joyce Campbell 1915–1994. *The Geographical Journal*, **160**: 361.

Wallis, H. (1995). Eila and her societies. *Imago Mundi*, **47**: 9–10.

Wallis, H. & Robinson, A.H. (1987) *Cartographical Innovations: an International Handbook of Mapping Terms.* Tring: Map Collector Publications /ICA.

Walsh, M. & Wrigley, C. (2001). Womanpower: the transformation of the labour force in the UK and the USA since 1945. ReFresh (Recent Findings in Research in Economic and Social History), **30.** www.ehs.org.uk/walsh30a.pdf, accessed online March 2008.

Waltham, A.C. & Sweeting, M.M. (eds) (1974). *The Limestones and Caves of NW England.* Newton Abbot: David and Charles.

Wanklyn, H. (1941). *The Eastern Marchlands of Europe.* London: George Philip.

Wanklyn, H. (1954). *Czechoslovakia.* London: George Philip.

Wanklyn, H. (1961). *Friedrich Ratzel.* Cambridge: Cambridge University Press.

Ward, E.M. (1920). The evolution of the Hastings coastline. *The Geographical Journal*, **56**: 107–20.

Ward, E.M. (1922). *English Coastal Evolution.* London: Methuen.

Ward, E.M. (1929/1948). *Days in Lakeland, Past and Present.* London: Methuen.

Ward, E.M. (1938). *Sea Wind.* London: Methuen.

Waters, R. (1991). Obituary: Alice Garnett 1903–1989. *The Geographical Journal*, **157**: 117.

Watts, R. (1981). The Unitarian contribution to the development of female education in the nineteenth century, unpublished paper. Harris and Manchester College, Oxford.

Whalley, B. (2004). Jean Mary Grove (née Clark). *Oxford Dictionary of National Biography.* Oxford: Oxford University Press.

Wheeler, P.T. (1967). The development and role of the geographical field group. *The East Midland Geographer*, 4: 185–95.

Wheeler, P.T. (1992). Gwendoline Joan Fuller BA, MA, PhD 1909–1992 (amended typescript). Nottingham University Archives.

Whewell, W. (1837). *History of the Inductive Sciences*. London: John Parker.

White, F.P. (1967). Review: E.G.R. Taylor, *The Mathematical Practitioners of Hanoverian England, 1714–1840. The Mathematical Gazette*, 51: 271.

Wilcox, D. (1975). *Explorers*. London: BBC.

Willatts, E.C. (1971). Planning and geography in the last three decades. *The Geographical Journal*, 137: 311–30.

Williams, P.W. (ed.) (1997). Tropical and sub-tropical karst – essays dedicated to the memory of Dr Marjorie Sweeting. *Zeitschrift fur Geomorphologie, Supplementbände*, Band 108.

Wise, M.J. (1980). Llewellyn Rodwell Jones. In T.W. Freeman & P. Pinchemel (eds), *Geographers: Biobibliographical Studies* 4. London: Mansell.

Wise, M.J. (1990). Obituary. Alice Garnett 1904–1989. *Transactions of the Institute of British Geographers*, NS 15: 113–16.

Withers, C.W. J. (2002). Constructing the geographical archive. *Area*, 34, 303–11.

Withers, C. W. J. (2004). Memory and the history of geographical knowledge: the commemoration of Mungo Park, African explorer. *Journal of Historical Geography*, 30: 316–39.

Withers, C. W. J. & Mayhew, R.J. (2002). Re-thinking 'disciplinary' history: geography in British universities, c. 1580–1887. *Transactions of the Institute of British Geographers*, NS27: 1–19.

Withers, C.W.J. (2006). Marion Newbigin, Catherine Snodgrass. In E. Ewan, S. Innes & S. Reynolds (eds), *Biogeographical Dictionary of Scottish Women* (pp. 282–3). Edinburgh: Edinburgh University Press.

Witz, A. (1992). *Professions and Patriarchy*. London: Routledge.

Witz, A. & Savage, M. (1992). The gender of organizations. In A. Witz & M. Savage (eds), *Gender and Bureaucracy*. Oxford:, Blackwell.

Women and Geography Study Group (1997) *Feminist Geographies. Explorations in Diversity and Difference*. London: Longman.

Women and Geography Study Group of the IBG (1984). *Geography and Gender: an Introduction to Feminist Geography*. London: Hutchinson and Explorations in Feminist Collective.

Woollacott, A. (1998). The fragmentary subject: feminist history, official records and self-representation. *Women's Studies International Forum*, 21: 329–39.

Wrigley, G.M. (1952). Adventures in serendipity: thirty years of the *Geographical Review. Geographical Review*, 42: 511–42.

Wylie Hutchison, I. (1923). *Original Companions*. London: Bodley Head.

Wylie Hutchison, I. (1930). *On Greenland's Closed Shore: the Fairyland of the Arctic*. Edinburgh: Blackwood.

Wylie Hutchison, I. (1934). *North to the Rime-Ringed Sun: being the Record of an Alaska-Canada Journey made in 1833–4*. London: Blackie.

Wylie Hutchison, I. (1937a). *Stepping Stones from Alaska to Asia*. London: Blackie.

Wylie Hutchison, I. (1937b). The discovery of a new reef near Attu Island. *The Geographical Journal*, 90: 541–4.

Younghusband, F. (1917). Geographical work in India for this society. *The Geographical Journal*, **49**: 401–12.

Zornlin, M. (1839). *Recreations in Physical Geography*. London: John W. Parker.

Zornlin, M. (1842). *Recreations in Hydrology*. London: John W. Parker.

ARCHIVE SOURCES*

British Library

Eila Campbell Papers (Map Room)
Eva Taylor Papers
Marie Stopes Papers

National Library of Scotland

Catherine Snodgrass Papers
Isobel Wylie Hutchison images

National Library of Wales

Elwyn and Margaret Davies Papers
Dorothy Sylvester Papers

RGS–IBG

Additional Papers 2
Application for election forms
A.R. Hinks Correspondence Files
Council Minute Books
E.G.R. Taylor Correspondence Files
E.M.J. Campbell Correspondence Block
Geographers in World War II Archive
Image Library
J.S. Keltie Correspondence Files
Marion Newbigin Correspondence Files
Picture Library
Violet Cressy-Marcks Correspondence Files
Women's Admission Files

* Full references are given in the notes at the end of the book. Note that these archive sources have been consulted over a near 20-year period and some may have been re-catalogued in the meantime.

RSGS

Council Minute Book 1884–
Isabella Bird File
Isobel Wylie Hutchison pictures

NORTHAMPTONSHIRE PUBLIC RECORD OFFICE

Dora Smee Papers
Ismay Papers

PUBLIC RECORD OFFICE ENGLAND AND WALES

Family Records: birth, marriage and death certificates
Geographical Association
Geographical Association Archives (in house), Sheffield
Geographical Association Papers, Sheffield City Archives

University Archives

London School of Economics: Hilda Ormsby Papers
Royal Holloway University of London: Blanche Hosgood, Monica Cole, Eunice
    Timberlake and Dora Smee
Queen Mary, University of London: Staff Files Betty Tunstall, Alice Mutton
University of Aberystwyth: Student Records
University of Bristol: Gladys Hickman
University of Cambridge: Newnham College Register, Girton College Register.
University of Edinburgh: Department of Geography, Catherine Snodgrass Papers
University of Glasgow: *Drumlin* Student Magazine
University of Hull: Staff Records
University of Liverpool: Staff Records
University of London Institute of Education: MI Long staff record
University of Manchester: Staff Records; Department of Geography: Staff Meeting
    Minutes Book 1960s
University of Newcastle: Gertrude Bell Papers (online)
University of Oxford: Alumni Records, staff Records, Mary Somerville Papers;
    Department of Geography: Mary Marshall Papers Marjorie Sweeting Papers,
    Geophil Student Magazine, Lady Margaret Hall Register
University of Sheffield: Alice Garnett curriculum vitae 1962.

Index

Page numbers in *italic* denotes as illustration/table
Page numbers in **bold** denotes a major section devoted to subject